ISACA®

CISM®

Review Questions, Answers & Explanations Manual

9th Edition

About ISACA

Nearing its 50th year, ISACA® (isaca.org) is a global association helping individuals and enterprises achieve the positive potential of technology. Technology powers today's world and ISACA equips professionals with the knowledge, credentials, education and community to advance their careers and transform their organizations. ISACA leverages the expertise of its half-million engaged professionals in information and cybersecurity, governance, audit, assurance, risk and innovation, as well as its enterprise performance subsidiary, CMMI® Institute, to help advance innovation through technology. ISACA has a presence in over 185 countries, including more than 215 chapters and offices in both the United States and China.

In addition, ISACA advances and validates business-critical skills and knowledge through the globally respected Certified Information Systems Auditor® (CISA®), Certified Information Security Manager® (CISM®), Certified in the Governance of Enterprise IT® (CGEIT®) and Certified in Risk and Information Systems Control™ (CRISC™) credentials.

Disclaimer

ISACA has designed and created *CISM® Review Questions, Answers & Explanations Manual 9th Edition* primarily as an educational resource to assist individuals preparing to take the CISA certification exam. It was produced independently from the CISA exam and the CISA Certification Committee, which has had no responsibility for its content. Copies of past exams are not released to the public and were not made available to ISACA for preparation of this publication. ISACA makes no representations or warranties whatsoever with regard to these or other ISACA publications assuring candidates' passage of the CISA exam.

ISACA
1700 E. Golf Road, Suite 400
Schaumburg, IL 60173, USA
Phone: +1.847.660.5505
Fax: +1.847.253.1755
Contact us: https://support.isaca.org
Website: www.isaca.org

Provide feedback: https://support.isaca.org
Participate in the ISACA Knowledge Center: www.isaca.org/knowledge-center

Twitter: www.twitter.com/ISACANews
LinkedIn: www.linkd.in/ISACAOfficial
Facebook: www.facebook.com/ISACAHQ
Instagram: www.instagram.com/isacanews

ISBN 978-1-60420-505-3
CISM® Review Questions, Answers & Explanations Manual 9th Edition
Printed in the United States of America

PREFACE

ISACA is pleased to offer the 1,000 questions in this *CISM® Review Questions, Answers & Explanations Manual 9th Edition*. The purpose of this manual is to provide the CISM candidate with sample questions and testing topics to help prepare and study for the CISM exam.

This manual consists of 1,000 multiple-choice study questions, answers and explanations, which are organized according to the newly revised (effective 2017) CISM job practice domains. These questions, answers and explanations are intended to introduce CISM candidates to the types of questions that may appear on the CISM exam. They are not actual questions from the exam. Many of these items appeared in previous editions of the *CISM® Review Questions, Answers & Explanations Manual* and/or supplements, but many have been rewritten or enhanced to be more representative of actual exam items and to provide further clarity or reflect a change in practice. The 1,000 questions are sorted by CISM domains. Additionally, 150 questions have been extracted to provide a sample exam with questions in the same proportion as the current CISM job practice. The candidate also may want to obtain a copy of the *CISM® Review Manual 15th Edition*, which provides the foundational knowledge of a CISM.

A job practice study is conducted at least every five years to ensure that the CISM certification is current and relevant. Further details regarding the new job practice can be found in the section titled New—CISM Job Practice.

ISACA has produced this publication as an educational resource to assist individuals preparing to take the CISM exam. It was produced independently from the CISM Certification Working Group, which has no responsibility for its content. Copies of past exams are not released to the public and are not made available to candidates. ISACA makes no representations or warranties whatsoever with regard to these or other ISACA or IT Governance Institute publications assuring candidates' passage of the CISM exam.

ISACA wishes you success with the CISM exam. Your commitment to pursuing the leading certification for information security managers is exemplary, and we welcome your comments and suggestions on the use and coverage of this manual. Once you have completed the exam, please take a moment to complete the online evaluation that corresponds to this publication *(www.isaca.org/studyaidsevaluation)*. Your observations will be invaluable as new questions, answers and explanations are prepared.

ACKNOWLEDGMENTS

The *CISM® Review Questions, Answers & Explanations Manual 9th Edition* is the result of the collective efforts of many volunteers over the past several years. ISACA members from throughout the global IS audit and control profession participated, generously offering their talents and expertise. This international team exhibited a spirit and selflessness that has become the hallmark of contributors to this valuable manual. Their participation and insight are truly appreciated.

NEW—CISM JOB PRACTICE

BEGINNING IN 2017, THE CISM EXAM WILL TEST THE NEW CISM JOB PRACTICE.

An international job practice analysis is conducted at least every five years or sooner to maintain the validity of the CISM certification program. A new job practice forms the basis of the CISM exam beginning in 2017.

The primary focus of the job practice is the current tasks performed and the knowledge used by CISMs. By gathering evidence of the current work practice of CISMs, ISACA is able to ensure that the CISM program continues to meet the high standards for the certification of professionals throughout the world.

The findings of the CISM job practice analysis are carefully considered and directly influence the development of new test specifications to ensure that the CISM exam reflects the most current best practices.

The new 2017 job practice reflects the areas of study to be tested and is compared below to the previous job practice. The complete CISM job practice can be found at *www.isaca.org/cismjobpractice*.

Previous CISM Job Practice	New 2017 CISM Job Practice
Domain 1: Information Security Governance (24%)	**Domain 1: Information Security Governance (24%)**
Domain 2: Information Risk Management and Compliance (33%)	**Domain 2: Information Risk Management (30%)**
Domain 3: Information Security Program Development and Management (25%)	**Domain 3: Information Security Program Development and Management (27%)**
Domain 4: Information Security Incident Management (18%)	**Domain 4: Information Security Incident Management (19%)**

Page intentionally left blank

TABLE OF CONTENTS

Page intentionally left blank

INTRODUCTION

OVERVIEW

The CISM exam evaluates a candidate's practical knowledge, including experience and application, of the job practice domains. We recommend that the exam candidate look to multiple resources to prepare for the exam, including the *CISM® Review Manual and Questions, Answers & Explanation (QAE) Manual* or database along with external publications. This section will cover some tips for studying for the exam and how best to use this QAE Manual in conjunction with other resources.

GETTING STARTED

Having adequate time to prepare for the CISM exam is critical. Most candidates spend between three and six months studying prior to taking the exam. Make sure you set aside a designated time each week to study, which you may wish to increase as your exam date approaches.

Developing a plan for your study efforts can also help you make the most effective use of your time prior to taking the exam.

CISM Self-assessment

In order to effectively study for the CISM exam, you should first identify the job practice areas in which you are weak. A good starting point is the CISM self-assessment, available at *www.isaca.org/Certification/CISM-Certified-Information-Security-Manager/Prepare-for-the-Exam/Pages/CISM-Self-Assessment.aspx.*

This 50-question sample exam is based on the question distribution of the CISM exam and can provide you with a high-level evaluation of your areas of need. When you complete the self-assessment, you will receive a summary of how you performed in each of the four job practice domains. You can use this summary to review the task and knowledge statements in the job practice and get an idea of where you should primarily focus your study efforts.

ABOUT THIS MANUAL

The *CISM QAE Manual* provides questions similar to those found on the CISM exam. They are developed using the task and knowledge statements as described in the CISM job practice.

This manual consists of 1,000 multiple-choice questions, answers and explanations (numbered S1-1, S1-2, etc.). These questions are selected and provided in two formats.

Questions Sorted by Domain

Questions, answers and explanations are provided (sorted) by the four CISM job practice domains. This allows the CISM candidate to refer to specific questions to evaluate comprehension of the topics covered within each domain. These questions are representative of CISM questions, although they are not actual exam items. They are provided to assist the CISM candidate in understanding the material in the *CISM® Review Manual 15th Edition* and to depict the type of question format typically found on the CISM exam. The numbers of questions, answers and explanations provided in the four domain chapters in this publication provide the CISM candidate with a maximum number of study questions.

Sample Exam

A random sample exam of 150 of the questions is also provided in this manual. **This exam is organized according to the domain percentages specified in the CISM job practice and used on the CISM exam:**

Information Security Governance..........24 percent
Information Risk Management..........30 percent
Information Security Program Development and Management..........27 percent
Information Security Incident Management..........19 percent

Candidates are urged to use this sample exam and the answer sheets provided to simulate an actual exam. There are two primary ways this sample exam may be used. The first is as a pretest, which is taken prior to any additional study. The sample exam in the QAE Manual is the same length as the actual CISM exam, as opposed to the CISM self-assessment, which is an abbreviated self-assessment tool. The pretest can help you to determine weaker domains. It can also help to orient you to the types of questions you may encounter in your study and during the exam.

The second way to use the sample exam is as a posttest. This will help you to determine the effectiveness of your study efforts as you approach the exam date. The results of this posttest can help you to focus on domains and task/knowledge statements that may require some additional review prior to taking the exam.

Sample exam answer sheets have been provided for both uses. In addition, a sample exam answer/reference key is included. These sample exam questions have been cross-referenced to the questions, answers and explanations by domain, so it is convenient to refer to the explanations of the correct answers. This publication is ideal to use in conjunction with the *CISM® Review Manual 15th Edition*.

It should be noted that the *CISM® Review Questions, Answers & Explanations Manual 9th Edition* has been developed to assist the CISM candidate in studying and preparing for the CISM exam. As you use this publication to prepare for the exam, please note that it covers a broad spectrum of information security management issues. Do not assume that reading and working the questions in this manual will fully prepare you for the exam. Because exam questions often relate to practical experience, it is recommended that you refer to your own experience and to other publications referred to in the *CISM® Review Manual 15th Edition*. These additional references are an excellent source of further detailed information and clarification. It is recommended that candidates evaluate the job practice domains in which they feel weak, or require a further understanding, and study accordingly.

Also, please note that this publication has been written using standard American English.

TYPES OF QUESTIONS ON THE CISM EXAM

CISM exam questions are developed with the intent of measuring and testing practical knowledge and the application of information security managerial principles and standards. As previously mentioned, all questions are presented in a multiple choice format and are designed for one best answer.

The candidate is cautioned to read each question carefully. Many times a CISM exam question will require the candidate to choose the appropriate answer that is **MOST** likely or **BEST**, or the candidate may be asked to choose a practice or procedure that would be performed **FIRST** related to the other answers. In every case, the candidate is required to read the question carefully, eliminate known wrong answers and then make the best choice possible. Knowing that these types of questions are asked and how to study to answer them will go a long way toward answering them correctly. The best answer is of the choices provided. There can be many potential solutions to the scenarios posed in the questions, depending on industry, geographical location, etc. It is advisable to consider the information provided in the question and to determine the best answer of the options provided.

Each CISM question has a stem (question) and four options (answer choices). The candidate is asked to choose the correct or best answer from the options. The stem may be in the form of a question or incomplete statement. In some instances, a scenario or description also may be included. These questions normally include a description of a situation and require the candidate to answer two or more questions based on the information provided.

A helpful approach to these questions includes the following:
- Read the entire stem and determine what the question is asking. Look for key words such as “BEST,” “MOST,” “FIRST,” etc., and key terms that may indicate what domain or concept that is being tested.
- Read all of the options, and then read the stem again to see if you can eliminate any of the options based on your immediate understanding of the question.
- Re-read the remaining options and bring in any personal experience to determine which is the best answer to the question.

Another condition the candidate should consider when preparing for the exam is to recognize that information security is a global profession, and individual perceptions and experiences may not reflect the more global position or circumstance. Because the exam and CISM manuals are written for the international information security community, the candidate will be required to be somewhat flexible when reading a condition that may be contrary to the candidate’s experience. It should be noted that CISM exam questions are written by experienced information security managers from around the world. Each question on the exam is reviewed by ISACA’s CISM Exam Item Development Working Group, which consists of international members. This geographic representation ensures that all exam questions are understood equally in every country and language.

> **Note:** ISACA study materials are living documents. As technology advances, these materials will be updated to reflect such advances. Further updates to this document before the date of the exam may be viewed at *www.isaca.org/studyaidupdates*.

Any suggestions to enhance the manual or questions related to the contents should be sent to *studymaterials@isaca.org*.

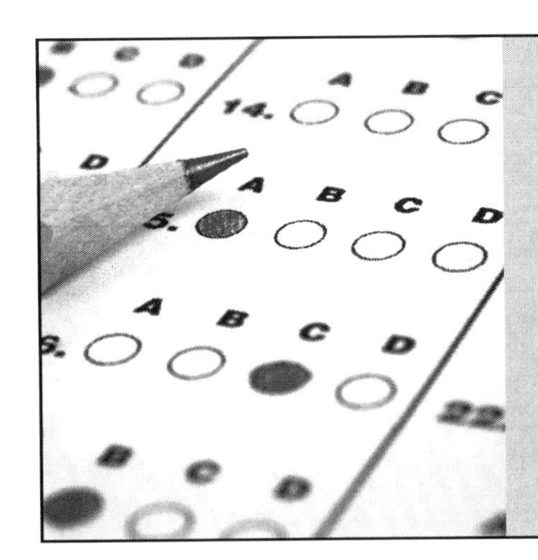

PRETEST

If you wish to take a pretest to determine strengths and weaknesses, the Sample Exam begins on page 409 and the pretest answer sheet begins on page 433. You can score your pretest with the Sample Exam Answer and Reference Key on page 431.

Page intentionally left blank

QUESTIONS, ANSWERS AND EXPLANATIONS BY DOMAIN

DOMAIN 1—INFORMATION SECURITY GOVERNANCE (24%)

S1-1 Which of the following steps should be **FIRST** in developing an information security plan?

A. Perform a technical vulnerabilities assessment.
B. Analyze the current business strategy.
C. Perform a business impact analysis.
D. Assess the current levels of security awareness.

B is the correct answer.

Justification:

A. Technical vulnerabilities as a component of risk will be most relevant in the context of threats to achieving the business objectives defined in the business strategy.
B. An information security manager needs to gain an understanding of the current business strategy and direction to understand the organization's objectives and the impact of the other answers on achieving those objectives.
C. A business impact analysis should be performed prior to developing a business continuity plan, but this would not be an appropriate first step in developing an information security plan because it focuses on availability, which is also primarily relevant in terms of the business objectives that are the basis of the strategy.
D. Without understanding the business strategy, it will not be possible to determine the current level of awareness because to be effective, awareness must include understanding the context and threats to the organization's business objectives.

S1-2 Senior management commitment and support for information security can **BEST** be obtained through presentations that:

A. use illustrative examples of successful attacks.
B. explain the technical risk to the organization.
C. evaluate the organization against good security practices.
D. tie security risk to key business objectives.

D is the correct answer.

Justification:

A. Senior management will not be as interested in examples of successful attacks if they are not tied to the impact on business environment and objectives.
B. Senior management will not be as interested in technical risk to the organization if it is not tied to the impact on business environment and objectives.
C. Industry good practices may be important to senior management to the extent they are relevant to the organization and its business objectives.
D. Senior management wants to understand the business justification for investing in security in relation to achieving key business objectives.

S1-3 The **MOST** appropriate role for senior management in supporting information security is the:

A. evaluation of vendors offering security products.
B. assessment of risk to the organization.
C. approval of policy statements and funding.
D. developing standards sufficient to achieve acceptable risk.

C is the correct answer.

Justification:

A. Evaluation of vendors is a day-to-day responsibility of the information security manager. In some organizations, business management may be involved in vendor evaluation, but their primary role is setting the organization's direction, oversight and governance.
B. Assessment of risk is a day-to-day responsibility of the information security manager.
C. Policies are a statement of senior management intent and direction. Therefore, senior management must approve them in addition to providing sufficient funding to achieve the organization's risk management objectives.
D. The development of standards that meet the policy intent is typically a function of the information security manager.

S1-4 Which of the following would be the **BEST** indicator of effective information security governance within an organization?

A. The steering committee approves security projects.
B. Security policy training is provided to all managers.
C. Security training is available to all employees on the intranet.
D. IT personnel are trained in testing and applying required patches.

A is the correct answer.

Justification:

A. The existence of a steering committee that approves all security projects would be an indication of the existence of a good governance program. To ensure that all stakeholders impacted by security considerations are involved, many organizations use a steering committee comprised of senior representatives of affected groups. This composition helps to achieve consensus on priorities and trade-offs and serves as an effective communication channel for ensuring the alignment of the security program with business objectives.
B. Security policy training is important at all levels of the organization and also an indicator of good governance. However, it must be guided and approved as a security project by the steering committee to ensure all parts of the organization are aware of the policies.
C. The availability of security training, while beneficial to the overall security program, does not ensure that employees are following the program and have the required level of awareness without a process to enforce awareness and compliance.
D. Even organizations with little overall governance may be effective in patching systems in a timely manner; this is not an indication of effective governance.

S1-5 Information security governance is **PRIMARILY** driven by:

A. technology constraints.
B. regulatory requirements.
C. litigation potential.
D. business strategy.

D is the correct answer.

Justification:

A. Strategy is the plan to achieve the business objectives of the organization that must be supported by governance. While technology constraints must be considered in developing governance and planning the strategy, it is not the driver.
B. Regulatory requirements must be addressed by governance and may affect how the strategy develops. However, regulatory requirements are not the driver of information security governance.
C. Litigation potential is usually an aspect of liability and is also a consideration for governance and when designing the strategy, but it may be a constraint, not a driver.
D. Business strategy is the main determinant of information security governance because security must align with the business objectives set forth in the business strategy.

S1-6 What is the **MOST** essential attribute of an effective key risk indicator (KRI)? The KRI:

A. is accurate and reliable.
B. provides quantitative metrics.
C. indicates required action.
D. is predictive of a risk event.

D is the correct answer.

Justification:

A. Key risk indicators (KRIs) usually signal developing risk but do not indicate what the actual risk is. In that context, they are neither accurate nor reliable.
B. KRIs typically do not provide quantitative metrics about risk.
C. KRIs will not indicate that any particular action is required other than to investigate further.
D. A KRI should indicate that a risk is developing or changing to show that investigation is needed to determine the nature and extent of a risk.

S1-7 Investments in information security technologies should be based on:

A. vulnerability assessments.
B. value analysis.
C. business climate.
D. audit recommendations.

B is the correct answer.

Justification:

A. Vulnerability assessments are useful, but they do not determine whether the cost of the technology is justified.
B. Investments in security technologies should be based on a value analysis and a sound business case.
C. Demonstrated value takes precedence over the current business because the climate is continually changing.
D. Basing decisions on audit recommendations alone would be reactive in nature and might not address the key business needs comprehensively.

S1-8 Determining which element of the confidentiality, integrity and availability (CIA) triad is **MOST** important is a necessary task when:

A. assessing overall system risk.
B. developing a controls policy.
C. determining treatment options.
D. developing a classification scheme.

B is the correct answer.

Justification:

A. Overall risk is not affected by determining which element of the triad is of greatest importance because overall risk is constructed from all known risk, regardless of the components of the triad to which each risk applies.

B. Because preventive controls necessarily must fail in either an open or closed state (i.e., fail safe or fail secure), and failing open favors availability while failing closed favors confidentiality—each at the expense of the other—a clear prioritization of the triad components is needed to develop a controls policy.

C. Although it is feasible that establishing a control that bolsters one component of the triad may diminish another, treatment options may be determined without a clear prioritization of the triad.

D. Classification is based on the potential impact of compromise and is not a function of prioritization within the confidentiality, integrity and availability (CIA) triad.

S1-9 Which of the following is characteristic of centralized information security management?

A. More expensive to administer
B. Better adherence to policies
C. More responsive to business unit needs
D. Faster turnaround of requests

B is the correct answer.

Justification:

A. Centralized information security management is generally less expensive to administer due to the economies of scale.

B. Centralization of information security management results in greater uniformity and better adherence to security policies.

C. With centralized information security management, information security is typically less responsive to specific business unit needs.

D. With centralized information security management, turnaround can be slower due to greater separation and more bureaucracy between the information security department and end users.

S1-10 Successful implementation of information security governance will **FIRST** require:

A. security awareness training.
B. updated security policies.
C. a computer incident management team.
D. a security architecture.

B is the correct answer.

Justification:

A. Security awareness training will promote the security policies, procedures and appropriate use of the security mechanisms but will not precede information security governance implementation.

B. Updated security policies are required to align management business objectives with security processes and procedures. Management objectives translate into policy; policy translates into standards and procedures.

C. An incident management team will not be the first requirement for the implementation of information security governance and can exist even if formal governance is minimal.

D. Information security governance provides the basis for architecture and must be implemented before a security architecture is developed.

S1-11 Which of the following individuals would be in the **BEST** position to sponsor the creation of an information security steering group?

A. Information security manager
B. Chief operating officer
C. Internal auditor
D. Legal counsel

B is the correct answer.

Justification:

A. Sponsoring the creation of the steering committee should be initiated by someone versed in the strategy and direction of the business. Because a security manager is looking to this group for direction, he/she is not in the best position to oversee the formation of this group.

B. The chief operating officer (COO) is highly placed within an organization and has the most knowledge of business operations and objectives. Sponsoring the creation of the steering committee should be initiated by someone versed in the strategy and direction of the business, such as the COO.

C. The internal auditor is an appropriate member of a steering group but would not oversee the formation of the committee.

D. Legal counsel is an appropriate member of a steering group but would not oversee the formation of the committee.

S1-12 Which of the following factors is the **MOST** significant in determining an organization's risk appetite?

A. The nature and extent of threats
B. Organizational policies
C. The overall security strategy
D. The organizational culture

D is the correct answer.

Justification:
A. Knowledge of the threat environment is constantly changing.
B. Policies are written in support of business objectives and parameters, including risk appetite.
C. Risk appetite is an input to the security strategy because the strategy is partly focused on mitigating risk to acceptable levels.
D. The extent to which the culture is risk adverse or risk aggressive, along with the objective ability of the organization to recover from loss, is the main factor in risk appetite.

S1-13 Which of the following attributes would be **MOST** essential to developing effective metrics?

A. Easily implemented
B. Meaningful to the recipient
C. Quantifiably represented
D. Meets regulatory requirements

B is the correct answer.

Justification:
A. Ease of implementation is valuable when developing metrics, but not essential. Metrics are most effective when they are meaningful to the person receiving the information.
B. Metrics will only be effective if the recipient can take appropriate action based upon the results.
C. Quantifiable representations can be useful, but qualitative measures are often just as useful.
D. Meeting legal and regulatory requirements may be important, but this is not always essential when developing metrics for meeting business goals.

S1-14 Which of the following is **MOST** appropriate for inclusion in an information security strategy?

A. Business controls designated as key controls
B. Security processes, methods, tools and techniques
C. Firewall rule sets, network defaults and intrusion detection system settings
D. Budget estimates to acquire specific security tools

B is the correct answer.

Justification:
A. Key business controls are only one part of a security strategy and must be related to business objectives.
B. A set of security objectives supported by processes, methods, tools and techniques together are the elements that constitute a security strategy.
C. Firewall rule sets, network defaults and intrusion detection system settings are technical details subject to periodic change and are not appropriate content for a strategy document.
D. Budgets will generally not be included in an information security strategy. Additionally, until the information security strategy is formulated and implemented, specific tools will not be identified and specific cost estimates will not be available.

S1-15 An information security manager can **BEST** attain senior management commitment and support by emphasizing:

A. organizational risk.
B. performance metrics.
C. security needs.
D. the responsibilities of organizational units.

A is the correct answer.

Justification:

A. **Information security exists to address risk to the organization that may impede achieving its objectives. Organizational risk will be the most persuasive argument for management commitment and support.**

B. Establishing metrics to measure security status will be viewed favorably by senior management after the overall organizational risk is identified.

C. The information security manager should identify information security needs based on organizational needs. Organizational or business risk should always take precedence.

D. Identifying organizational responsibilities will be most effective if related directly to addressing organizational risk.

S1-16 Which of the following roles would represent a conflict of interest for an information security manager?

A. Evaluation of third parties requesting connectivity
B. Assessment of the adequacy of disaster recovery plans
C. Final approval of information security policies
D. Monitoring adherence to physical security controls

C is the correct answer.

Justification:

A. Evaluation of third parties requesting connectivity is an acceptable practice and does not present any conflict of interest.

B. Assessment of disaster recovery plans is an acceptable practice and does not present any conflict of interest.

C. **Because senior management is ultimately responsible for information security, it should approve information security policy statements; the information security manager should not have final approval.**

D. Monitoring of adherence to physical security controls is an acceptable practice and does not present any conflicts of interest.

X **S1-17** Which of the following situations must be corrected **FIRST** to ensure successful information security governance within an organization?

A. The information security department has difficulty filling vacancies.
B. The chief operating officer approves security policy changes.
C. The information security oversight committee only meets quarterly.
D. The data center manager has final sign-off on all security projects.

D is the correct answer.

Justification:

A. Difficulty in filling vacancies is not uncommon due to the shortage of qualified information security professionals.
B. It is important to have senior management, such as the chief operating officer, approve security policies to ensure they meet management intent and direction.
C. It is not inappropriate for an oversight or steering committee to meet quarterly.
D. A steering committee should be in place to approve all security projects. The fact that the data center manager has final sign-off for all security projects indicates that a steering committee is not being used and that information security is relegated to a subordinate place in the organization. This would indicate a failure of information security governance.

S1-18 Which of the following requirements would have the **LOWEST** level of priority in information security?

A. Technical
B. Regulatory
C. Privacy
D. Business

A is the correct answer.

Justification:

A. Information security priorities may, at times, override technical specifications, which then must be rewritten to conform to minimum security standards.
B. Regulatory requirements are government-mandated and, therefore, not subject to override.
C. Privacy requirements are usually government-mandated and, therefore, not subject to override.
D. The needs of the business should always take precedence in deciding information security priorities.

S1-19 Which of the following is **MOST** likely to be discretionary?

A. Policies
B. Procedures
C. Guidelines
D. Standards

C is the correct answer.

Justification:

A. Policies define management's security goals and expectations for an organization. These are defined in more specific terms within standards and procedures.
B. Procedures describe how work is to be done.
C. Guidelines provide recommendations that business management must consider in developing practices within their areas of control; as such, they are discretionary.
D. Standards establish the allowable operational boundaries for people, processes and technology.

S1-20 Security technologies should be selected **PRIMARILY** on the basis of their:

A. ability to mitigate business risk.
B. evaluations in trade publications.
C. use of new and emerging technologies.
D. benefits in comparison to their costs.

D is the correct answer.

Justification:

A. The most fundamental evaluation criterion for the appropriate selection of any security technology is its ability to reduce or eliminate business risk but only if the cost is acceptable.
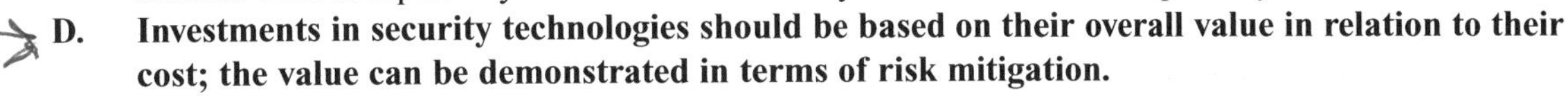
B. The technology's ability to cost-effectively mitigate risk for a particular organization takes precedence over how it is evaluated in trade publications.
C. While new or emerging technologies may offer potential benefits, the lack of being time tested reduces their acceptability in critical areas and by itself will not be the primary selection basis.
D. Investments in security technologies should be based on their overall value in relation to their cost; the value can be demonstrated in terms of risk mitigation.

S1-21 Which of the following are seldom changed in response to technological changes?

A. Standards
B. Procedures
C. Policies
D. Guidelines

C is the correct answer.

Justification:

A. Security standards must be revised and updated based on the impact of technology changes.
B. Procedures must be revised and updated based on the impact of technology or standards changes.
C. Policies are high-level statements of management intent and direction, which is not likely to be affected by technology changes.
D. Guidelines must be revised and updated based on the impact of technology changes.

S1-22 When creating an effective data-protection strategy, the information security manager must understand the flow of data and its protection at various stages. This is **BEST** achieved with:

A. a third-party vulnerability assessment.
B. a tailored methodology based on exposure.
C. an insurance policy for accidental data losses.
D. a tokenization system set up in a secure network environment.

B is the correct answer.

Justification:

A. Vulnerability assessments, third-party or otherwise, do not take into account threat and other factors that influence risk treatment.
B. Organizations classify data according to their value and exposure. The organization can then develop a sensible plan to invest budget and effort where they matter most.
C. An insurance policy is a risk treatment option for the transfer/sharing of risk. Whether it is an appropriate action requires a cost-benefit analysis and a more complete understanding of the risk involved.
D. Tokenization is a technique used to protect data, but whether it is appropriate cannot be known without an understanding of the various exposures to which the data are subject.

S1-23 Which of the following is characteristic of decentralized information security management across a geographically dispersed organization?

A. More uniformity in quality of service
B. Better adherence to policies
C. Better alignment to business unit needs
D. More savings in total operating costs

C is the correct answer.

Justification:

A. Uniformity in quality of service tends to vary from unit to unit.
B. Adherence to policies is likely to vary considerably between various business units.
C. Decentralization of information security management generally results in better alignment to business unit needs because security management is closer to the end user.
D. Decentralization of information security management is generally more expensive to administer due to the lack of economies of scale.

X **S1-24** Which of the following is the **MOST** appropriate position to sponsor the design and implementation of a new security infrastructure in a large global enterprise?

A. Chief security officer
B. Chief operating officer
C. Chief privacy officer
D. Chief legal counsel

B is the correct answer.

Justification:

A. Although the chief security officer knows what is needed, the sponsor for this task should be someone with far-reaching influence across the organization.
B. The chief operating officer is most knowledgeable of business operations and objectives.
C. The chief privacy officer may not have the knowledge of the day-to-day business operations and overall security requirements to ensure proper guidance.
D. The chief legal counsel will typically have a narrow legal focus on contracts and stock and other regulatory requirements and have little knowledge of overall organizational security requirements.

X **S1-25** The **MOST** important element(s) to consider when developing a business case for a project is the:

A. feasibility and value proposition.
B. resource and time requirements.
C. financial analysis of benefits.
D. alignment with organizational objectives.

A is the correct answer.

Justification:

A. Feasibility and whether the value proposition makes sense will be major considerations of whether a project will proceed.
B. Resources and time needed are important but will be a component of the value proposition in terms of costs.
C. Financial analysis of benefits is a component of the value proposition, but there would typically be other benefits that should be proposed.
D. The value proposition would, as a matter of course, have to include alignment with the organization's objectives.

S1-26 Acceptable levels of information security risk should be determined by:

A. legal counsel.
B. security management.
C. external auditors.
D. the steering committee.

D is the correct answer.

Justification:

A. Legal counsel is not in a position to determine what levels of business risk the organization is willing to assume.
B. An acceptable level of risk in an organization is a business decision, not a security decision.
C. External auditors can point out areas of risk but are not in a position to determine what levels of risk the organization is willing to assume.
D. Senior management, represented in the steering committee, has ultimate responsibility for determining what levels of risk the organization is willing to assume.

S1-27 The **PRIMARY** goal of developing an information security strategy is to:

A. establish security metrics and performance monitoring.
B. educate business process owners regarding their duties.
C. ensure that legal and regulatory requirements are met.
D. support the business objectives of the organization.

D is the correct answer.

Justification:

A. Establishing metrics and performance monitoring is very important to the extent they indicate the achievement of business objectives, but this is only one aspect of the primary requirement to support business objectives.
B. Educating business process owners is subordinate to supporting the business objectives and is only incidental to developing an information security strategy.
C. Meeting legal and regulatory requirements is just one of the objectives of the strategy needed to support business objectives.
D. The purpose of information security in an organization is to assist the organization in achieving its objectives, and it is the primary goal of an information security strategy.

S1-28 Senior management commitment and support for information security can **BEST** be enhanced through:

A. a formal security policy sponsored by the chief executive officer.
B. regular security awareness training for employees.
C. periodic review of alignment with business management goals.
D. senior management sign-off on the information security strategy.

C is the correct answer.

Justification:

A. Although having the chief executive officer sign-off on the security policy makes for good visibility and demonstrates good tone at the top, it is a one-time discrete event that may be quickly forgotten by senior management.
B. Security awareness training for employees will not have as much effect on senior management commitment as alignment with business goals.
C. Ensuring that security activities continue to be aligned and support business goals is critical to obtaining management support.
D. Although having senior management sign-off on the security strategy makes for good visibility and demonstrates good tone at the top, it is a one-time discrete event that may be quickly forgotten by senior management.

S1-29 Which of the following activities **MOST** commonly falls within the scope of an information security governance steering committee?

A. Interviewing candidates for information security specialist positions
B. Developing content for security awareness programs
C. Prioritizing information security initiatives
D. Approving access to critical financial systems

C is the correct answer.

Justification:

A. Interviewing specialists should be performed by the information security manager.
B. Development of program content should be performed by the information security staff.
C. Prioritizing information security initiatives falls within the scope of an information security governance committee.
D. Approving access to critical financial systems is the responsibility of individual system data owners.

S1-30 Which of the following is the **MOST** important factor when designing information security architecture?

A. Technical platform interfaces
B. Scalability of the network
C. Development methodologies
D. Stakeholder requirements

D is the correct answer.

Justification:

A. Interoperability is important but without merit if a technologically elegant solution is achieved that does not meet the needs of the business.
B. Scalability is important but only to the extent the architecture meets stakeholder requirements.
C. There are a number of viable developmental methodologies, and the choice of which is used is not particularly important as long as it meets the needs of the organization.
D. The most important factor for information security is that it advances the interests of the business, as defined by stakeholder requirements.

S1-31 An information security manager receives a report showing an increase in the number of security events. The **MOST** likely explanation is:

A. exploitation of a vulnerability in the information system.
B. threat actors targeting the organization in greater numbers.
C. failure of a previously deployed detective control.
D. approval of a new exception for noncompliance by management.

A is the correct answer.

Justification:

A. Exploitation of a vulnerability is likely to generate security events.

B. Absent a change in vulnerability, an increase in the number of threat actors targeting the organization would not explain an increase in security events.

C. An increase in the number of security events that appear on reports suggests that detective controls are likely working properly.

D. Exceptions approved by management may result in a higher number of security events on reports if notice of the exceptions is not provided to information security to allow updates to monitoring. However, exceptions are typically communicated to the information security manager, so this is an unlikely explanation for the increase.

S1-32 Which of the following is the **MOST** appropriate task for a chief information security officer to perform?

A. Update platform-level security settings.
B. Conduct disaster recovery test exercises.
C. Approve access to critical financial systems.
D. Develop an information security strategy.

D is the correct answer.

Justification:

A. Updating platform-level security settings would typically be performed by lower-level personnel because this is a basic administrative task.

B. Conducting recovery test exercises would typically be performed by operational personnel.

C. Approving access would be the job of the data owner.

D. Developing a strategy for information security would be the most appropriate task for the chief information security officer.

S1-33 When an information security manager is developing a strategic plan for information security, the time line for the plan should be:

A. aligned with the IT strategic plan.
B. based on the current rate of technological change.
C. three to five years for both hardware and software.
D. aligned with the business strategy.

D is the correct answer.

Justification:

A. Any planning for information security should be properly aligned with the needs of the business, not necessarily the IT strategic plan.

B. Technology needs should not come before the needs of the business.

C. Planning should not be done on an artificial timetable that ignores business needs.

D. Any planning for information security should be properly aligned with the needs of the business.

S1-34 Which of the following is the **MOST** important information to include in a strategic plan for information security?

A. Information security staffing requirements
B. Current state and desired future state
C. IT capital investment requirements
D. Information security mission statement

B is the correct answer.

Justification:

A. Staffing requirements stem from the implementation time lines and requirements of the strategic plan.
B. It is most important to present a vision for the future and then create a road map from the current state to the desired future state based on a gap analysis of the requirements to achieve the desired or future state.
C. IT capital investment requirements are generally not determined at the strategic plan level but rather as a result of gap analysis and the options on how to achieve the objectives of the strategic plan.
D. The mission statement is typically a short, high-level aspirational statement of overall organizational objectives and only directly affects the information security strategy in a very limited way.

S1-35 Information security projects should be prioritized on the basis of:

A. time required for implementation.
B. impact on the organization.
C. total cost for implementation.
D. mix of resources required.

B is the correct answer.

Justification:

A. Time required for implementation is potentially one impact on the organization but is subordinate to the overall impact of the project on the organization.
B. Information security projects should be assessed on the basis of the positive impact that they will have on the organization.
C. Total cost for implementation is just one aspect of the impact of the project on the organization.
D. A mix of resources required is not particularly relevant to prioritizing security projects.

X **S1-36** Which of the following would **BEST** prepare an information security manager for regulatory reviews?

A. Assign an information security administrator as regulatory liaison.
B. Perform self-assessments using regulatory guidelines and reports.
C. Assess previous regulatory reports with process owners input.
D. Ensure all regulatory inquiries are sanctioned by the legal department.

B is the correct answer.

Justification:

A. Directing regulators to a specific person or department is not as effective as performing self-assessments.
B. Self-assessments provide the best feedback on readiness and permit identification of items requiring remediation.
C. Assessing previous regulatory reports is not as effective as performing self-assessments because conditions may have changed.
D. The legal department should review all formal inquiries, but this does not help prepare for a regulatory review.

S1-37 From an information security manager perspective, what is an immediate benefit of clearly defined roles and responsibilities?

A. Enhanced policy compliance
B. Improved procedure flows
C. Segregation of duties
D. Better accountability

D is the correct answer.

Justification:
A. Defining roles and responsibilities does not by itself improve policy compliance without proper monitoring and enforcement of accountability.
B. Procedure flows are not necessarily affected by defining roles and responsibilities.
C. Segregation of duties is more likely to occur as a result of policy compliance enforcement than simply defining roles and responsibilities, although that is a necessary first step.
D. Defining roles and responsibilities makes it clear who is accountable for performance and outcomes.

S1-38 Which of the following roles is responsible for legal and regulatory liability?

A. Chief security officer
B. Chief legal counsel
C. Board of directors and senior management
D. Information security steering group

C is the correct answer.

Justification:
A. The chief security officer is not individually liable for failures of security in the organization.
B. The chief legal counsel is not individually liable for failures of security in the organization.
C. The board of directors and senior management are ultimately responsible for ensuring regulations are appropriately addressed.
D. The information security steering group is not individually liable for failures of security in the organization.

S1-39 While implementing information security governance, an organization should **FIRST**:

A. adopt security standards.
B. determine security baselines.
C. define the security strategy.
D. establish security policies.

C is the correct answer.

Justification:
A. Adopting suitable security standards that implement the intent of the policies follows the development of policies that support the strategy.
B. Security baselines are established as a result of determining acceptable risk, which should be defined as a requirement prior to strategy development.
C. Security governance must be developed to meet and support the objectives of the information security strategy.
D. Policies are a primary instrument of governance and must be developed or modified to support the strategy.

S1-40 The **MOST** basic requirement for an information security governance program is to:

A. be aligned with the corporate business strategy.
B. be based on a sound risk management approach.
C. provide adequate regulatory compliance.
D. provide good practices for security initiatives.

A is the correct answer.

Justification:

A. To be effective and receive senior management support, an information security program must be aligned with the corporate business strategy.
B. An otherwise sound risk management approach may be of little benefit to an organization unless it specifically addresses and is consistent with the organization's business strategy.
C. The governance program must address regulatory requirements that affect that particular organization to an extent determined by management, but this is not the most basic requirement.
D. Good practices are generally a substitute for specific knowledge of the organization's requirements and may be excessive for some and inadequate for others.

S1-41 Information security policy enforcement is the responsibility of the:

A. security steering committee.
B. chief information officer.
C. chief information security officer.
D. chief compliance officer.

C is the correct answer.

Justification:

A. The security steering committee should ensure that a security policy is in line with corporate objectives but typically is not responsible for enforcement.
B. The chief information officer may to some extent be involved in the enforcement of the policy but is not directly responsible for it.
C. Information security policy enforcement is generally the responsibility of the chief information security officer.
D. The chief compliance officer is usually involved in determining the level of compliance but is usually not directly involved in the enforcement of the policy.

S1-42 An information security manager at a global organization has to ensure that the local information security program will initially be in compliance with the:

A. corporate data privacy policy.
B. data privacy policy where data are collected.
C. data privacy policy of the headquarters' country.
D. data privacy directive applicable globally.

B is the correct answer.

Justification:

A. The corporate data privacy policy, being internal, cannot supersede the local law.
B. As a subsidiary, the local entity will have to comply with the local law for data collected in the country. Senior management will be accountable for compliance.
C. In case of data collected locally (and potentially transferred to a country with a different data privacy regulation), the local law applies, not the law applicable to the head office. The data privacy laws are country-specific.
D. With local regulations differing from the country in which the organization is headquartered, it is improbable that a groupwide policy will address all the local legal requirements. The data privacy laws are country-specific.

S1-43 Segregation of duties (SoD) has been designed and introduced into an accounts payable system. Which of the following should be in place to **BEST** maintain the effectiveness of SoD?

A. A strong password rule is assigned to disbursement staff.
B. Security awareness is publicized by the compliance department.
C. An operational role matrix is aligned with the organizational chart.
D. Access privilege is reviewed when an operator's role changes.

D is the correct answer.

Justification:

A. Password strength is important for each staff member, but complexity of passwords does not ensure segregation of duties (SoD).
B. Effective SoD is not based on self-governance, so security awareness is inadequate.
C. It is not uncommon for staff to have ancillary roles beyond what is shown on the organizational chart, so aligning a role matrix with the organizational chart is not sufficiently granular to maintain the effectiveness of SoD.
D. In order to maintain the effectiveness of SoD established in an application system, user access privilege must be reviewed whenever an operator's role changes. If this effort is neglected, there is a risk that a single staff member could acquire excessive operational capabilities. For instance, if a cash disbursement staff member accidentally acquires a trade input role, this person is technically able to accomplish illegal payment operation.

S1-44 Information security frameworks can be **MOST** useful for the information security manager because they:

A. provide detailed processes and methods.
B. are designed to achieve specific outcomes.
C. provide structure and guidance.
D. provide policy and procedure.

C is the correct answer.

Justification:
A. Frameworks are general structures rather than detailed processes and methods.
B. Frameworks do not specify particular outcomes but may provide the structure to assess outcomes against requirements.
C. Frameworks are like a skeleton; they provide the outlines and basic structure but not the specifics of process and outcomes.
D. Frameworks do not specify policies and procedures. Their creation is left to the implementer.

S1-45 Business goals define the strategic direction of the organization. Functional goals define the tactical direction of a business function. Security goals define the security direction of the organization. What is the **MOST** important relationship between these concepts?

A. Functional goals should be derived from security goals.
B. Business goals should be derived from security goals.
C. Security goals should be derived from business goals.
D. Security and business goals should be defined independently from each other.

C is the correct answer.

Justification:
A. Functional goals and security goals need to be aligned at the operational level, but neither is derived from the other.
B. Security is not an end in itself, but it should serve the overall business goals.
C. Security goals should be developed based on the overall business strategy. The business strategy is the most important steering mechanism for directing the business and is defined by the highest management level.
D. If security goals are defined independently from business goals, the security function would not support the overall business strategy or it might even hinder the achievement of overall business objectives.

S1-46 A security manager is preparing a report to obtain the commitment of executive management to a security program. Inclusion of which of the following items would be of **MOST** value?

A. Examples of genuine incidents at similar organizations
B. Statement of generally accepted good practices
C. Associating realistic threats to corporate objectives
D. Analysis of current technological exposures

C is the correct answer.

Justification:
A. While examples of incidents to other organizations may help obtain senior management buy-in, it should be based on realistic threats to the organization's corporate objectives.
B. Good practices are rarely useful, although they may enhance senior management buy-in. However, this is not as substantial as realistic threats to the organization's corporate objectives.
C. Linking realistic threats to key business objectives will direct executive attention to them.
D. Analysis of current technological exposures may enhance senior management buy-in but is not as substantial as realistic threats to the organization's corporate objectives.

S1-47 The **PRIMARY** concern of an information security manager documenting a formal data retention policy is:

A. generally accepted industry good practices.
B. business requirements.
C. legislative and regulatory requirements.
D. storage availability.

B is the correct answer.

Justification:
A. Good practices are rarely the most effective answer for a particular organization. They may be a useful guide but not a primary concern.
B. The primary concern will be business requirements that may include the regulatory issues management has decided to address.
C. Legislative and regulatory requirements are only relevant if compliance is a business need.
D. Storage is irrelevant because whatever is needed must be provided.

S1-48 Who in an organization has the responsibility for classifying information?

A. Data custodian
B. Database administrator
C. Information security officer
D. Data owner

D is the correct answer.

Justification:
A. The data custodian is responsible for securing the information in alignment with the data classification.
B. The database administrator carries out the technical administration of the database and handling requirements that apply to data in storage and transit in accordance with the protection standards for each classification.
C. The information security officer oversees the overall data classification and handling process to ensure conformance to policy and standards.
D. The data owner has responsibility for data classification consistent with the organization's classification criteria.

X **S1-49** What is the **PRIMARY** role of the information security manager related to the data classification and handling process within an organization?

A. Defining and ratifying the organization's data classification structure
B. Assigning the classification levels to the information assets
C. Securing information assets in accordance with their data classification
D. Confirming that information assets have been properly classified

A is the correct answer.

Justification:
A. Defining and ratifying the data classification structure consistent with the organization's risk appetite and the business value of information assets is the primary role of the information security manager related to the data classification and handling process within the organization.
B. The final responsibility for assigning the classification levels to information assets rests with the data owners.
C. The job of securing information assets is the responsibility of the data custodians.
D. Confirming proper classification of information assets may be a role of an information security manager performing security reviews or of IS auditors based on the organization's classification criteria.

S1-50 Which of the following is **MOST** important in developing a security strategy?

A. Creating a positive security environment
B. Understanding key business objectives
C. Having a reporting line to senior management
D. Allocating sufficient resources to information security

B is the correct answer.

Justification:
A. A positive security environment (culture) enables successful implementation of the security strategy but is not as important as alignment with business objectives during the development of the strategy.
B. Alignment with business strategy is essential in determining the security needs of the organization; this can only be achieved if key business objectives driving the strategy are understood.
C. A reporting line to senior management may be helpful in developing a strategy but does not ensure an understanding of business objectives necessary for strategic alignment.
D. Allocation of resources is not likely to be effective if the business objectives are not well understood.

S1-51 Who is ultimately responsible for an organization's information?

A. Data custodian
B. Chief information security officer
C. Board of directors
D. Chief information officer

C is the correct answer.

Justification:
A. The data custodian is responsible for the maintenance and protection of data. This role is usually filled by the IT department.
B. The chief information security officer is responsible for security and carrying out senior management's directives.
C. Responsibility for all organizational assets, including information, falls to the board of directors, which is tasked with responding to issues that affect the information's protection.
D. The chief information officer is responsible for information technology within the organization but is not ultimately legally responsible for an organization's information.

S1-52 An information security manager mapping a job description to types of data access is **MOST** likely to adhere to which of the following information security principles?

A. Ethics
B. Proportionality
C. Integration
D. Accountability

B is the correct answer.

Justification:

A. Ethics have the least to do with mapping a job description to types of data access.
B. Information security controls, including access, should be proportionate to the criticality and/or sensitivity of the asset (i.e., the potential impact of compromise).
C. Principles of integration are not relevant to mapping a job description to types of data access.
D. The principle of accountability would be the second most adhered to principle because people with access to data may not always be accountable.

S1-53 Which of the following is the **MOST** important prerequisite for establishing information security management within an organization?

A. Senior management commitment
B. Information security framework
C. Information security organizational structure
D. Information security policy

A is the correct answer.

Justification:

A. Senior management commitment is necessary in order for each of the other elements to succeed. Without senior management commitment, the other elements will likely be ignored within the organization.
B. Without senior management commitment, an information security framework is not likely to be implemented.
C. Without senior management commitment, it is not likely that there is support for developing an information security organizational structure.
D. The development of effective policies as a statement of management intent and direction is likely to be inadequate without senior management commitment to information security.

S1-54 What will have the **HIGHEST** impact on standard information security governance models?

A. Number of employees
B. Distance between physical locations
C. Complexity of organizational structure
D. Organizational budget

C is the correct answer.

Justification:
A. The number of employees has little or no effect on standard information security governance models.
B. The distance between physical locations has little or no effect on standard information security governance models.
C. Information security governance models are highly dependent on the overall organizational structure. Some of the elements that impact organizational structure are multiple missions and functions across the organization, leadership, and lines of communication.
D. Organizational budget may have some impact on suitable governance models depending on the one chosen because some models will be more costly to implement.

S1-55 In order to highlight to management the importance of integrating information security in the business processes, a newly hired information security officer should **FIRST**:

A. prepare a security budget.
B. conduct a risk assessment.
C. develop an information security policy.
D. obtain benchmarking information.

B is the correct answer.

Justification:
A. Preparing a security budget follows risk assessment to determine areas of concern.
B. Risk assessment, analysis, evaluation and impact analysis will be the starting point for driving management's attention to information security.
C. Developing an information security policy is based on and follows risk assessment.
D. Benchmarking information will only be relevant after a risk assessment has been performed for comparison purposes.

S1-56 How should an information security manager balance the potentially conflicting requirements of an international organization's security standards with local regulation?

A. Give organizational standards preference over local regulations.
B. Follow local regulations only.
C. Make the organization aware of those standards where local regulations causes conflicts.
D. Negotiate a local version of the organization standards.

D is the correct answer.

Justification:
A. Organizational standards must be subordinate to local regulations.
B. It would be incorrect to follow local regulations only because there must be recognition of organizational requirements.
C. Making an organization aware of standards is a sensible step but is not a complete solution.
D. Negotiating a local version of the organization's standards is the most effective compromise in this situation.

S1-57 The **FIRST** step in developing an information security management program is to:

A. identify business risk that affects the organization.
B. establish the need for creating the program.
C. assign responsibility for the program.
D. assess adequacy of existing controls.

B is the correct answer.

Justification:
A. The task of identifying business risk that affects the organization is assigned and acted on after establishing the need for creating the program.
B. In developing an information security management program, the first step is to establish the need for creating the program. This is a business decision based more on judgment than on any specific quantitative measures. The other choices are assigned and acted on after establishing the need.
C. The task of assigning responsibility for the program is assigned and acted on after establishing the need for creating the program.
D. The task of assessing the adequacy of existing controls is assigned and acted on after establishing the need for creating the program.

S1-58 Which of the following should an information security manager **PRIMARILY** use when proposing the implementation of a security solution?

A. Risk assessment report
B. Technical evaluation report
C. Business case
D. Budgetary requirements

C is the correct answer.

Justification:
A. The risk assessment report provides the rationale for the business case for implementing a particular security solution.
B. The technical evaluation report provides supplemental information for the business case.
C. The information security manager needs to have knowledge of the development of business cases to illustrate the costs and benefits, or value proposition, of the various security solutions.
D. Budgetary requirements provides part of the information required in the business case.

S1-59 To justify its ongoing information security budget, which of the following would be of **MOST** use to the information security department?

A. Security breach frequency
B. Annual loss expectancy
C. Cost-benefit analysis
D. Peer group comparison

C is the correct answer.

Justification:
A. The frequency of information security breaches may assist in justifying the budget but is not the key tool because it does not address the benefits from the budget expenditure.
B. Annual loss expectancy does not address the potential benefit of information security investment.
C. Cost-benefit analysis is the best way to justify budget.
D. Peer group comparison would provide support for the necessary information security budget, but it would not take into account the specific needs and activities of the organization.

S1-60 Which of the following situations would **MOST** inhibit the effective implementation of security governance?

A. The complexity of technology
B. Budgetary constraints
C. Conflicting business priorities
D. Lack of high-level sponsorship

D is the correct answer.

Justification:
A. Complexity of technology should be factored into the governance model of the organization but is not likely to have a major effect on security governance.
B. Budget constraints will inhibit effective implementation of security governance but is likely to be a consequence of the lack of high-level sponsorship and, therefore, secondary.
C. Conflicting business priorities must be addressed by senior management in order to implement effective security governance, which will be more likely to be accomplished with high-level sponsorship.
D. The need for senior management involvement and support is a key success factor for the implementation of appropriate security governance.

S1-61 To achieve effective strategic alignment of information security initiatives, it is important that:

A. steering committee leadership rotates among members.
B. major organizational units provide input and reach a consensus.
C. the business strategy is updated periodically.
D. procedures and standards are approved by all departmental heads.

B is the correct answer.

Justification:
A. Rotation of steering committee leadership does not help in achieving strategic alignment.
B. It is important to achieve consensus on risk and controls and obtain inputs from various organizational entities because security must be aligned with the needs of the various parts of the organization.
C. Updating business strategy does not lead to strategic alignment of security initiatives.
D. Procedures and standards do not need to be approved by ALL departmental heads.

S1-62 In implementing information security governance, the information security manager is **PRIMARILY** responsible for:

A. developing the security strategy.
B. reviewing the security strategy.
C. communicating the security strategy.
D. approving the security strategy.

A is the correct answer.

Justification:

A. **The information security manager is responsible for developing a security strategy based on business objectives with the help of business process owners.**
B. Reviewing the security strategy is the responsibility of a steering committee or management.
C. The information security manager is not necessarily responsible for communicating the security strategy.
D. Management must approve and fund the security strategy implementation.

S1-63 The **MOST** useful way to describe the objectives in the information security strategy is through:

X

A. attributes and characteristics of the desired state.
B. overall control objectives of the security program.
C. mapping the IT systems to key business processes.
D. calculation of annual loss expectations.

A is the correct answer.

Justification:

A. **Security strategy will typically cover a wide variety of issues, processes, technologies and outcomes that can best be described by a set of characteristics and attributes that are desired.**
B. Control objectives are a function of acceptable risk determination and one part of strategy development but at a high level, best described in terms of desired outcomes.
C. Mapping IT to business processes must occur as one part of strategy implementation but is too specific to describe general strategy objectives.
D. Calculation of annual loss expectations would not describe the objectives in the information security strategy.

S1-64 Which of the following choices is the **MOST** likely cause of significant inconsistencies in system configurations?

A. A lack of procedures
B. Inadequate governance
C. Poor standards
D. Insufficient training

B is the correct answer.

Justification:

A. A lack of proper procedures may well be the issue, but that is a failure of governance. Good governance would ensure that procedures are consistent with standards that meet policy intent. Procedures for configuration that meet standards for a particular security domain will be consistent.
B. Governance is the rules the organization operates by and the oversight to ensure compliance as well as feedback mechanisms that provide assurance that the rules are followed. A failure of one or more of those processes is likely to be the reason that system configurations are inconsistent.
C. Poor standards are also a sign of inadequate governance and likely to result in poor consistency in configurations.
D. Insufficient training indicates that there are no requirements, the requirements are not being met or the trainers are not competent in the subject matter, which is also a lack of effective governance resulting in a lack of oversight, clear requirements for training or a lack of suitable metrics.

S1-65 The **MOST** important characteristic of good security policies is that they:

A. state expectations of IT management.
B. state only one general security mandate.
C. are aligned with organizational goals.
D. govern the creation of procedures and guidelines.

C is the correct answer.

Justification:

A. Stating expectations of IT management omits addressing overall organizational goals and objectives.
B. Stating only one general security mandate is the next best option because policies should be clear; otherwise, policies may be confusing and difficult to understand and enforce.
C. The most important characteristic of good security policies is that they are aligned with organizational goals. Failure to align policies and goals makes them ineffective and potentially misleading in governing the creation of standards and procedures.
D. Policies govern the creation of standards, which in turn, govern the development of procedures.

S1-66 An information security manager must understand the relationship between information security and business operations in order to:

A. support organizational objectives.
B. determine likely areas of noncompliance.
C. assess the possible impacts of compromise.
D. understand the threats to the business.

A is the correct answer.

Justification:

A. Security exists to provide a level of predictability for operations, support for the activities of the organization and to ensure preservation of the organization. Business operations must be the driver for security activities in order to set meaningful objectives, determine and manage the risk to those activities, and provide a basis to measure the effectiveness of and provide guidance to the security program.

B. Regulatory compliance may or may not be an organizational requirement. If compliance is a requirement, some level of compliance must be supported, but compliance is only one aspect.

C. It is necessary to understand the business goals in order to assess potential impacts, but this is just one way in which security supports organizational objectives.

D. It is necessary to understand the business goals in order to evaluate threats, but this is just one way in which security supports organizational objectives.

S1-67 The **MOST** effective approach to address issues that arise between IT management, business units and security management when implementing a new security strategy is for the information security manager to:

A. escalate issues to an external third party for resolution.
B. ensure that senior management provide authority for security to address the issues.
C. insist that managers or units not in agreement with the security solution accept the risk.
D. refer the issues to senior management along with any security recommendations.

D is the correct answer.

Justification:

A. Senior management would be responsible for escalating issues to relevant external third parties.

B. Security is only one aspect of issues that may arise between parts of the organization, and authority to address issues must be at a higher level to arbitrate between conflicting requirements.

C. It is unlikely that the security manager is in a position to insist that business units accept risk, and the matter should be escalated to senior management to arbitrate conflicting requirements.

D. Senior management is in the best position to arbitrate issues between organizational units because they will look at the overall needs of the business in reaching a decision.

S1-68 Obtaining senior management support for establishing a warm site can **BEST** be accomplished by:

A. establishing a periodic risk assessment.
B. promoting regulatory requirements.
C. developing a business case.
D. developing effective metrics.

C is the correct answer.

Justification:

A. A risk assessment should be included in the business case but by itself will not be as effective in gaining management support.
B. Informing management of regulatory requirements may help gain support for initiatives, but given that many organizations are not in compliance with regulations, it is unlikely to be sufficient.
C. A complete business case, including a cost-benefit analysis, will be most persuasive to management.
D. Good metrics that provide assurance that initiatives are meeting organizational goals will also be useful but are likely to be insufficient in gaining management support.

S1-69 Which of the following elements is **MOST** important when developing an information security strategy?

A. Defined objectives
B. Time frames for delivery
C. Adoption of a control framework
D. Complete policies

A is the correct answer.

Justification:

A. Without defined objectives, a strategy—the plan to achieve objectives—cannot be developed.
B. Time frames for delivery are important but not critical for inclusion in the strategy document.
C. The adoption of a control framework is not critical prior to developing an information security strategy.
D. Policies are developed subsequent to, and as a part of, implementing a strategy.

X1

S1-70 There is a concern that lack of detail in the recovery plan may prevent an organization from meeting its required time objectives when a security incident strikes. Which of the following is **MOST** likely to ensure the recovery time objectives would be met?

A. Establishment of distributed operation centers
B. Delegation of authority in recovery execution
C. Outsourcing of the business restoration process
D. Incremental backup of voluminous databases

B is the correct answer.

Justification:

A. Establishment of distributed operation centers does not compensate for a lack of detail in the recovery plan.
B. When recovery is underway in response to an incident, there are many cases where decisions need to be made at each management level. This may take up considerable time due to escalation procedures. Therefore, it is desirable that delegation of authority becomes effective during the recovery process. Scope of delegation of authority in recovery execution may be assessed beforehand and documented in business continuity policies and procedures.
C. Outsourcing will not resolve any failure to meet recovery time objectives, unless the recovery strategy includes a clear line of authority and adequate detail in the plan.
D. Incremental backup of voluminous databases may be recommended to expedite the data backup process. However, it generally increases the time needed to recover.

S1-71 Which of the following is the **BEST** justification to convince management to invest in an information security program?

A. Cost reduction
B. Compliance with company policies
C. Protection of business assets
D. Increased business value

D is the correct answer.

Justification:
A. Cost reduction by itself is rarely the motivator for implementing an information security program.
B. Compliance is secondary to business value.
C. Increasing business value may include protection of business assets.
D. Investing in an information security program should increase business value as a result of fewer business disruptions, fewer losses and increased productivity.

S1-72 Achieving compliance with a particular information security standard selected by management would **BEST** be described as a:

A. key goal indicator.
B. critical success factor.
C. key performance indicator.
D. business impact analysis.

C is the correct answer.

Justification:
A. A key goal indicator defines a clear objective sought by an organization. A key goal indicator is defined as a measure that tells management, after the fact, whether an IT process has achieved its business requirements; usually expressed in terms of information criteria.
B. Critical success factors are steps that must be achieved to accomplish high-level goals. A critical success factor is defined as the most important issue or action for management to achieve control over and within its IT processes.
C. A key performance indicator indicates how well a process is progressing according to expectations. Another definition for a key performance indicator is a measure that determines how well the process is performing in enabling the goal to be reached.
D. A business impact analysis defines risk impact; its main purpose is not to achieve compliance. It is defined as an exercise that determines the impact of losing the support of any resource to an enterprise, establishes the escalation of that loss over time, identifies the minimum resources needed to recover, and prioritizes the recovery of processes and the supporting system.

S1-73 Which of the following choices is the **MOST** important consideration when developing the security strategy of a company operating in different countries?

A. Diverse attitudes toward security by employees and management
B. Time differences and the ability to reach security officers
C. A coherent implementation of security policies and procedures in all countries

D. Compliance with diverse laws and governmental regulations

D is the correct answer.

Justification:

A. Attitudes among employees and managers may vary by country, and this will impact implementation of a security policy. However, the impact is not nearly as significant as the variance in national laws.
B. Time differences and reachability are not significant considerations when developing a security strategy.
C. Implementation occurs after a security strategy has been developed.
D. In addition to laws varying from one country to another, they can also conflict, making it difficult for an organization to create an overarching enterprise security policy that adequately addresses the requirements in each nation. The repercussions of failing to adhere to multiple legal frameworks at the same time go well beyond the impacts of the other considerations listed.

S1-74 Which of the following **BEST** contributes to the development of an information security governance framework that supports the maturity model concept?

A. Continuous analysis, monitoring and feedback
B. Continuous monitoring of the return on security investment
C. Continuous risk reduction
D. Key risk indicator setup to security management processes

A is the correct answer.

Justification:

A. To improve the governance framework and achieve a higher level of maturity, an organization needs to conduct continuous analysis, monitoring and feedback comparing the desired state of maturity to the current state.
B. Return on security investment may show the performance result of the security-related activities in terms of cost-effectiveness; however, this is not an indication of maturity level.
C. Continuous risk reduction would demonstrate the effectiveness of the security governance framework but does not indicate a higher level of maturity.
D. Key risk indicator setup is a tool to be used in internal control assessment, and it presents a threshold to alert management when controls are being compromised in business processes. This is a control tool rather than a maturity model support tool.

S1-75 The **MOST** complete business case for security solutions is one that:

A. includes appropriate justification.
B. explains the current risk profile.
C. details regulatory requirements.
D. identifies incidents and losses.

A is the correct answer.

Justification:

A. There are a number of possible justifications for implementing a security solution. The key is to choose the most appropriate justification, which could be any of the other choices.
B. The current risk profile may be one of a number of possible justifications.
C. Regulatory requirements may be one of many possible justifications.
D. Incidents and losses are possible justifications for implementing security solutions.

S1-76 Which of the following choices is a necessary attribute of an effective information security governance framework?

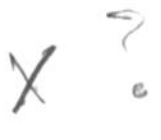

A. An organizational structure with minimal conflicts of interest, with sufficient resources and defined responsibilities
B. Organizational policies and guidelines in line with predefined procedures
C. Business objectives aligned with a predefined security strategy
D. Security guidelines that address multiple facets of security such as strategy, regulatory compliance and controls

A is the correct answer.

Justification:

A. An information security framework will help ensure the protection of information assets from confidentiality, integrity and availability perspectives. Organizational structures that minimize conflicts of interest are important for this to work effectively.
B. Organizational procedures and guidelines must be aligned with policies.
C. The security strategy must be aligned with business objectives.
D. The security policy addresses multiple facets of security.

S1-77 Which of the following actions would help to change an organization's security culture?

A. Develop procedures to enforce the information security policy.
B. Obtain strong management support.
C. Implement strict technical security controls.
D. Periodically audit compliance with the information security policy.

B is the correct answer.

Justification:

A. Procedures will support an information security policy, but this is not likely to have much impact on the culture of the organization.
B. Because culture in an organization is a reflection of senior management whether intentional or accidental, only management support and pressure will help to change an organization's culture.
C. Technical controls will provide more security to an information system and staff; however, this does not mean the culture will be changed.
D. Auditing will help to ensure the effectiveness of the information security policy; however, auditing is not effective in changing the culture of the company.

S1-78 An enterprise has been recently subject to a series of denial-of-service attacks due to a weakness in security. The information security manager needs to present a business case for increasing the investment in security. The **MOST** significant challenge in obtaining approval from senior management for the proposal is:

A. explaining technology issues of security.
B. demonstrating value and benefits.
C. simulating various risk scenarios.
D. obtaining benchmarking data for comparison.

B is the correct answer.

Justification:
A. In a business case, there is no need to explain technology issues of security.
B. Business cases are prepared on the basis of business value and benefits to the organization.
C. Simulating various risk scenarios must be done as part of risk assessment, but these are not to be included in a business case.
D. Benchmarking data is an approach to value demonstration.

? X **S1-79** The enactment of policies and procedures for preventing hacker intrusions is an example of an activity that belongs to:

A. risk management.
B. compliance.
C. IT management.
D. governance.

D is the correct answer.

Justification:
A. Risk management is about identifying risk and adequate countermeasures and would be concerned if such policies and procedures are necessary based on a risk analysis. However, the enactment would not fall into the area of risk management.
B. Compliance would be concerned with the adequacy of the policies and procedures to achieve the control objectives and whether employees act according to the policies and procedures.
C. IT management would be concerned about setting the policies into operation (e.g., by providing training and resources).
D. Governance is concerned with implementing adequate mechanisms for ensuring that organizational goals and objectives can be achieved. Policies and procedures are common governance mechanisms.

S1-80 An IS manager has decided to implement a security system to monitor access to the Internet and prevent access to numerous sites. Immediately upon installation, employees flood the IT help desk with complaints of being unable to perform business functions on Internet sites. This is an example of:

A. conflicting security controls with organizational needs.
B. strong protection of information resources.
C. implementing appropriate controls to reduce risk.
D. proving information security's protective abilities.

A is the correct answer.

Justification:

A. The needs of the organization were not taken into account, so there is a conflict.
B. This example is not strong protection.
C. A control that significantly restricts the ability of users to do their job is not appropriate.
D. Proving protection abilities at an unacceptable cost or performance is a poor strategy. This does not prove the ability to protect, but proves the ability to interfere with business.

S1-81 An organization's information security strategy should be based on:

A. managing risk relative to business objectives.
B. managing risk to a zero level and minimizing insurance premiums.
C. avoiding occurrence of risk so that insurance is not required.
D. transferring most risk to insurers and saving on control costs.

A is the correct answer.

Justification:

A. Organizations must manage risk to a level that is acceptable for their business model, goals and objectives.
B. A zero-level approach may be costly and not provide the effective benefit of additional revenue to the organization. Long-term maintenance of this approach may not be cost-effective.
C. Risk varies as business models and geography, regulatory and operational processes change.
D. Insurance is generally used to protect against low-probability high-impact events and requires that the organization have certain operational controls to mitigate risk in place in addition to generally high deductibles. Therefore, transferring most risk is not cost-effective.

S1-82 Which of the following should be included in an annual information security budget that is submitted for management approval?

A. A cost–benefit analysis of budgeted resources
B. All of the resources that are recommended by the business
C. Total cost of ownership
D. Baseline comparisons

A is the correct answer.

Justification:

A. A brief explanation of the benefit of expenditures in the budget helps to convey the context of how the purchases that are being requested meet goals and objectives, which in turn helps build credibility for the information security function or program. Explanations of benefits also help engage senior management in the support of the information security program.

B. While the budget should consider all inputs and recommendations that are received from the business, the budget that is ultimately submitted to management for approval should include only those elements that are intended for purchase.

C. Total cost of ownership may be requested by management and may be provided in an addendum to a given purchase request, but it is not usually included in an annual budget.

D. Baseline comparisons (cost comparisons with other companies or industries) may be useful in developing a budget or providing justification in an internal review for an individual purchase but would not be included with a request for budget approval.

S1-83 Which of the following choices is the **BEST** attribute of key risk indicators?

A. High flexibility and adaptability
B. Consistent methodologies and practices
C. Robustness and resilience
D. The cost-benefit ratio

B is the correct answer.

Justification:

A. High flexibility and adaptability are commendable attributes but do not provide a consistent baseline on which to determine significant deviations.

B. Effective key risk indicators result from the deviation from baselines, and consistent methodologies and practices establish the baseline.

C. Robustness and resilience are commendable attributes but do not provide a consistent baseline on which to determine significant deviations.

D. The cost-benefit ratio does not indicate risk.

S1-84 The data access requirements for an application should be determined by the:

A. legal department.
B. compliance officer.
C. information security manager.
D. business owner.

D is the correct answer.

Justification:

A. The legal department can advise but does not have final responsibility.

B. The compliance officer can advise but does not have final responsibility.

C. The information security manager can advise but does not have final responsibility.

D. Business owners are ultimately responsible for their applications.

S1-85 The **PRIMARY** purpose of an information security program is to:

A. provide protection to information assets consistent with business strategy and objectives.
B. express the results of an operational risk assessment in terms of business impact.
C. protect the confidentiality of business information and technology resources.
D. develop information security policy and procedures in line with business objectives.

A is the correct answer.

Justification:
A. The information security program must be primarily aligned with the business's strategy and objectives.
B. The risk assessment program is focused on identifying risk scenarios based on the business's strategy and objectives. It must consider all risk, not just operational risk.
C. The protection of the information security program needs to address confidentiality, integrity and availability, not just confidentiality.
D. Security policy and procedures will be developed as part of the security program to achieve protection for information assets consistent with business strategy and objectives.

S1-86 Effective governance of enterprise security is **BEST** ensured by:

A. using a bottom-up approach.
B. management by the IT department.
C. referring the matter to the organization's legal department.
D. using a top-down approach.

D is the correct answer.

Justification:
A. Focus on the regulatory issues and management priorities may not be reflected effectively by a bottom-up approach.
B. Governance of enterprise security affects the entire organization and is not a matter concerning only the management of IT.
C. The legal department is part of the overall governance process and may provide useful input but cannot take full responsibility.
D. Effective governance of enterprise security needs to be a top-down initiative, with the board and executive management setting clear policies, goals and objectives and providing for ongoing monitoring of the same.

S1-87 The **FIRST** step to create an internal culture that embraces information security is to:

A. implement stronger controls.
B. conduct periodic awareness training.
C. actively monitor operations.
D. gain endorsement from executive management.

D is the correct answer.

Justification:
A. The implementation of stronger controls may lead to circumvention.
B. Awareness training is important but must be based on policies and supported by management.
C. Actively monitoring operations will not directly affect culture.
D. Endorsement from executive management in the form of policy approval provides intent, direction and support.

S1-88 Which of the following recommendations is the **BEST** one to promote a positive information security governance culture within an organization?

A. Strong oversight by the audit committee
B. Organizational governance transparency
C. Collaboration across business lines
D. Positive governance ratings by stock analysts

C is the correct answer.

Justification:
A. Supervision by the audit committee is unlikely to occur and would be of little help.
B. Governance transparency may contribute to the security management practice but is not directly linked to the establishment of a positive governance culture.
C. To promote a positive governance culture, it is essential to establish collaboration across business lines. In this way, line management will speak a common language and share the same goals.
D. Positive governance ratings by stock analysts may be useful for investors but will have little effect on internal organizational culture.

S1-89 A risk assessment and business impact analysis (BIA) have been completed for a major proposed purchase and new process for an organization. There is disagreement between the information security manager and the business department manager who will be responsible for evaluating the results and identified risk. Which of the following would be the **BEST** approach of the information security manager?

A. Acceptance of the business manager's decision on the risk to the corporation
B. Acceptance of the information security manager's decision on the risk to the corporation
C. Review of the risk assessment with executive management for final input
D. Create a new risk assessment and BIA to resolve the disagreement

C is the correct answer.

Justification:
A. The business manager is likely to be focused on getting the business done as opposed to the risk posed to the organization.
B. The typical information security manager is focused on risk, and on average, he/she will overestimate risk by about 100 percent—usually considering worst case scenarios rather than the most probable events.
C. Executive management will be in the best position to consider the big picture and the trade-offs between security and functionality in the entire organization.
D. There is no indication that the assessments are inadequate or defective in some way; therefore, repeating the exercise is not warranted.

S1-90 Who is accountable for ensuring that information is categorized and that specific protective measures are taken?

A. The security officer
B. Senior management
C. The end user
D. The custodian

B is the correct answer.

Justification:
A. The security officer supports and implements information security to achieve senior management objectives.
B. Routine administration of all aspects of security is delegated, but top management must retain overall accountability.
C. The end user does not perform categorization.
D. The custodian supports and implements information security measures as directed.

S1-91 Which of the following is the **MOST** appropriate as a means of obtaining commitment from senior management for implementation of the information security strategy?

A. Educational material discussing the importance of good information security practices
B. Regular group meetings to review the challenges and requirements of daily operations
C. A cost-benefit analysis detailing how the requested implementation budget will be used
D. A formal presentation highlighting the relationship between security and business goals

D is the correct answer.

Justification:

A. Education regarding good information security practices is a part of implementing an effective information security program and is best distributed across the workforce.
B. Senior managers are unlikely to be able to accommodate regular group meetings, and daily operations are better addressed at the level of business process owners.
C. A detailed cost-benefit analysis of the implementation budget is a supporting document that can be referenced to answer questions and establish credibility, not the means by which an information security manager obtains commitment from senior management to implementation of the information security strategy.
D. A formal presentation to senior management is used as a means to educate and communicate key aspects of the overall security program and how security is a necessary enabler for the achievement of business goals.

S1-92 Information security should:

A. focus on eliminating all risk.
B. balance technical and business requirements.
C. be driven by regulatory requirements.
D. be defined by the board of directors.

B is the correct answer.

Justification:

A. It is not practical or feasible to eliminate all risk.
B. Information security should ensure that business objectives are met given available technical capabilities, resource constraints and compliance requirements.
C. The extent of compliance with regulatory requirements is a business decision and must be defined by management.
D. Defining information security is an executive and operational function, not a board function.

S1-93 Which of the following choices would **BEST** align information security objectives to business objectives?

A. A capability maturity model
B. A process assessment model
C. A risk assessment and analysis
D. A business balanced scorecard

D is the correct answer.

Justification:

A. A capability maturity model is not as inclusive as a scorecard, does not provide as complete a perspective and is more focused on process.
B. While providing greater detail into processes and capabilities, a process assessment model still only provides a process-focused view rather than a multidimensional one.
C. A risk assessment is used to identify vulnerabilities and controls but does not address alignment.
D. A business balanced scorecard will align information security goals with the business goals and provides a multidimensional view of both quantitative and qualitative factors.

S1-94 What is the **MAIN** risk when there is no user management representation on the information security steering committee?

A. Functional requirements are not adequately considered.
B. User training programs may be inadequate.
C. Budgets allocated to business units are not appropriate.
D. Information security plans are not aligned with business requirements.

D is the correct answer.

Justification:

A. Functional requirements and user training programs are considered to be part of project development but are not the main risk.
B. Specifics of training programs are not normally under the purview of the steering committee.
C. The steering committee does not approve budgets for business units.
D. The steering committee usually controls the execution of the information security strategy, and lacking representation of user management, the committee may fail to consider impact on productivity and adequate user controls.

S1-95 Senior management is reluctant to budget for the acquisition of an intrusion prevention system. The chief information security officer should do which of the following activities?

A. Develop and present a business case for the project.
B. Seek the support of the users and information asset custodians.
C. Invite the vendor for a proof of concept demonstration.
D. Organize security awareness training for management.

A is the correct answer.

Justification:

A. Senior management needs to understand the link between the acquisition of an intrusion prevention system (IPS) and the organization's business objectives. A business case is the best way to present this information.
B. Stakeholder buy-in is an important part of the acquisition and implementation process, but senior management needs to see the value of budgeting for the purchase before anything else can happen.
C. Senior management probably believes that the IPS will do what it promises. What is lacking is an understanding of how and what the IPS does to provide value to the organization.
D. Security awareness training may provide some insight into the value of security tools in general, but the decision to allocate funds for an IPS will be made only on the basis of the specific value that the IPS provides.

X **S1-96** An organization that appoints a chief information security officer:

A. improves collaboration among the ranks of senior management.
B. acknowledges a commitment to legal responsibility for information security.
C. infringes on the governance role of the board of directors.
D. enhances the financial accountability of technology projects.

B is the correct answer.

Justification:

A. Whether senior managers collaborate is not substantially influenced by the presence or absence of a chief information security officer (CISO).
B. Appointing a CISO creates a clear line of responsibility for information security. The scope and breadth of information security today is such that the authority required and responsibility taken will inevitably end up with a senior manager, so not having a CISO does not prevent someone from being responsible, but it increases the potential for confusion over precisely who is responsible and may result in unrecognized liability. Accountability lies with the board of directors.
C. The board of directors retains its governance role regardless of whether a CISO is formally designated.
D. The CISO is typically not associated with financial accountability for technology projects.

S1-97 The director of auditing has recommended a specific information security monitoring solution to the information security manager. What should the information security manager do **FIRST**?

A. Obtain comparative pricing bids and complete the transaction with the vendor offering the best deal.
B. Add the purchase to the budget during the next budget preparation cycle to account for costs.
C. Perform an assessment to determine correlation with business goals and objectives.
D. Form a project team to plan the implementation.

C is the correct answer.

Justification:

A. Comparative pricing bids and completing the transaction with the vendor offering the best deal is not necessary until a determination has been made regarding whether the product fits the goals and objectives of the business.
B. Adding the purchase to the budget is not necessary until a determination has been made regarding whether the product fits the goals and objectives of the business.
C. An assessment must be made first to determine that the proposed solution is aligned with business goals and objectives.
D. Forming a project team for implementation is not necessary until a determination has been made regarding whether the product fits the goals and objectives of the business.

S1-98 When setting up an information classification scheme, the role of the information owner is to:

A. ensure that all data on an information system are protected according to the classification policy.
B. determine the classification of information across his/her scope of responsibility.
C. identify all information that requires backup according to its criticality and classification.
D. delegate the classification of information to responsible information custodians.

B is the correct answer.

Justification:

A. The information system owner is responsible for protecting information on an information system according to the policy and mandates of the information owner.
B. The role of the information owner is usually broader than one department or one system and requires that information be classified in a consistent manner across the owner's area of responsibility.
C. Ensuring backup of data is the role of the information custodian and operations group.
D. The information owner may delegate the classification to another responsible manager but not to the data custodian. The information owner remains accountable for the information classification even if the actual work effort is delegated.

S1-99 Which of the following choices would influence the content of the information security strategy to the **GREATEST** extent?

A. Emerging technology
B. System compromises
C. Network architecture
D. Organizational goals

D is the correct answer.

Justification:

A. Emerging technology may help bring an organization up to current standards, but it may not be part of the organization's goals or mission.
B. Handling or preventing system compromises is important but should not influence the overall strategy, and it is most likely to affect control design and implementation.
C. The network architecture must support the security strategy in support of organizational goals.
D. If the security strategy supports organizational goals, it is more likely to obtain funding and upper management support.

S1-100 Which of the following metrics will provide the **BEST** indication of organizational risk?

A. Annual loss expectancy
B. The number of information security incidents
C. The extent of unplanned business interruptions
D. The number of high-impact vulnerabilities

C is the correct answer.

Justification:

A. Annual loss expectancy is the quantification of loss exposure based on probability and frequency of outages with a known or estimated cost. It is part of a business impact analysis and may be calculated at the organization and/or system level, but it is based on projections rather than on observed data.
B. The number of recorded or recognized incidents does not reveal impact.
C. An unplanned business interruption is a standard measure because it provides a quantifiable measure of how much business may be lost due to the inability to acquire, process and produce results that affect the customer(s).
D. The number of high-impact vulnerabilities provides an indication of weakness within the information network and/or systems but is not by itself an indicator of risk.

S1-101 The formal declaration of organizational information security goals and objectives should be found in the:

A. information security procedures.
B. information security principles.
C. employee code of conduct.
D. information security policy.

D is the correct answer.

Justification:
A. Security procedures are usually detailed as step-by-step actions to ensure that activities meet a given standard.
B. Security principles are not always organizationally specific.
C. A code of conduct is a standard requirement that encompasses more than security.
D. The information security policy is management's formal declaration of security goals and objectives.

S1-102 Which of the following would be the **FIRST** step when developing a business case for an information security investment?

A. Defining the objectives
B. Calculating the cost
C. Defining the need
D. Analyzing the cost-effectiveness

C is the correct answer.

Justification:
A. Without a clear definition of the needs to be addressed, the objectives cannot be determined.
B. The objectives cannot be determined without a definition of the needs to be addressed; therefore, the costs to achieve the objectives cannot be determined before the needs are defined.
C. Without a clear definition of the needs to be fulfilled, the rest of the components of the business case cannot be determined.
D. Without a need requiring a solution, cost-effectiveness cannot be determined.

S1-103 Effective information security requires a combination of management, administrative and technical controls because:

A. technical controls alone are unable to adequately compensate for faulty processes.
B. senior management is unlikely to fund adequate deployment of technical controls.
C. the approach to addressing or treating specific risk has a significant impact on costs.
D. development of the right strategy needs to be iterative to achieve the desired state.

A is the correct answer.

Justification:
A. In addition to typically being less costly, processes are considerably more effective when flaws in a process are the source of risk. Attempting to counteract process flaws using technical controls will generally impose substantial restrictions on business operations and burden the organization with disproportionate cost without addressing the root cause of the problem.
B. Cost is always a consideration, and technical controls tend to be more costly than other types of controls. However, even with unlimited funding, the information security manager is unlikely to be able to adequately compensate for faulty processes solely by deploying technical controls.
C. While the approach to addressing or treating specific risk has a significant impact on cost, it does not explain why deploying technical controls alone cannot create and maintain information security.
D. Regardless of how many iterations of examination and deployment may occur, deployment solely of technical controls will not create and maintain information security.

X

S1-104 Which of the following is an indicator of effective governance?

A. A defined information security architecture
B. Compliance with international security standards
C. Periodic external audits
D. An established risk management program

D is the correct answer.

Justification:
A. A defined information security architecture is helpful but by itself is not a strong indicator of effective governance.
B. Compliance with international standards is not an indication of the use of effective governance.
C. Periodic external audits may serve to provide an opinion on effective governance.
D. A risk management program is a key component of effective governance.

S1-105 Which of the following **BEST** indicates senior management commitment toward supporting information security?

A. Assessment of risk to the assets
B. Approval of risk management methodology
C. Review of inherent risk to information assets
D. Review of residual risk for information assets

B is the correct answer.

Justification:
A. An assessment of risk to assets by itself does not indicate commitment and support.
B. Management sign-off on risk management methodology indicates support and commitment to effective information security.
C. A review of inherent risk is not an indication of commitment and support.
D. Reviewing residual risk may be a step in gaining commitment and support but by itself is not sufficient.

S1-106 Which of the following is the **MOST** effective approach to identify events that may affect information security across a large multinational enterprise?

A. Review internal and external audits to indicate anomalies.
B. Ensure that intrusion detection sensors are widely deployed.
C. Develop communication channels throughout the enterprise.
D. Conduct regular enterprisewide security reviews.

C is the correct answer.

Justification:
A. Audits are performed periodically, and a number of events can occur between audits that might not be detected and responded to in a timely manner.
B. Intrusion detection sensors are useful in detecting intrusion events but not other types of events.
C. Good communication channels can provide timely reporting of events across a large enterprise as well as providing channels for dissemination of security information.
D. Enterprisewide security reviews are an enormous task and will be, at best, periodic; therefore, they are unlikely to provide timely information on events.

S1-107 Maturity levels are an approach to determine the extent that sound practices have been implemented in an organization based on outcomes. Another approach that has been developed to achieve essentially the same result is:

A. controls applicability statements.
B. process performance and capabilities.
C. probabilistic risk assessment.
D. factor analysis of information risk.

B is the correct answer.

Justification:

A. A controls applicability statement identifies which risk controls are applied but is not directly related to performance or maturity.
B. The process performance and capabilities approach provides a more detailed perspective of maturity levels and serves essentially the same purpose.
C. Probabilistic risk assessment provides quantitative results of probability and magnitude of risk not related to performance or capabilities.
D. Factor analysis of information risk is an approach to assessing risk that does not address performance.

S1-108 The **PRIMARY** objective for information security program development should be:

A. creating an information security strategy.
B. establishing incident response procedures.
C. implementing cost-effective security solutions.
D. reducing the impact of risk on the business.

D is the correct answer.

Justification:

A. An information security strategy is important, but it is one of the ways to achieve the objective of reducing risk and impact.
B. Establishing incident response procedures is important, but it is one of the ways to achieve the objective of reducing risk and impact.
C. Cost-effective security solutions are essential but not the objective of security program development.
D. Reducing risk to and impact on the business is the most important objective of an information security program.

S1-109 During a stakeholder meeting, a question was asked regarding who is ultimately accountable for the protection of sensitive data. Assuming all of the following roles exist in the enterprise, which would be the **MOST** appropriate answer?

A. Security administrators
B. The IT steering committee
C. The board of directors
D. The information security manager

C is the correct answer.

Justification:

A. Security administrators are responsible for implementing, monitoring and enforcing security rules established and authorized by management, but they are not ultimately accountable for the protection of sensitive data.
B. The IT steering committee assists in the delivery of the IT strategy, oversees management of IT service delivery and IT projects, and focuses on implementation aspects, but it is not ultimately accountable for the protection of sensitive data.
C. The board of directors is ultimately accountable for information security as it is for all organizational assets.
D. The information security manager is responsible for identifying and explaining to stakeholders the risk to the organization's information, presenting alternatives for mitigation, and then implementing an approach supported by the organization. While authority is delegated by senior management, ultimate accountability for the protection of sensitive data rests with the board.

S1-110 Which of the following is the **MOST** important objective of an information security strategy review?

A. Ensuring that risk is identified, analyzed and mitigated to acceptable levels
B. Ensuring the information security strategy is aligned with organizational goals
C. Ensuring the best return on information security investments
D. Ensuring the efficient utilization of information security resources

B is the correct answer.

Justification:

A. Without alignment with business goals, the risk identified and mitigated as part of the information security strategy may not be the most significant to the business.
B. The most important part of an information security strategy is that it supports the business objectives and goals of the enterprise.
C. Maximizing return on information security investment can only be achieved if the information security strategy is aligned with the business strategy.
D. Efficient utilization of resources at the enterprise level can only be achieved if the information security strategy is aligned with the business strategy.

S1-111 Information security governance must be integrated into all business functions and activities **PRIMARILY** to:

A. maximize security efficiency.
B. standardize operational activities.
C. achieve strategic alignment.
D. address operational risk.

D is the correct answer.

Justification:

A. Efficiency is not necessarily an attribute of the integration of governance throughout the organization, but the effectiveness of the governance program to address and reduce business risk is such an attribute.
B. Standardization will help create a more efficient program, but it will not necessarily establish a risk mitigation process that will address operational risk to assist business in better managing risk functions and processes.
C. While good governance may help promote strategic alignment, the main reason to ensure integration of governance in all organizational functions is to prevent gaps in the management of risk and maintain acceptable risk levels throughout the organization.
D. All aspects of organizational activities pose risk that is mitigated through effective information security governance and the development and implementation of policies, standards and procedures.

X **S1-112** Which of the following choices is the **BEST** indicator of the state of information security governance?

A. A defined maturity level
B. A developed security strategy
C. Complete policies and standards
D. Low numbers of incidents

A is the correct answer.

Justification:

A. A defined maturity level is the best overall indicator of the state of information security governance. The maturity level indicates how mature a process is on a scale from 0 (incomplete process) to 5 (optimizing process).
B. A developed security strategy is an important first step, but it must be implemented properly to be effective and by itself is not an indication of the state of governance.
C. Complete policies and standards are required for effective governance but are only one part of the requirement. By themselves, they are not an indicator of the effectiveness of governance.
D. The number of incidents is relatively unconnected to the effectiveness of information security governance. Trends in incidents would be a better indicator.

S1-113 From an information security perspective, which of the following will have the **GREATEST** impact on a financial enterprise with offices in various countries and involved in transborder transactions?

A. Current and future technologies
B. Evolving data protection regulations
C. Economizing the costs of network bandwidth
D. Centralization of information security

B is the correct answer.

Justification:
A. Current and future technologies would be considered but will not generally be affected by operational regions or countries.
B. Information security laws vary from country to country and an enterprise must be aware of and comply with the applicable laws from each country.
C. Economizing the costs of network bandwidth would be considered but will not directly be affected by global operations.
D. Centralization of information security is not a significant factor in multinational operations.

S1-114 Strategic alignment is **PRIMARILY** achieved when services provided by the information security department:

A. reflect the requirements of key business stakeholders.
B. reflect the desires of the IT executive team.
C. reflect the requirements of industry good practices.
D. are reliable and cost-effective.

A is the correct answer.

Justification:
A. Information security exists to minimize business disruptions and support the achievement of organizational objectives. When the services provided reflect the requirements of key business stakeholders, there is alignment.
B. The IT executive team is just one of the stakeholders, and their desires may not reflect the requirements of the rest of the organization.
C. Good practices may be excessive or insufficient for a particular organization and do nothing to ensure alignment of information security and business.
D. Services should be reliable and cost-effective, but this does not ensure alignment with business objectives.

S1-115 The security responsibility of data custodians in an organization will include:

A. assuming overall protection of information assets.
B. determining data classification levels.
C. implementing security controls in products they install.
D. ensuring security measures are consistent with policy.

D is the correct answer.

Justification:
A. Senior management has overall responsibility for protection of the information assets.
B. Data owners determine data classification levels for information assets so that appropriate levels of controls can be provided to meet the requirements relating to confidentiality, integrity and availability.
C. Implementation of security controls in products is the responsibility of the IT developers.
D. Security responsibilities of data custodians within an organization include ensuring that appropriate security measures are maintained and are consistent with organizational policy and standards.

S1-116 Who can **BEST** approve plans to implement an information security governance framework?

A. Internal auditor
B. Information security management
C. Steering committee
D. Infrastructure management

C is the correct answer.

Justification:

A. An internal auditor is secondary to the authority and influence of senior management.
B. Information security management should not have the authority to approve the security governance framework.
C. Senior management that is part of the security steering committee is in the best position to approve plans to implement an information security governance framework.
D. Infrastructure management will not be in the best position because it focuses more on the technologies than on the business.

S1-117 What is the **MOST** important item to be included in an information security policy?

A. The definition of roles and responsibilities
B. The scope of the security program
C. The key objectives of the security program
D. Reference to procedures and standards of the security program

C is the correct answer.

Justification:

A. The definition of roles and responsibilities is part of implementing an information security governance framework.
B. The scope of the security program should be defined in the charter of the information security program.
C. Stating the objectives of the security program is the most important element to ensure alignment with business goals.
D. Reference to standards that interpret the policy may be included, but the multitude of procedures controlled by those standards would not normally be referenced.

S1-118 In an organization, information systems security is the responsibility of:

A. all personnel.
B. information systems personnel.
C. information systems security personnel.
D. functional personnel.

A is the correct answer.

Justification:

A. All personnel of the organization have the responsibility of ensuring information systems security—this can include indirect personnel such as those dealing with physical security.
B. Information systems security cannot be the sole responsibility of information systems personnel because they cannot ensure security across the entire enterprise.
C. Information systems security cannot be the sole responsibility of information systems security personnel alone because they must rely on numerous other departments to ensure security.
D. Information systems security cannot be the responsibility of functional personnel alone because they also have limited authority and must count on other personnel to collectively ensure security.

S1-119 The corporate information security policy should:

A. address corporate network vulnerabilities.
B. address the process for communicating a violation.
C. be straightforward and easy to understand.
D. be customized to specific target audiences.

C is the correct answer.

Justification:
A. Information security policies are high level and do not address network vulnerabilities directly.
B. Information security policies are high level and do not address the process for communicating a violation.
C. As high-level statements, information security policies should be straightforward and easy to understand.
D. As policies, information security policies should provide a uniform message to all groups and user roles.

S1-120 Requiring all employees and contractors to meet personnel security/suitability requirements commensurate with their position's sensitivity level and subject to personnel screening is an example of a security:

A. policy.
B. strategy.
C. guideline.
D. baseline.

A is the correct answer.

Justification:
A. A security policy is a general statement to define management objectives with respect to security.
B. The security strategy is the plan to achieve security objectives.
C. Guidelines are optional actions and helpful narrative.
D. A security baseline is a set of minimum security requirements that is acceptable to an organization.

S1-121 The **MOST** important aspect in establishing good information security policies is to ensure that they:

A. have the consensus of all concerned groups.
B. are easy to access by all employees.
C. capture the intent of management.
D. have been approved by the internal audit department.

C is the correct answer.

Justification:
A. Having the consensus of all concerned groups may be desirable but is not a requirement of good policies, which are the intent and direction of senior management.
B. Availability of policies is important but not an indicator of good information security content.
C. Policies should reflect the intent and direction of senior management.
D. The internal audit department tests compliance with policy, but it does not write the policies.

S1-122 Which of the following is the **MOST** important consideration when developing an information security strategy?

A. Resources available to implement the program
B. Compliance with legal and regulatory constraints
C. Effectiveness of risk mitigation
D. Resources required to implement the strategy

C is the correct answer.

Justification:

A. The availability of resources is a factor in developing and implementing the program but is not the main consideration.
B. Legal and regulatory requirements must be considered in the strategy to the extent management determines the appropriate level of compliance.
C. Effectively managing information risk to acceptable levels (in alignment with the business objectives) is the most important overall consideration of an information security strategy.
D. The requirements for resources in implementing the strategy is a consideration but a secondary one.

S1-123 Which of the following is the **MOST** effective way to measure strategic alignment of an information security program?

A. Survey business stakeholders
B. Track audits over time
C. Evaluate incident losses
D. Analyze business cases

A is the correct answer.

Justification:

A. The best indicator of strategic alignment is the opinion of the business stakeholders—and the best way to obtain this information is to periodically ask them.
B. Audits might indicate something is amiss, but audits do not have a direct correlation with the effectiveness of the information security program to support business goals and objectives.
C. Incident losses may indicate the overall effectiveness of the program but may have more to do with inadequate budgets or staffing than with alignment.
D. Business cases for security projects may indicate where alignment went astray. However, business cases are indirect and analysis of them would be too late to be useful as an indicator of alignment.

S1-124 Business objectives should be evident in the security strategy by:

A. inferred connections.
B. standardized controls.
C. managed constraints.
D. direct traceability.

D is the correct answer.

Justification:

A. Inferred connections to business objectives are not as good as traceable connections.
B. Standardized controls may or may not be relevant to a particular business objective.
C. Addressing and managing constraints alone is not as useful as also defining explicit benefits.
D. The security strategy will be most useful if there is a direct traceable connection with business objectives.

S1-125 It is **MOST** important that information security architecture be aligned with which of the following?

A. Industry good practices
B. Business goals and objectives
C. Information technology plans
D. International information security frameworks

B is the correct answer.

Justification:

A. Industry good practices may serve as a guideline but may be excessive or insufficient for a particular organization.
B. A security architecture is based on policies and both must be aligned with business goals and objectives.
C. Information technology plans must be aligned with business goals and objectives.
D. International frameworks can serve as a general guide to the extent it supports business goals and objectives.

S1-126 Which of the following is **MOST** likely to remain constant over time? An information security:

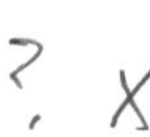

A. policy.
B. standard.
C. strategy.
D. procedure.

C is the correct answer.

Justification:

A. Policies do not change as frequently as procedures and standards; however, security policies do change to adjust to new regulations or laws or to address emerging technology trends, and these changes do not typically require adjustments to the information security strategy.
B. Standards change more frequently because they must often be adjusted to compensate for changes in technology and business processes.
C. Of the choices provided, the information security strategy is the least likely to change. An information security strategy is a reflection of high-level objectives and the direction of the security program, as dictated by business leadership. All information security policies, standards and procedures are derived from the information security strategy.
D. Procedures change more frequently because they must often be adjusted to compensate for changes in technology and business processes.

S1-127 An organization's board of directors is concerned about recent fraud attempts that originated over the Internet. What action should the board take to address this concern?

A. Direct information security regarding specific resolutions that are needed to address the risk.
B. Research solutions to determine appropriate actions for the organization.
C. Take no action; information security does not report to the board.
D. Direct management to assess the risk and to report the results to the board.

D is the correct answer.

Justification:

A. The board does not direct the security operations, which is delegated to executive management.
B. The board would not research solutions but might direct executive management to do so.
C. Taking no action would not be a responsible course of action.
D. The board would typically direct executive management to assess the risk and report results.

S1-128 New regulatory and legal compliance requirements that will have an effect on information security will **MOST** likely come from the:

A. corporate legal officer.
B. internal audit department.
C. affected departments.
D. compliance officer.

C is the correct answer.

Justification:

A. Corporate legal departments are often focused on contractual matters and disclosure requirements for reporting to the agencies regulating publicly held corporations.
B. Internal auditors would typically be concerned with review of existing compliance requirements rather than with new legal or regulatory requirements.
C. The departments affected by legal and regulatory requirements (such as the human resources department) are typically advised by their respective associations of new or changing regulations and the probable impacts on various organizations.
D. Compliance officers are typically charged with determining compliance with internal policies and standards.

X **S1-129** Which of the following choices would be the **MOST** significant key risk indicator?

A. A deviation in employee turnover
B. The number of packets dropped by the firewall
C. The number of viruses detected
D. The reporting relationship of IT

A is the correct answer.

Justification:

A. Significant changes in employee turnover indicate that something significant is impacting the workforce, which deserves the attention of the information security manager. If a large number of senior developers are leaving the research and development group, for instance, it may indicate that a competitor is attempting to obtain the organization's development plans or proprietary technology.
B. An increase in the number of packets being dropped may indicate a change in the threat environment, but there is no impact unless legitimate traffic is being impacted. Therefore, the number of packets dropped is not an effective key risk indicator (KRI).
C. An increase in the number of viruses detected may indicate a change in the threat environment, but the increase in detected viruses also indicates that the threat is adequately countered by existing controls.
D. Changes in reporting relationships come about as a result of intentional business decisions, so the reporting relationship of IT is not a KRI.

S1-130 The **BEST** approach to developing an information security program is to use a:

A. process.
B. framework.
C. model.
D. guideline.

B is the correct answer.

Justification:
A. Generally, a framework is more flexible than a process and avoids the rigidity of process approaches.
B. Frameworks, such as International Organization for Standardization (ISO) 27001 and COBIT, are the most common approach to program development.
C. A reference model is one approach to architecture, but it has less flexibility than a framework.
D. Guidelines can provide useful suggestions but by themselves are not as useful as a framework.

S1-131 The **FIRST** step in developing a business case is to:

A. determine the probability of success.
B. calculate the return on investment.
C. analyze the cost-effectiveness.
D. define the issues to be addressed.

D is the correct answer.

Justification:
A. Without a clear definition of the issues to be addressed, the probability of success is low.
B. Without a clear definition of the issues to be addressed, the solutions proposed and the results expected, return on investment cannot be determined.
C. Without a clear definition of the issues to be addressed, the solutions proposed and the results expected, cost-effectiveness cannot be determined.
D. Without a clear definition of the issues to be addressed, solutions and expected results determined, the other components of a business case cannot be determined.

S1-132 Which of the following choices would provide the **BEST** measure of the effectiveness of the security strategy?

A. Minimizing risk across the enterprise
B. Countermeasures existing for all known threats
C. Losses consistent with annual loss expectations
D. The extent to which control objectives are met

D is the correct answer.

Justification:
A. Minimizing risk is not the objective. The objective is achieving control objectives and thereby achieving acceptable risk levels. Risk reduction beyond the acceptable level is likely to not be cost-effective and to be a waste of resources.
B. There are some threats for which no countermeasures exist (e.g., comet strikes).
C. The extent of losses is not a reliable indication of the effectiveness of the strategy. Losses may or may not exceed expectations for a variety of reasons and relate to impacts rather than to risk levels.
D. Control objectives are developed to achieve acceptable levels of risk. To the extent that is achieved is a good measure of the effectiveness of the strategy.

S1-133 A regulatory authority has just introduced a new regulation pertaining to the release of quarterly financial results. The **FIRST** task that the security officer should perform is to:

A. identify whether current controls are adequate.
B. communicate the new requirement to audit.
C. implement the requirements of the new regulation.
D. conduct a cost-benefit analysis of implementing the control.

A is the correct answer.

Justification:

A. If current security practices and procedures already meet the new regulation, then there is no need to implement new controls.
B. It is likely that audit is already aware of the new regulation, and this is not the first thing to do.
C. The new regulation should only be implemented after determining existing controls do not meet requirements.
D. A cost-benefit analysis would be useful after determining current controls are not adequate.

S1-134 The **MOST** important requirement for gaining management commitment to the information security program is to:

A. benchmark a number of successful organizations.
B. demonstrate potential losses and other impacts that can result from a lack of support.
C. inform management of the legal requirements of due care.
D. demonstrate support for desired outcomes.

D is the correct answer.

Justification:

A. While benchmarking similar organizations can be helpful in some instances to make a case for management support of the information security program, benchmarking by itself is not likely to be sufficient.
B. Management often considers security to be a financial drain and over reactive. Showing probable outcomes can help build a case, but demonstrating how the program will materially assist in achieving the desired business outcomes will be more effective.
C. Legal requirements are best presented by the legal department and are just another risk.
D. The most effective approach to gain support from management for the information security program is to persuasively demonstrate how the program will help achieve the desired outcomes. This can be done by providing specific business support in areas of operational predictability and regulatory compliance, and by improving resource allocation and meaningful performance metrics.

S1-135 Serious security incidents typically lead to renewed focus on information security by management. To **BEST** use this attention, the information security manager should make the case for:

A. improving integration of business and information security processes.
B. increasing information security budgets and staffing levels.
C. developing tighter controls and stronger compliance efforts.
D. acquiring better supplemental technical security controls.

A is the correct answer.

Justification:
A. Close integration of information security governance with overall organization governance is likely to provide better long-term security by institutionalizing its activities and increasing visibility in all organization activities.
B. Increased budgets and staff may improve security, but they will not have the same beneficial impact as incorporating security into the strategic levels of the organization's operations.
C. Control strength and compliance efforts must be balanced against business requirements, culture and other organization factors that are best accomplished at governance levels.
D. While technical security controls may improve some aspects of security, they will not address management issues nor provide enduring changes that are needed for an overall improvement of the enterprise security posture.

S1-136 Which person or group should have final approval of an organization's information technology (IT) security policies?

A. Business unit managers
B. Chief information security officer
C. Senior management
D. Chief information officer

C is the correct answer.

Justification:
A. Business unit managers should have input into information technology (IT) security policies, but they should not have authority to give final approval.
B. The chief information security officer would more than likely be the primary author of the policies and, therefore, is not the appropriate individual to approve the policies.
C. Senior management should have final approval of all organization policies, including IT security policies.
D. The chief information officer should provide input into the IT security policies but should not have the authority to give final approval.

S1-137 Which one of the following groups has final responsibility for the effectiveness of security controls?

A. The security administrator who implemented the controls
B. The organization's chief information security officer
C. The organization's senior management
D. The information systems auditor who recommended the controls

C is the correct answer.

Justification:

A. Senior management, not the security administrator, holds ultimate responsibility for the effectiveness of security controls. Although the authority to implement is delegated, responsibility cannot be delegated.
B. The chief information security officer may have been delegated authority to verify the effectiveness of security controls, final responsibility still rests with senior management.
C. Senior management holds ultimate responsibility for the effectiveness of security controls.
D. The information systems auditor may be assigned testing of the effectiveness of security controls but is not responsible for their effectiveness.

S1-138 For an organization's information security program to be highly effective, who should have final responsibility for authorizing information system access?

A. Information owner
B. Security manager
C. Chief information officer
D. System administrator

A is the correct answer.

Justification:

A. Because the information owner best understands the nature of the information in the system and who needs access to the information, the information owner or manager should provide authorization for users to access the information systems under their control.
B. The security manager will be responsible for ensuring the access controls are functioning properly and assessing the risk of unauthorized access but is not responsible for granting access rights.
C. The chief information officer will only have responsibility for authorizing access to information related to systems and networks within his or her domain.
D. System administrators are primarily custodians and should only have access to their required operational systems information.

S1-139 An organization has consolidated global operations. The chief information officer has asked the chief information security officer to develop a new organization information security strategy. Which of the following actions should be taken **FIRST**?

A. Identify the assets.
B. Conduct a risk assessment.
C. Define the scope.
D. Perform a business impact analysis.

C is the correct answer.

Justification:

A. The scope of the program must be determined before asset identification can be performed.
B. The scope of the program must be determined before a risk assessment can be performed.
C. The scope of the program must be determined before any of the other steps can be performed.
D. The scope of the program must be determined before a business impact analysis can be performed.

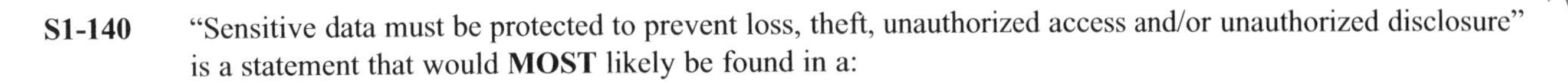

S1-140 "Sensitive data must be protected to prevent loss, theft, unauthorized access and/or unauthorized disclosure" is a statement that would **MOST** likely be found in a:

A. guideline.
B. policy.
C. procedure.
D. standard.

B is the correct answer.

Justification:

A. A guideline is a suggested action that is not mandatory.
B. A policy is a principle that is used to set direction in an organization. It can be a course of action to steer and influence decisions. The wording of the policy must make the course of action mandatory and it must set the direction.
C. A procedure is a particular way of accomplishing something.
D. A standard sets the allowable boundaries for people, processes and technologies that must be met to meet the intent of the policy.

S1-141 To improve the security of an organization's human resources system, an information security manager was presented with a choice to either implement an additional packet filtering firewall **OR** a heuristics-based intrusion detection system. How should the security manager with a limited budget choose between the two technologies?

A. Risk analysis
B. Business impact analysis
C. Return on investment analysis
D. Cost-benefit analysis

D is the correct answer.

Justification:

A. Risk analysis identifies the risk and treatment options.
B. A business impact analysis identifies the impact from the loss of systems or organization functions.
C. Return on investment analysis compares the magnitude and timing of investment gains directly with the magnitude and timing of investment costs.
D. Cost-benefit analysis measures the cost of a safeguard versus the benefit it provides and includes risk assessment. The cost of a control should not exceed the benefit to be derived from it. The degree of control employed is a matter of good business judgment.

S1-142 Which of the following is the **PRIMARY** reason to change policies during program development?

A. The policies must comply with new regulatory and legal mandates.
B. Appropriate security baselines are no longer set in the policies.
C. The policies no longer reflect management intent and direction.
D. Employees consistently ignore the policies.

C is the correct answer.

Justification:
A. Regulatory requirements typically are better addressed with standards and procedures than with high-level policies.
B. Standards set security baselines, not policies.
C. Policies must reflect management intent and direction. Policies should be changed only when management determines that there is a need to address new legal and regulatory or contractual or business requirements.
D. Employees not abiding by policies is a compliance and enforcement issue rather than a reason to change the policies.

X

S1-143 The concept of governance, risk and compliance serves **PRIMARILY** to:

A. align organization assurance functions.
B. ensure that all three activities are addressed by policy.
C. present the correct sequence of security activities.
D. define the responsibilities of information security.

A is the correct answer.

Justification:
A. Governance, risk and compliance (GRC) is an effort to integrate assurance activities across an organization to achieve greater efficiency and effectiveness.
B. It is unlikely that all three activities would not be covered by policies, but GRC may unify existing policies to reduce complexity and any differences that exist.
C. GRC deals directly with sequence of security activities and all three may occur concurrently.
D. GRC is about integration of these activities, not specific responsibilities of various groups.

S1-144 Which of the following factors is **MOST** important for the successful implementation of an organization's information security program?

A. Senior management support
B. Budget for security activities
C. Regular vulnerability assessments
D. Knowledgeable security administrators

A is the correct answer.

Justification:
A. Senior management support is critical to the implementation of any security program.
B. An appropriate budget for security activities is not likely without the support of senior management.
C. Vulnerability assessments are an important element of a successful security program but will be of little use without management support for addressing issues that arise.
D. Knowledgeable security administrators are important for a successful security program, but they are not likely to be effective without management support.

S1-145 An information security strategy presented to senior management for approval **MUST** incorporate:

A. specific technologies.
B. compliance mechanisms.
C. business priorities.
D. detailed procedures.

C is the correct answer.

Justification:

A. The strategy is a forward-looking document that reflects awareness of technological baselines and developments in general, but specific technologies are typically addressed at lower levels based on the strategy.

B. Mechanisms for compliance with legal and regulatory requirements are generally controls implemented at the tactical level based on direction from the strategy.

C. Strategy is the high-level approach by which priorities and goals can be met. The information security strategy necessarily must incorporate the priorities of the business to be meaningful.

D. Detailed procedures are inappropriate at the strategic level.

S1-146 Which of the following will require the **MOST** effort when supporting an operational information security program?

A. Reviewing and modifying procedures
B. Modifying policies to address changing technologies
C. Writing additional policies to address new regulations
D. Drafting standards to address regional differences

A is the correct answer.

Justification:

A. When an information security program is operational, few changes to policies or standards will be needed. Procedures, however, are designed at a more granular level and will require reasonably frequent modification. Because procedures are more detailed and can be technology specific, there are generally far more procedures than standards or policies. Consequently, review and modification of procedures will consume the majority of effort.

B. While technology does change, it is relatively rare for a technology shift to be so disruptive as to require a modification of policy. Most technological changes can be addressed at lower levels (e.g., in standards or procedures).

C. New regulations may require the creation of a new policy, but this does not happen nearly as often or consume as much time in an operational program as the review and modification of procedures.

D. Global organizations may need to customize policy through the use of regional standards, but an operational program will already have most of these standards in place. Even where they need to be drafted, the level of effort required to customize policy by region is less than what will be needed to review and modify the vast body of procedures that change more frequently.

X **S1-147** A newly hired information security manager notes that existing information security practices and procedures appear *ad hoc*. Based on this observation, the next action should be to:

A. assess the commitment of senior management to the program.
B. assess the maturity level of the organization.
C. review the corporate standards.
D. review corporate risk management practices.

C is the correct answer.

Justification:
A. While management may not be exercising due care, it is concerned enough to engage a new information security manager. Assessing the commitment of senior management will not address the immediate concern of *ad hoc* practices and procedures.
B. It is evident from the initial review that maturity is very low and efforts required for a complete assessment are not warranted. It may be better to address the immediate problem of *ad hoc* practices and procedures.
C. The absence of current, effective standards is a concern that must be addressed promptly.
D. It is apparent that risk management is not being practiced; establishing an effective program will take time. A more prudent initial activity is to implement basic controls.

S1-148 Compliance with security policies and standards is the responsibility of:

A. the information security manager.
B. executive management.
C. the compliance officer.
D. all organizational units.

D is the correct answer.

Justification:
A. The information security manager usually has some responsibility for monitoring and assessing security compliance, but enforcement is typically the responsibility of the unit management or human resources.
B. Executive management normally monitors compliance.
C. Compliance officers are usually concerned with legal and regulatory compliance issues such as privacy.
D. Compliance responsibilities are usually shared across organizational units and the results shared with executive management and the board of directors' audit or compliance committee.

S1-149 Which of the following is a risk that would **MOST** likely be overlooked by an information security review during an onsite inspection of an offshore provider?

A. Cultural differences
B. Technical skills
C. Defense in depth
D. Adequate policies

A is the correct answer.

Justification:
A. Individuals in different cultures often have a different perspective on what information is considered sensitive or confidential and how it should be handled that may not be consistent with the organization's requirements. Cultural norms are not usually an area of consideration in a security review or during an onsite inspection.
B. Technical skills are common scope areas for a security review to ensure that the offshore provider meets acceptable standards.
C. Controls design and operational effectiveness are common scope areas for a security review to ensure that the offshore provider meets acceptable standards.
D. Information security policies are common scope areas for a security review to ensure that the offshore provider meets acceptable standards.

S1-150 Which of the following is the **MOST** important component of information security governance? 

A. Appropriate monitoring and metrics
B. An established strategy for moving forward
C. An information security steering committee
D. Senior management involvement

D is the correct answer.

Justification:
A. Monitoring and metrics can determine progress but are effective only if there is management support.
B. Strategy is only one building block of information security governance and cannot work without management support.
C. A steering committee cannot exist without management support.
D. Senior management must champion the process and information security spokespersons to create an effective information security governance framework.

S1-151 Which of the following is the **MOST** important outcome of an information security strategy?

A. Consistent policies and standards
B. Ensuring that residual risk is at an acceptable level
C. An improvement in the threat landscape
D. Controls consistent with international standards

B is the correct answer.

Justification:

A. Consistency of document design facilitates maintenance, while consistency of document content across units and entities ensures that documents are applied uniformly; consistency does not ensure alignment with business objectives.
B. Residual risk is the remaining risk after management has implemented a risk response or treatment. An important objective of a security strategy is to implement cost-effective controls that ensure that residual risk remains within the organization's acceptable risk and tolerance levels.
C. Most threats cannot be affected by policy; however, risk likelihood and impact can be affected.
D. Standard controls may or may not be relevant to a particular business objective.

?, X **S1-152** Obtaining senior management support for an information security initiative can **BEST** be accomplished by:

A. developing and presenting a business case.
B. defining the risk that will be addressed.
C. presenting a financial analysis of benefits.
D. aligning the initiative with organizational objectives.

A is the correct answer.

Justification:

A. A business case is inclusive of the other options and includes and specifically addresses them.
B. A business case must enumerate the risk that the initiative will address.
C. The value proposition is an essential part of the business case that addresses the financial aspects of the initiative.
D. The business case must show how the initiative will align with and support organizational objectives.

S1-153 The **PRIMARY** focus of information security governance is to:

A. adequately protect the information and knowledge base of the organization.
B. provide assurance to senior management that the security posture is adequate.
C. safeguard the IT systems that store and process business information.
D. optimize the information security strategy to achieve business objectives.

D is the correct answer.

Justification:

A. While adequately protecting information and the knowledge base is important, governance is ultimately about achieving business objectives.
B. Unless information security strategy is aligned with business objectives, there is no basis to determine the adequacy of the security posture.
C. Information security governance is more than IT systems.
D. Governance ensures that business objectives are achieved by evaluating stakeholder needs, conditions and options; setting direction through prioritization and decision making; and monitoring performance, compliance and progress against plans.

S1-154 The **MOST** important basis for developing a business case is the:

A. risk that will be addressed.
B. financial analysis of benefits.
C. alignment with organizational objectives.
D. feasibility and value proposition.

D is the correct answer.

Justification:
A. Risk that will be addressed is a part of what determines feasibility and whether the benefits are sufficient for the cost.
B. Benefits analysis is a part of what determines feasibility and whether the benefits are sufficient for the cost.
C. Alignment with organizational objectives is a part of what determines feasibility and whether the benefits are sufficient for the cost.
D. The feasibility and value proposition are the primary factors in determining whether a project will proceed.

S1-155 Which of the following is the **MOST** important consideration when developing an information security strategy?

A. Supporting business objectives
B. Maximizing the effectiveness of available resources
C. Ensuring that legal and regulatory constraints are addressed
D. Determining the effect on the organizational roles and responsibilities

A is the correct answer.

Justification:
A. The overall objective of an information security strategy is to support business objectives and activities and minimize disruptions.
B. Maximizing the effectiveness of resources is one of the factors in developing a strategy but is secondary to supporting organizational activities.
C. The strategy must consider legal and regulatory requirements, but they are just one of the potential impact considerations.
D. Organizational structure affects the approaches to developing a strategy but is just one of the considerations.

S1-156 The **MOST** important outcome of aligning information security governance with corporate governance is to:

A. show that information security understands the rules.
B. provide regulatory compliance.
C. maximize the cost-effectiveness of controls.
D. minimize the number of rules and regulations required.

C is the correct answer.

Justification:
A. While it is important that information security understands the corporate rules, that is not the main reason for alignment.
B. Regulatory compliance is not a primary driver in governance alignment.
C. Corporate governance includes a structure and rules that in most cases are related to managing various types of risk. A lack of alignment can result in potentially duplicate or contradictory controls, which negatively impacts cost-effectiveness.
D. Minimizing the number of rules is helpful; however, it is just one element of achieving cost-effectiveness.

S1-157 Which of the following is the **MOST** important consideration for a control policy?

A. Data protection
B. Life safety
C. Security strategy
D. Regulatory factors

B is the correct answer.

Justification:

A. Protecting data is not as important as protecting life.
B. For physical controls, such as electrically controlled doors with swipe card access, the most important consideration is safety, such as ensuring that the doors fail open in case of fire.
C. The control policy is part of the information security strategy.
D. Compliance with regulatory requirements, where relevant, is important, but ultimately, the safety of people has the highest priority.

S1-158 Senior management has expressed some concern about the effectiveness of the information security program. What can the information security manager do to gain the support of senior management for the program?

A. Rebuild the program on the basis of a recognized, auditable standard.
B. Calculate the cost-benefit analysis of the existing controls that are in place.
C. Interview senior managers to address their concerns with the program.
D. Present a report from the steering committee supporting the program.

C is the correct answer.

Justification:

A. The key to gaining support from senior management is understanding their concerns and making sure that those concerns are addressed. Replacing the entire program as a response to general concerns would not be appropriate without more information.
B. A cost-benefit analysis of controls demonstrates that the controls that have been put in place were preferable to alternative methods of risk treatment, but this evidence does not address the question of overall program effectiveness.
C. It is not uncommon for senior managers to have concerns. An effective information security manager will discuss these concerns and make changes as needed to address them.
D. The steering committee generally reports to senior management, so if senior managers express concern regarding the effectiveness of the program, the concern may be directed in part at the steering committee.

S1-159 Which of the following indicators is **MOST** likely to be of strategic value?

A. Number of users with privileged access
B. Trends in incident frequency
C. Annual network downtime
D. Vulnerability scan results

B is the correct answer.

Justification:

A. The number of users with privileged access, if excessive, can pose unnecessary risk but is more of an operational metric.
B. Trends in incident frequency will show whether the information security program is improving and heading in the right direction.
C. Network downtime is a relevant operational metric in terms of service level agreements but, without trends over time, is not a useful strategic metric.
D. Vulnerability scans are an operational metric.

S1-160 Which of the following is the **MOST** cost-effective approach to achieve strategic alignment?

X

A. Periodically survey management
B. Implement a governance framework
C. Ensure that controls meet objectives
D. Develop an enterprise architecture

A is the correct answer.

Justification:

A. Achieving and maintaining strategic alignment means that business process owners and managers believe that information security is effectively supporting their organizational activities. This can most easily and inexpensively be determined by periodic surveys, which will also indicate improvement or degradation over time.
B. Implementing an appropriate governance framework may improve strategic alignment in addition to a number of other benefits, but is far more complex, time-consuming and expensive and may not capture as directly business owners' perceptions or show changes over time.
C. While important, controls meeting objectives may not be perceived by managers as helpful to the business and may, in fact, be seen as an impediment to their activities.
D. An enterprise architecture should consider business objectives during design and development, but in an effort to balance many other requirements such as security and functionality, it may or may not be perceived as supporting business activities.

S1-161 Which of the following is **PRIMARILY** related to the emergence of governance, risk and compliance?

A. The increasing need for controls
B. The policy development process
C. The integration of assurance-related activities
D. A model for information security program development

C is the correct answer.

Justification:

A. One of the outcomes of governance, risk and compliance (GRC) is the increased attention on general controls, because they are more pervasive and cost-effective than application-level controls. However, the **PRIMARY** driver for GRC has been the increased complexity and diversity of assurance requirements and the need to address these through one integrated process.
B. As with most information security activities, appropriate policy support is needed for effective GRC implementation, but that is only one aspect of achieving integration.
C. GRC is a process to integrate multiple disparate but related activities to improve effectiveness, reduce or eliminate conflicting approaches, and reduce costs.
D. GRC is not a model, but an approach to achieving greater assurance process integration.

S1-162 Which of the following is **MOST** likely to be responsible for establishing the information security requirements over an application?

A. IT steering committee
B. Data owner
C. System owner
D. IS auditor

B is the correct answer.

Justification:

A. The IT steering committee is an executive management–level committee that assists in the delivery of the IT strategy, oversees day-to-day management of IT service delivery and IT projects, and focuses on implementation aspects.
B. Data owners determine the level of controls deemed necessary to secure data and the applications that store or process the data.
C. System owners are responsible for platforms, rather than for applications or data.
D. The IS auditor evaluates the adequacy, efficiency and effectiveness of controls.

S1-163 Which of the following **BEST** supports continuous improvement of the risk management process?

A. Regular review of risk treatment options
B. Classification of assets in order of criticality
C. Adoption of a maturity model
D. Integration of assurance functions

C is the correct answer.

Justification:

A. Risk treatment is an element of the risk management process. Other elements such as risk identification, risk communication and acceptance also need to be considered.
B. Classification of assets is important, but is an element of the risk management process and is not sufficient to ensure continuous improvement.
C. A maturity model such as the capability maturity model (CMM) can be used to classify an organization as initial, repeatable, defined, managed or optimized. As a result, an organization can easily know where it falls and then start working to reach the optimized state.
D. There are many benefits from integrating assurance functions. However, this is not a holistic approach because the best of assurance functions will be reactive if risk management does not cascade through the entire organization. Measures must be taken to ensure that the entire staff, rather than only the assurance functions, is risk conscious.

S1-164 Which of the following is **MOST** important in the development of information security policies?

A. Adopting an established framework
B. Using modular design for easier maintenance
C. Using prevailing industry standards
D. Gathering stakeholder requirements

D is the correct answer.

Justification:

A. A framework will not be effective without including the management intent and direction provided by policies.
B. While using a modular design should be a key consideration, it is not as important as considering stakeholder input. Stakeholder input not only promotes policy completeness, it also facilitates stakeholder buy-in.
C. Prevailing industry standards are important but may not be appropriate or suitable to address unique or specific issues in an organization.
D. The primary stakeholders in policies are management, and policies are the primary governance tool employed in an organization; therefore, the policies must reflect management intent and direction.

S1-165 An organization has recently developed and approved an access control policy. Which of the following will be **MOST** effective in communicating the access control policy to the employees?

A. Requiring employees to formally acknowledge receipt of the policy
B. Integrating security requirements into job descriptions
C. Making the policy available on the intranet
D. Implementing an annual retreat for employees on information security

A is the correct answer.

Justification:

A. Requiring employees to formally acknowledge receipt of the policy does not guarantee that the policy has been read or understood but establishes employee acknowledgement of the existence of the new policy. Each communication should identify a point of contact for follow-up questions.

B. Current employees do not necessarily reread job descriptions that would contain the new policy.

C. Making the policy available on the intranet does not ensure that the document has been read, nor does it create an audit trail that establishes that employees have been made aware of the policy.

D. An annual event may not be timely and may not rectify significant gaps in awareness.

S1-166 Which of the following is the **MOST** important step in developing a cost-effective information security strategy that is aligned with business requirements?

A. Identification of information assets and resource ownership
B. Valuation of information assets
C. Determination of clearly defined objectives
D. Classification of assets as to criticality and sensitivity

C is the correct answer.

Justification:

A. Identification of information assets and asset ownership is a good starting point for implementing an information security strategy. However, having a clear objective is essential.

B. Valuation of the information assets is best performed after the asset inventory has been compiled and the asset owners are assigned. Asset owners generally classify assets according to the organization's asset classification scheme. Asset classification represents the business value of the asset to the organization and is the basis for the required protection levels.

C. Determining the objectives of information security provides the basis for a plan to achieve those objectives, which is the definition of a strategy.

D. Asset classification represents the business value of the asset to the organization and is the basis for the required protection levels.

S1-167 In a mature organization, it would be expected that the security baseline could be approximated by which of the following?

A. Organizational policies are in place.
B. Enterprise architecture is documented.
C. Control objectives are being met.
D. Compliance requirements are addressed.

C is the correct answer.

Justification:
A. Policies, as a statement of management intent and direction, will only indicate the security baseline in general sense.
B. Enterprise architecture may or may not provide an indication of some of the controls implemented.
C. The control objectives, when achieved, set the security baselines.
D. Compliance requirements will indicate some of the controls required indicative of what the baseline should be but only in the areas related to specific regulations.

S1-168 Systems thinking as it relates to information security is:

A. a prescriptive methodology for designing the systems architecture.
B. an understanding that the whole is greater than the sum of its parts.
C. a process that ensures alignment with business objectives.
D. a framework for information security governance.

B is the correct answer.

Justification:
A. While systems thinking is essential to developing a sound systems architecture, it is not a prescriptive approach.
B. A systems approach for developing information security includes the understanding that the whole is more than the sum of its parts and changes in any one part affect the rest.
C. Alignment with business objectives is one of the desired outcomes, but systems thinking does not ensure it.
D. Systems thinking is not a framework for information security governance, although the systems approach can be helpful in implementing an effective information security governance framework as well as an information security management program.

S1-169 Which of the following choices **BEST** justifies an information security program?

A. The impact on critical IT assets
B. A detailed business case
C. Steering committee approval
D. User acceptance

B is the correct answer.

Justification:

A. The impact on IT assets is an important component but by itself is insufficient to justify the information security program.
B. A business case contains documentation of the rationale for making a business investment, used both to support a business decision on whether to proceed with the investment and as an operational tool to support management of the investment through its full economic life cycle. This will provide the justification for the information security program by demonstrating the benefits of implementation.
C. Approval by the steering committee validates the justification contained in the business case.
D. User acceptance is highly relevant during the system development life cycle; however, it is ill-advised to rely on user acceptance as justification for a security program, particularly because security and performance often clash.

S1-170 Information security policy development should **PRIMARILY** be based on:

A. vulnerabilities.
B. exposures.
C. threats.
D. impacts.

C is the correct answer.

Justification:

A. Absent a threat, vulnerabilities do not pose a risk. Vulnerability is defined as a weakness in the design, implementation, operation or internal control of a process that could expose the system to adverse impacts from threat events.
B. Exposure is only important if there is a threat. Exposure is defined as the potential loss to an area due to the occurrence of an adverse event.
C. Policies are developed in response to perceived threats. If there is no perceived threat, there is no need for a policy. A threat is defined as anything (e.g., object, substance, human) that is capable of acting against an asset in a manner that can result in harm.
D. Impact is not an issue if no threat exists. The impact is generally quantified as a direct financial loss in the short term or an ultimate (indirect) financial loss in the long term.

S1-171 The **FIRST** action for an information security manager to take when presented with news that new regulations are being applied to how organizations handle sensitive data is to determine:

A. processes and activities that may be affected.
B. how senior management would prefer to respond.
C. whether the organization qualifies for an exemption.
D. the approximate cost of compliance.

A is the correct answer.

Justification:

A. Changes to information security are best made on the basis of risk. To determine the risk associated with the new regulations, the information security manager must first know what processes and activities may be affected.

B. Senior management will not have a basis for preference until potential effects are determined and compliance requirements identified.

C. Requesting exemptions comes at a cost, at least in terms of time and potentially with reputational consequences. Also, if there is little or no effect on the organization, there will be no need to request an exemption even if one is available.

D. Until the scope of potential effects and the changes that may be needed to comply are both understood, the cost of compliance cannot be reasonably approximated.

S1-172 Which of the following requirements is the **MOST** important when developing information security governance?

A. Complying with applicable corporate standards
B. Achieving cost-effectiveness of risk mitigation
C. Obtaining consensus of business units
D. Aligning with organizational goals

D is the correct answer.

Justification:

A. Corporate standards are established on the basis of policies that support organizational strategy. Complying with corporate standards is only one aspect of developing governance.

B. While cost-effectiveness of risk mitigation approaches is an important consideration, aspects of information security governance cannot be implemented if contrary to organizational goals.

C. Consensus is valuable, but not required.

D. Information security governance is the set of responsibilities and practices exercised by the board and executive management with the goal of providing strategic direction, ensuring that objectives are achieved, ascertaining that risk is managed appropriately and verifying that the enterprise's resources are used responsibly. It should support and reflect the goals of the organization.

X **S1-173** What is the **MOST** important consideration when developing a business case for an information security investment?

A. The impact on the risk profile of the organization
B. The acceptability to the board of directors
C. The implementation benefits
D. The affordability to the organization

C is the correct answer.

Justification:

A. The impact on the risk profile can be one component of the business case but does not include all the areas the business case would cover.
B. The basis for acceptance among the directors should be the impact on the risk profile.
C. **A business case is defined as documentation of the rationale for making a business investment, used both to support a business decision on whether to proceed with the investment and as an operational tool to support management of the investment through its full economic life cycle. A business case covers not only long-term benefits, but short-term ones as well as the costs.**
D. While cost is important to consider, if the benefits outweigh the costs, it will be in the best interests of the organization to go ahead with the investment.

S1-174 An organization has decided to implement governance, risk and compliance processes into several critical areas of the enterprise. Which of the following objectives is the **MAIN** one?

A. To reduce governance costs
B. To improve risk management
C. To harmonize security activities
D. To meet or maintain regulatory compliance

B is the correct answer.

Justification:

A. Governance costs may or may not be reduced, but that is not the primary objective.
B. The overarching objective of governance, risk and compliance (GRC) is improved risk management achieved by integrating these interrelated activities across the enterprise, primarily focused on finance, legal and IT domains.
C. Convergence of security activities would be just one element of GRC.
D. Achieving an appropriate level of regulatory compliance is likely to be one of the goals, but with the overall objective of more effective and efficient management of risk.

S1-175 The acceptable limits defined by organizational standards are **PRIMARILY** determined by:

A. likelihood and impact.
B. risk appetite.
C. relevant policies.
D. the defined strategy.

B is the correct answer.

Justification:
A. The likelihood of an adverse event and the consequences of such an event will provide the risk, but only the risk appetite will indicate if the risk is acceptable.
B. Risk appetite is the amount of risk, on a broad level, that an entity is willing to accept in pursuit of its mission. This would set the acceptable limits for organizational standards.
C. Standards interpret policy, but the boundaries are set based on risk.
D. Standards are a part of implementing strategy, but the boundaries are set based on risk.

S1-176 What is the **MOST** likely reason that an organizational policy can be eliminated?

A. There is no credible threat.
B. The policy is ignored by staff.
C. Underlying standards are obsolete.
D. The policy is not required by regulatory requirements.

A is the correct answer.

Justification:
A. If it is certain that there is no threat, then there is no risk and a policy is not needed to address it.
B. Noncompliance is not a good reason to eliminate a policy.
C. If the standards are obsolete, then they should be brought current, but that is not a reason to eliminate the policy.
D. If there is a potential risk, then there is a reason to have the policy, independent of whether regulation mandates that particular control.

S1-177 Which of the following factors is the **MOST** important for determining the success of an information security strategy?

A. It is approved by the chief technology officer.
B. It is aligned with the long-term IT plan.
C. It is aligned with goals set by the board of directors.
D. It is supported by key performance indicators.

C is the correct answer.

Justification:
A. The chief technology officer will support the strategy but is not the person who would generally approve it.
B. The long-term IT plan in part implements the information security strategy.
C. The strategy is the plan to achieve objectives. The board of directors sets these objectives.
D. Key performance indicators are used to measure the milestones necessary to achieve the goals.

S1-178 It is essential for the board of directors to be involved with information security activities primarily because of concerns regarding:

A. technology.
B. liability.
C. compliance.
D. strategy.

B is the correct answer.

Justification:
A. The board is typically not essential in selecting particular technical solutions.
B. The insurance policies that organizations typically obtain to shield owners and key stakeholders from liability frequently require a good-faith effort on the part of the board to exercise due care as a precondition for coverage. If the board is not involved, this liability protection may be lost.
C. Compliance is addressed as part of the risk management program.
D. The board sets goals, for which strategies are then developed by senior management or subordinate steering committees.

S1-179 Which of the following would be the **BEST** indicator that an organization has good governance?

A. Risk assessments
B. Maturity level
C. Audit reports
D. Loss history

B is the correct answer.

Justification:
A. While it is likely that good results on risk assessments will align with good governance, they are only indirectly correlated with good governance, and many other factors are involved such as industry sector, exposure, etc.
B. A high score on the capability maturity model (CMM) scale is a good indicator of good governance.
C. Audit reports generally deal with specifics of compliance and specific risk rather than overall governance.
D. Loss history will be affected by many factors other than governance.

S1-180 Which of the following ways is the **BEST** to establish a basis on which to build an information security governance program?

A. Align the business with an information security framework.
B. Understand the objectives of the various business units.
C. Direct compliance with regulatory and legal requirements.
D. Meet with representatives of the various security functions.

B is the correct answer.

Justification:

A. Frameworks are beneficial as a means of tracking what functions should be performed by effective governance, but the establishment of governance is primarily a matter of understanding business objectives.
B. The governance program needs to be a comprehensive security strategy intrinsically linked with business objectives. It is impossible to build an effective program for governance without understanding the objectives of the business units, and the objectives of the business units can best be understood by examining their processes and functions.
C. Meeting regulatory and legal requirements may be included among the objectives of the business, but compliance with laws and regulations is not the primary function of information security governance. Depending on the cost associated with doing so, businesses may, in some cases, even opt to accept the risk of noncompliance.
D. Governance reflects the approach to achieving the objectives of the business. Meeting with the security functions can only provide insight with regard to the technical posture and goals as they currently exist; it does not provide a basis on which to build a program.

S1-181 An organization has decided to implement bring your own device (BYOD) for laptops and mobile phones. What should the information security manager focus on **FIRST**?

A. Advising against implementing BYOD because of a security risk
B. Preparing a business case for new security tools for BYOD
C. Updating the security awareness program to include BYOD
D. Determining an information security strategy for BYOD

D is the correct answer.

Justification:

A. The organization has already made the decision to implement bring your own device (BYOD). The security manager's role is to identify and communicate the risk and determine how to implement this decision in the most secure way.
B. A business case can be prepared if new tools are required for implementing BYOD; however, this requirement will be based on the security strategy.
C. The security strategy must take into account BYOD before the security awareness program may be updated to include it.
D. The information security manager should determine whether the existing strategy can accommodate BYOD and, if not, then what changes are needed. A risk assessment and other tools may be part of this process.

S1-182 Which of the following choices is **MOST** likely to ensure that responsibilities are carried out?

A. Signed contracts
B. Severe penalties
C. Assigned accountability
D. Clear policies

C is the correct answer.

Justification:

A. Contracts can define responsibilities, but it is essential that individuals are accountable.
B. Penalties can reinforce accountability and are a deterrent control but will not ensure that responsibilities are always discharged properly.
C. Assigning accountability to individuals is most likely to ensure that duties are properly carried out.
D. Policies generally record a high-level principle or course of action that has been decided on; they are advantageous, but it is more effective to establish direct accountability to ensure that responsibilities are performed.

S1-183 The purpose of an information security strategy is to:

A. express the goals of an information security program and the plan to achieve them.
B. outline the intended configuration of information system security controls.
C. mandate the behavior and acceptable actions of all information system users.
D. authorize the steps and procedures necessary to protect critical information systems.

A is the correct answer.

Justification:

A. The purpose of the strategy is to set out the goals of the information security program and the plan to achieve those objectives.
B. A strategy is usually too high level to deal specifically with control configuration.
C. Some elements of strategy may deal with required behaviors and actions, but it will not be a mandate, rather part of a process to achieve a particular objective.
D. Strategy will not deal with authorizing specific actions.

X **S1-184** What should be the **PRIMARY** basis of a road map for implementing information security governance?

A. Policies
B. Architecture
C. Legal requirements
D. Strategy

D is the correct answer.

Justification:

A. Policies are developed or modified after a strategy is defined and are one of the controls to implement it.
B. Logical security architecture will be a reflection of the road map and may serve as the road map after a strategy has been developed.
C. While legal and regulatory requirements must be considered, the road map is based on the strategy, which in turn is based on the organization's objectives.
D. The road map detailing the steps, resources and time lines for development of the strategy is developed after the strategy is determined.

S1-185 Which of the following choices will **MOST** influence how the information security program will be designed and implemented?

A. Type and nature of risk
B. Organizational culture
C. Overall business objectives
D. Lines of business

B is the correct answer.

Justification:

A. The specific risk faced by the organization will affect the security program, but how this risk is perceived and dealt with depends on the organizational culture.
B. The organizational culture generally influences risk appetite and risk tolerance in addition to how issues are perceived and dealt with and many other aspects, which has significant influence over how an information security program should be designed and implemented.
C. Business objectives will determine the specific kinds of risk to be addressed but will not greatly influence the actual program development and implementation.
D. The lines of business will affect the specific kinds of risk to be addressed but will not greatly influence the actual program development and implementation.

S1-186 While governance, risk and compliance (GRC) can be applied to any area of an organization, it is **MOST** often focused on which of the following areas?

A. Operations and marketing
B. IT, finance and legal
C. Audit, risk and regulations
D. Information security and risk

B is the correct answer.

Justification:

A. Governance, risk and compliance (GRC) is generally not used in support of operations and marketing.
B. GRC is largely concerned with ensuring that processes in IT, finance and legal are in compliance with regulatory requirements; that proper rules are in place; and that risk is appropriately addressed.
C. Audit, risk and regulations are support functions to IT, finance and legal.
D. Information security and risk can be a part of GRC and interrelate to audit, risk and regulations, but are primarily in support of IT, finance and legal.

S1-187 Responsibility for information security and related activities involves multiple departments. What is the **PRIMARY** reason the information security manager should develop processes that integrate these roles and responsibilities?

A. To mitigate the tendency for security gaps to exist between assurance functions
B. To reduce manpower requirements for providing effective information security
C. To ensure effective business continuity and disaster recovery
D. To simplify specification development and acquisition processes

A is the correct answer.

Justification:

A. Wherever multiple departments have shared responsibility for related activities, gaps tend to emerge or there can be unneeded duplication of activities. Integrating the roles and responsibilities is the best way to mitigate these gaps, minimize duplication and ensure consistent risk management.
B. Integrating roles and responsibilities may allow for reductions in manpower, but this is not the driving consideration of integration.
C. Integrating roles and responsibilities may allow for more effective business continuity and disaster recovery, but this is just one of the considerations for integration.
D. Specifications and acquisition may not be affected depending on the particular needs of each department involved.

S1-188 An information security manager wants to implement a security information and event management (SIEM) system not funded in the current budget. Which of the following choices is **MOST** likely to persuade management of this need?

A. A comprehensive risk assessment
B. An enterprisewide impact assessment
C. A well-developed business case
D. Computing the net present value of future savings

C is the correct answer.

Justification:

A. A risk assessment is a process used to identify and evaluate risk and its potential effects. This may be part of a business case but alone is less likely to persuade management.
B. An enterprisewide impact assessment would review the possible consequences of a risk. This may be part of a business case but alone is less likely to persuade management.
C. A business case demonstrating the need and the value proposition is most likely to be persuasive to management. All of the other options could be part of a well-developed business case.
D. The net present value would be calculated by using an after-tax discount rate of an investment and a series of expected incremental cash outflows (the initial investment and operational costs) and cash inflows (cost savings or revenues) that occur at regular periods during the life cycle of the investment. This may be part of a business case but alone is less likely to persuade management.

S1-189 Which of the following reasons is the **MOST** important to develop a strategy before implementing an information security program?

A. To justify program development costs
B. To integrate development activities
C. To gain management support for an information security program
D. To comply with international standards

B is the correct answer.

Justification:

A. Justification for program costs will need to be achieved prior to developing the strategy and is more likely based on a business case than on the strategy.
B. A strategy is a plan to achieve an objective that serves to align and integrate program activities to achieve the defined outcomes.
C. Management support will need to be achieved prior to developing the strategy and is more likely based on a business case than on the strategy.
D. Compliance with international standards, such as International Organization for Standardization (ISO) 27001, does not necessarily require a cohesive plan of action or strategy and can be done piecemeal. If meeting the standard is one of the objectives, a strategy should encompass the actions needed to meet those requirements.

S1-190 Decisions regarding information security are **BEST** supported by:

A. statistical analysis.
B. expert advice.
C. benchmarking.
D. effective metrics.

D is the correct answer.

Justification:

A. A statistical analysis of metrics can be helpful but only if the underlying metrics are sound.
B. Expert advice may be useful, but effective metrics are a better indication.
C. Other organizations would typically only provide some guidance, but decisions should be based on effective metrics.
D. Effective metrics are essential to provide information needed to make decisions. Metrics are a quantifiable entity that allows the measurement of the achievement of a process goal.

S1-191 Which of the following tasks should information security management undertake **FIRST** while creating the information security strategy of the organization?

A. Understand the IT service portfolio.
B. Investigate the baseline security level.
C. Define the information security policy.
D. Assess the risk associated with IT.

A is the correct answer.

Justification:

A. While defining the information security strategy, it is essential to align it with the business and the IT strategy. In order to do that, the security manager must first focus on understanding the business and the IT strategy.
B. Investigating baseline security is a task associated with strategy implementation.
C. Defining the information security policy is performed after defining security strategy.
D. Risk assessment is performed to determine the control objectives, which is generally performed after the security strategy is defined.

S1-192 Which of the following elements are the **MOST** essential to develop an information security strategy?

A. Complete policies and standards
B. An appropriate governance framework
C. Current state and objectives
D. Management intent and direction

C is the correct answer.

Justification:

A. Policies and standards are some of the primary tools to implement a strategy and are subsequent steps in the process.
B. Implementing the information security strategy is the activity that populates or develops the governance framework.
C. Because a strategy is essentially a plan to achieve an objective, it is essential to know the current state of information security and the desired future state or objectives.
D. Management intent and direction is essential to developing objectives; the current state is also required.

S1-193 Requirements for an information security program should be based **PRIMARILY** on which of the following choices?

A. Governance policies
B. Desired outcomes
C. Specific objectives
D. The security strategy

B is the correct answer.

Justification:

A. Policies are one of the resources used to develop the strategy, which is based on specific objectives that meet the requirements.
B. The desired outcomes for the security program will be high-level achievements related to acceptable risk across the enterprise and will determine the requirements that must be met to achieve those outcomes.
C. Objectives are the steps required to achieve the desired outcomes.
D. The security strategy is the road map to achieve the objectives that result in the desired outcomes.

S1-194 Which of the following choices is the **BEST** indication that the information security manager is achieving the objective of value delivery?

A. Having a high resource utilization
B. Reducing the budget requirements
C. Utilizing the lowest cost vendors
D. Minimizing the loaded staff cost

A is the correct answer.

Justification:

A. Value delivery means that good rates of return and a high utilization of resources are achieved.
B. The budget level is not an indication of value delivery.
C. The lowest cost vendors may not present the best value.
D. Staff-associated overhead costs by themselves are not an indicator of value delivery.

S1-195 Which of the following internal or external influences on an organization is the **MOST** difficult to estimate?

A. Vulnerability posture
B. Compliance requirements
C. Outsourcing expenses
D. Threat landscape

D is the correct answer.

Justification:

A. The vulnerability posture of an organization can be estimated with a high degree of accuracy through systematic, iterative review of systems, data flows, people and processes.
B. Compliance requirements may be ambiguous at first, but as requirements are reviewed and narrowed, their influence on an organization becomes more predictable until the requirements change or expand over time.
C. The long-term costs of outsourcing are difficult to predict, but the cost is generally clear for defined periods of time (e.g., contract periods). In contrast, the threat landscape is always difficult to estimate.
D. Threats originate from independent sources that may be natural or human-directed. Neither can be positively predicted in all cases. Human-directed threats in particular are extremely difficult to estimate in an information security context because very small numbers of threat actors (including individuals with no assistance) may be ready and able to initiate threat events for any reason at all, including reasons that are not sensible to the individual or an impartial observer.

S1-196 Which of the following challenges associated with information security documentation is **MOST** likely to affect a large, established organization?

A. Standards change more slowly than the environment.
B. Policies change faster than they can be distributed.
C. Procedures are ignored to meet operational requirements.
D. Policies remain unchanged for long periods of time.

A is the correct answer.

Justification:

A. Large, established organizations tend to have numerous layers of review and approval associated with changes to standards. These review mechanisms are likely to be outpaced by changes in technology and the risk environment.

B. Policies are meant to reflect strategic goals and objectives. In small or immature organizations, the policy model may be poorly implemented, resulting in rapid changes to policies that are treated more like standards, but this situation is unlikely to arise in a large, established organization.

C. Large, established organizations typically have formal training programs and internal controls that keep activities substantially in line with published procedures.

D. Although policies should be subject to periodic review and not be regarded as static, properly written policies should require significant changes only when there are substantial changes in strategic goals and objectives. It is reasonable that a large, established organization would experience policy changes only rarely.

S1-197 An information security manager is **PRIMARILY** responsible for:

A. managing the risk to the information infrastructure.
B. implementing a standard configuration for IT assets.
C. conducting a business impact analysis (BIA).
D. closing identified technical vulnerabilities.

A is the correct answer.

Justification:

A. An information security manager is primarily responsible and accountable for managing the information security risk management plan by involving various asset and risk owners to identify and implement appropriate responses.

B. An information security manager may help in standardizing the baseline configuration for IT assets, but implementing the configuration is the responsibility of the asset owners.

C. The information security manager may facilitate a business impact analysis (BIA) conducted by business process owners.

D. The information security manager monitors closure of vulnerabilities by asset owners.

DOMAIN 2—INFORMATION RISK MANAGEMENT (30%)

S2-1 An effective risk management program should reduce risk to:

A. zero.
B. an acceptable level.
C. an acceptable percent of revenue.
D. an acceptable probability of occurrence.

B is the correct answer.

Justification:
A. Reducing risk to zero is impossible, and the attempt would be cost prohibitive.
B. An effective risk management program reduces the risk to an acceptable level; this is achieved by reducing the probability of a loss event through preventive measures as well as reducing the impact of a loss event through corrective measures.
C. Tying risk to a percentage of revenue is inadvisable because there is no direct correlation between the two.
D. Reducing the probability of risk occurrence may not always be possible, as in the case of natural disasters.

S2-2 Which of the following **BEST** indicates a successful risk management practice?

A. Overall risk is quantified.
B. Inherent risk is eliminated.
C. Residual risk is acceptable.
D. Control risk is tied to business units.

C is the correct answer.

Justification:
A. The fact that overall risk has been quantified does not necessarily indicate the existence of a successful risk management practice.
B. Eliminating inherent risk is virtually impossible.
C. A successful risk management practice reduces residual risk to acceptable levels.
D. Although the tying of control risk to business may improve accountability, this is not as desirable as achieving acceptable residual risk levels.

S2-3 Which of the following should a successful information security management program use to determine the amount of resources devoted to mitigating exposures?

A. Risk analysis results
B. Audit report findings
C. Penetration test results
D. Amount of IT budget available

A is the correct answer.

Justification:
A. Risk analysis results are the most useful and complete source of information for determining the amount of resources to devote to mitigating exposures.
B. Audit report findings may not address all risk and do not address annual loss frequency.
C. Penetration test results provide only a limited view of exposures.
D. The IT budget is not tied to the exposures faced by the organization.

S2-4 Which of the following will **BEST** protect an organization from insider security attacks?

A. Static Internet Protocol addressing
B. Internal address translation
C. Prospective employee background checks
D. Employee awareness certification program

C is the correct answer.

Justification:
A. Static Internet Protocol addressing does little to prevent an insider attack.
B. Internal address translation using nonroutable addresses is useful against external attacks but not against insider attacks.
C. Because past performance is a strong predictor of future performance, background checks of prospective employees best prevents attacks from originating within an organization.
D. Employees who certify that they have read security policies are desirable, but this does not guarantee that the employees behave honestly.

S2-5 For risk management purposes, the value of a physical asset should be based on:

A. original cost.
B. net cash flow.
C. net present value.
D. replacement cost.

D is the correct answer.

Justification:
A. Original cost may be significantly different than the current cost of replacing the asset.
B. Net cash flow does not accurately reflect the true value of the asset.
C. Net present value does not accurately reflect the true value of the asset.
D. The value of a physical asset should be based on its replacement cost because this is the amount that would be needed to replace the asset if it were to become damaged or destroyed.

S2-6 In a business impact analysis, the value of an information system should be based on the overall:

A. cost of recovery.
B. cost to recreate.
C. opportunity cost.
D. cost of emergency operations.

C is the correct answer.

Justification:
A. The cost of recovering the system is not the basis for determining the value of the system to the organization, rather the loss of revenues and/or other costs is the primary basis.
B. The cost to recreate is also not a basis for valuing the system; the cost to the organization of the loss of the function is the basis.
C. Opportunity cost reflects the cost to the organization resulting from the loss of a function.
D. Cost of emergency operations is unrelated to the value of an information system.

S2-7 Which of the following is the **BEST** source for determining the value of information assets?

A. Individual business managers
B. Business systems analysts
C. Information security management
D. Industry benchmarking results

A is the correct answer.

Justification:

A. Individual business managers are in the best position to determine the value of information assets since they are most knowledgeable of the assets' impact on the business.

B. Business systems analysts are not as knowledgeable as individual business managers regarding the impact on the business.

C. Information security managers are not as knowledgeable as individual business managers regarding the impact on the business.

D. Peer companies' industry averages do not necessarily provide information that is detailed enough nor are they as relevant to the unique aspects of the business as information from individual business managers.

S2-8 During which phase of development is it **MOST** appropriate to begin assessing the risk of a new application system?

A. Feasibility
B. Design
C. Development
D. Testing

A is the correct answer.

Justification:

A. Risk should be addressed as early in the development of a new application system as possible. The projected risk associated with a new system may make it not feasible.

B. In some cases, identified risk could be mitigated through design changes. If needed changes are not identified until design has already commenced, such changes become more expensive. For this reason, beginning risk assessment during the design phase is not the best solution.

C. The development phase is too late in the system development life cycle (SDLC) for effective risk mitigation.

D. Waiting to assess risk until testing can result in having to start over on the project.

S2-9 Which of the following would be **MOST** useful in developing a series of recovery time objectives?

A. Gap analysis
B. Regression analysis
C. Risk analysis
D. Business impact analysis

D is the correct answer.

Justification:

A. A gap analysis is useful in assessing the differences between the current state and a future state.

B. Regression analysis is used to retest earlier program abends or logical errors that occurred during the initial testing phase.

C. Risk analysis is a process by which frequency and magnitude of IT risk scenarios are estimated.

D. Recovery time objectives (RTOs) are a primary deliverable of a business impact analysis. RTOs define the amount of time allowed for the recovery of a business function or resource after a disaster occurs.

S2-10 Risk acceptance is a component of which of the following?

A. Risk assessment
B. Risk mitigation
C. Risk identification
D. Risk monitoring

B is the correct answer.

Justification:

A. Risk assessment includes identification and analysis to determine the likelihood and potential consequences of a compromise, which is not when risk is to be considered for acceptance or requires mitigation.
B. If after risk evaluation a risk is unacceptable, acceptability is determined after risk mitigation efforts.
C. Risk identification is the process during assessment during which viable risk is identified through developing a series of potential risk scenarios.
D. Monitoring is unrelated to risk acceptance.

S2-11 Risk management programs are designed to reduce risk to:

A. a level that is too small to be measurable.
B. the point at which the benefit exceeds the expense.
C. a level that the organization is willing to accept.
D. a rate of return that equals the current cost of capital.

C is the correct answer.

Justification:

A. Reducing risk to a level too small to measure is impractical and is often cost prohibitive.
B. Depending on the risk preference of an organization, it may or may not choose to pursue risk mitigation to the point at which the benefit equals or exceeds the expense. Therefore, choice C is a more precise answer.
C. Risk should be reduced to a level that an organization is willing to accept.
D. To tie risk to a specific rate of return ignores the qualitative aspects of risk that must also be considered.

S2-12 At what interval should a risk assessment **TYPICALLY** be conducted?

A. Once a year for each business process and subprocess
B. Every three to six months for critical business processes
C. On a continuous basis
D. Annually or whenever there is a significant change

D is the correct answer.

Justification:

A. Conducting a risk assessment once a year is insufficient if important changes take place.
B. Conducting a risk assessment every three to six months for critical processes is not typical and may not be necessary, or it may not address important changes in a timely manner.
C. Performing risk assessments on a continuous basis is generally financially not feasible; it is more cost-effective to conduct risk assessments annually or whenever there is a significant change. Continuous risk monitoring is different from continuous risk assessment and should be conducted on a continuous basis to help the enterprise identify when risk events occur.
D. Risk is constantly changing. Conducting a risk assessment annually or whenever there is a significant change offers the best alternative because it takes into consideration a reasonable time frame and allows flexibility to address significant change.

S2-13 Which of the following risk scenarios would **BEST** be assessed using qualitative risk assessment techniques?

A. Theft of purchased software
B. Power outage lasting 24 hours
C. Permanent decline in customer confidence
D. Temporary loss of email services

C is the correct answer.

Justification:
A. Theft of software can be quantified into monetary amounts.
B. Power outages can be quantified into monetary amounts more precisely than they can be with qualitative techniques.
C. A permanent decline in customer confidence does not lend itself well to measurement by quantitative techniques. Qualitative techniques are more effective in evaluating things such as customer loyalty and goodwill.
D. Temporary loss of email can be easily quantified into monetary amounts.

S2-14 A business impact analysis is the **BEST** tool for determining:

A. total cost of ownership.
B. priority of restoration.
C. annual loss expectancy.
D. residual risk.

B is the correct answer.

Justification:
A. A business impact analysis (BIA) is not used to determine total cost of ownership to the organization.
B. A BIA is the best tool for determining the priority of restoration for applications.
C. A BIA is not used to determine annual loss expectancy to the organization.
D. A BIA is not used to determine residual risk to the organization.

S2-15 Quantitative risk analysis is **MOST** appropriate when assessment results:

A. include customer perceptions.
B. contain percentage estimates.
C. lack specific details.
D. contain subjective information.

B is the correct answer.

Justification:
A. Qualitative analysis is a more appropriate approach for customer perceptions, which are difficult to express in a purely quantitative manner.
B. Percentage estimates are characteristic of quantitative risk analysis.
C. Qualitative analysis is a more appropriate approach when there is a lack of specific details.
D. Qualitative analysis is a more appropriate approach for subjective information.

S2-16 Which of the following is the **MOST** appropriate use of gap analysis?

A. Evaluating a business impact analysis
B. Developing a business balanced scorecard
C. Demonstrating the relationship between controls
D. Measuring current state versus desired future state

D is the correct answer.

Justification:
A. A gap analysis is not as appropriate for evaluating a business impact analysis.
B. A gap analysis is not as appropriate for developing a business balanced scorecard.
C. A gap analysis is not as appropriate for demonstrating the relationship between controls.
D. A gap analysis is most useful in addressing the differences between the current state and future state.

S2-17 Which of the following situations presents the **GREATEST** information security risk for an organization with multiple, but small, domestic processing locations?

A. Systems operation guidelines are not enforced.
B. Change management procedures are poor.
C. Systems development is outsourced.
D. Systems capacity management is not performed.

B is the correct answer.

Justification:
A. Because guidelines are generally not mandatory, their lack of enforcement is not a primary concern.
B. The lack of effective oversight is likely to result in inconsistent change management activities, which can present a serious security risk.
C. Systems that are developed by third-party vendors are becoming common and do not represent an increase in security risk as much as poor change management.
D. Poor capacity management may not necessarily represent a major security risk.

S2-18 The decision as to whether an IT risk has been reduced to an acceptable level should be determined by:

A. organizational requirements.
B. information systems requirements.
C. information security requirements.
D. international standards.

A is the correct answer.

Justification:
A. Organizational requirements should determine when a risk has been reduced to an acceptable level.
B. The acceptability of a risk is ultimately a management decision, which may or may not be consistent with information systems requirements.
C. The acceptability of a risk is ultimately a management decision, which may or may not be consistent with information security requirements.
D. Because each organization is unique, international standards may not represent the best solution for specific organizations and are primarily a guideline.

S2-19 Which of the following is the **PRIMARY** reason for implementing a risk management program? A risk management program:

A. allows the organization to eliminate risk.
B. is a necessary part of management's due diligence.
C. satisfies audit and regulatory requirements.
D. assists in increasing the return on investment.

B is the correct answer.

Justification:
A. The elimination of risk is not possible.
B. The key reason for performing risk management is that it is an essential part of management's due diligence.
C. Satisfying audit and regulatory requirements is of secondary importance.
D. A risk management program may or may not increase the return on investment.

S2-20 Which of the following groups would be in the **BEST** position to perform a risk analysis for a business?

A. External auditors
B. A peer group within a similar business
C. Process owners
D. A specialized management consultant

C is the correct answer.

Justification:
A. External parties, including auditors, do not have the necessary level of detailed knowledge on the inner workings of the business.
B. Peer groups would not have the detailed understanding of the business to be effective at analyzing a particular organization's risk.
C. Process owners have the most in-depth knowledge of risk and compensating controls within their environment.
D. Management consultants are expected to have the necessary skills in risk analysis techniques but would still have to rely on the group with intimate knowledge of the business.

S2-21 Which of the following types of risk is **BEST** assessed using quantitative risk assessment techniques?

A. Stolen customer data
B. An electrical power outage
C. A defaced web site
D. Loss of the software development team

B is the correct answer.

Justification:
A. The effect of the theft of customer data could lead to a permanent decline in customer confidence, which does not lend itself to measurement by quantitative techniques.
B. The loss of electrical power for a short duration is more easily measurable than the other choices and can be quantified into monetary amounts that can be assessed with quantitative techniques.
C. The risk of web site defacement by hackers is nearly impossible to quantify but could also lead to a permanent decline in customer confidence, which does not lend itself to measurement by quantitative techniques.
D. Loss of a majority of the software development team would be impossible to quantify.

S2-22 Which of the following **BEST** helps calculate the impact of losing frame relay network connectivity for 18 to 24 hours?

A. Hourly billing rate charged by the carrier
B. Value of the data transmitted over the network
C. Aggregate compensation of all affected business users
D. Financial losses incurred by affected business units

D is the correct answer.

Justification:
A. Presumably the carrier would not charge if connectivity were lost, and this would not be useful in calculating impact.
B. The value of data is not affected by lost connectivity and would not help calculate impact.
C. Compensation of affected business users is not based on connectivity and would be useless in calculating impact.
D. Financial losses incurred by the business units would be a major factor in calculating impact of lost connectivity.

S2-23 Which of the following is the **MOST** usable deliverable of an information security risk analysis?

A. Business impact analysis report
B. List of action items to mitigate risk
C. Assignment of risk to process owners
D. Quantification of organizational risk

B is the correct answer.

Justification:
A. The business impact analysis report is a useful report primarily for future incident response and business continuity purposes but does not mitigate current risk.
B. List of action items to mitigate risk is the most useful in presenting direct, actionable items to address organizational risk.
C. Assigning risk is useful but does not by itself result in risk mitigation activities.
D. Quantification of risk does not directly result in risk mitigation activities.

S2-24 Ongoing tracking of remediation efforts to mitigate identified risk can **BEST** be accomplished through the use of which of the following approaches?

A. Tree diagrams
B. Venn diagrams
C. Heat charts
D. Bar charts

C is the correct answer.

Justification:
A. Tree diagrams are useful for decision analysis.
B. Venn diagrams show the connection between sets but are not useful in indicating status.
C. Heat charts, sometimes referred to as stoplight charts, quickly and clearly show the current status of remediation efforts.
D. Bar charts show relative size but are a less direct presentation approach to tracking status of remediation efforts.

S2-25 Which two components **PRIMARILY** must be assessed in an effective risk analysis?

A. Visibility and duration
B. Likelihood and impact
C. Probability and frequency
D. Financial impact and duration

B is the correct answer.

Justification:
A. Visibility and duration are not the primary elements of a risk analysis.
B. Likelihood and impact are the primary elements that are determined in a risk analysis.
C. Probability is the same as likelihood, and frequency is considered when determining annual loss expectancy, but this is a secondary analysis element.
D. Financial impact is one of the primary considerations, but duration is a secondary element of the analysis.

S2-26 Information security managers should use risk assessment techniques to:

A. justify selection of risk mitigation strategies.
B. maximize the return on investment.
C. provide documentation for auditors and regulators.
D. quantify risk that would otherwise be subjective.

A is the correct answer.

Justification:
A. Information security managers should use risk assessment techniques as one of the main basis to justify and implement a risk mitigation strategy as efficiently as possible.
B. Risk assessment is only one part of determining return on investment.
C. Providing documentation for auditors and regulators is a secondary aspect of using risk assessment techniques.
D. If assessed risk is subjective, risk assessment techniques will not meaningfully quantify them.

S2-27 Which of the following is **MOST** essential when assessing risk?

A. Providing equal coverage for all asset types
B. Benchmarking data from similar organizations
C. Considering both monetary value and likelihood of loss
D. Focusing on valid past threats and business losses

C is the correct answer.

Justification:
A. Providing equal coverage for all asset types when assessing risk may not be relevant, depending on the significance the asset type has to the organization (e.g., automobile fleet is not likely to have as much significance as the data center).
B. Benchmarking other organizations when assessing risk is of relatively little value.
C. The likelihood of loss and the monetary value of those losses are the most essential elements to consider in assessing risk.
D. Past threats and losses may be instructive of potential future events but is not the most essential consideration when assessing risk.

S2-28 In which of the following areas are data owners **PRIMARILY** responsible for establishing risk mitigation?

A. Platform security
B. Entitlement changes
C. Intrusion detection
D. Antivirus controls

B is the correct answer.

Justification:
A. Platform security is usually the responsibility of the information security manager.
B. Data owners are responsible for assigning user entitlements and approving access to the systems for which they are responsible.
C. Intrusion detection is the responsibility of the information security manager.
D. Antivirus controls are the responsibility of the information security manager.

S2-29 The **PRIMARY** goal of a corporate risk management program is to ensure that an organization's:

A. IT assets in key business functions are protected.
B. business risk is addressed by preventive controls.
C. stated objectives are achieved.
D. IT facilities and systems are always available.

C is the correct answer.

Justification:
A. Protecting IT assets is one goal among many others included in the stated objectives. However, these should be put in the perspective of achieving an organization's objectives.
B. Preventive controls are not always possible or necessary; risk management will address issues with an appropriate mix of preventive and corrective controls to achieve the stated objectives.
C. Risk management's primary goal is to ensure an organization maintains the ability to achieve its objectives.
D. Ensuring infrastructure and systems availability is one typical goal included in the stated objectives.

S2-30 When performing a quantitative risk analysis, which of the following is **MOST** important to estimate the potential loss?

A. Evaluate productivity losses
B. Assess the impact of confidential data disclosure
C. Calculate the value of the information or asset
D. Measure the probability of occurrence of each threat

C is the correct answer.

Justification:
A. Determining how much productivity could be lost and how much it would cost is a step in the estimation of potential risk process.
B. Knowing the impact if confidential information is disclosed is also a step in the estimation of potential risk.
C. Calculating the value of the information or asset is the first step in a risk analysis process to determine the impact to the organization, which is the ultimate goal.
D. Measuring the probability of occurrence for each threat identified is a step in performing a threat analysis and, therefore, a partial answer.

S2-31 What is the **PRIMARY** objective of a risk management program?

A. Minimize inherent risk.
B. Eliminate business risk.
C. Implement effective controls.
D. Achieve acceptable risk.

D is the correct answer.

Justification:
A. Inherent risk may already be acceptable and require no remediation. Minimizing below the acceptable level is not the objective and usually raises costs.
B. Elimination of business risk is not possible.
C. Effective controls are naturally a clear objective of a risk management program to the extent of achieving the primary goal of achieving acceptable risk across the organization.
D. The goal of a risk management program is to ensure that acceptable risk levels are achieved and maintained.

S2-32 After completing a full IT risk assessment, who is in the **BEST** position to decide which mitigating controls should be implemented?

A. Senior management
B. The business manager
C. The IT audit manager
D. The information security officer

B is the correct answer.

Justification:
A. Senior management will have to ensure that the business manager has a clear understanding of the risk assessed, but it will not be in a position to decide on specific controls.
B. The business manager will be in the best position, based on the risk assessment and mitigation proposals, to decide which controls should/could be implemented, in line with the business strategy and with budget.
C. The IT audit manager will take part in the process to identify threats and vulnerabilities and make recommendations for mitigations.
D. The information security officer could make some decisions regarding implementation of controls. However, the business manager will have a broader business view and better understanding of control impact on the business goals and, therefore, will be in a better position to make strategic decisions.

S2-33 What is the **FIRST** step of performing an information risk analysis?

A. Establish the ownership of assets.
B. Evaluate the risk to the assets.
C. Take an asset inventory.
D. Categorize the assets.

C is the correct answer.

Justification:
A. Assets must be inventoried before the ownership of the assets can be established.
B. Assets must be inventoried before risk to the assets can be evaluated.
C. Assets must be inventoried before any of the other choices can be performed.
D. Assets must be inventoried before they can be categorized.

S2-34 What is the **PRIMARY** benefit of performing an information asset classification?

A. It links security requirements to business objectives.
B. It identifies controls commensurate with impact.
C. It defines access rights.
D. It establishes asset ownership.

B is the correct answer.

Justification:
A. Asset classification indirectly links security to business objectives on the basis of business value of assets.
B. Classification levels are based on the business value (or potential impact) of assets and the stronger controls needed for higher classification.
C. Classification does not define access rights.
D. Classification does not establish ownership.

S2-35 Which of the following is **MOST** essential for a risk management program to be effective?

A. Flexible security budget
B. Sound risk baseline
C. Detection of new risk
D. Accurate risk reporting

C is the correct answer.

Justification:
A. A flexible security budget is essential for implementing risk management. However, without identifying new risk, other procedures will only be useful for a limited period.
B. A sound risk baseline is essential for implementing risk management. However, without identifying new risk, other procedures will only be useful for a limited period.
C. All of these procedures are essential for implementing risk management. However, without identifying new risk, other procedures will only be useful for a limited period.
D. Accurate risk reporting is essential for implementing risk management. However, without identifying new risk, other procedures will only be useful for a limited period.

S2-36 Which of the following attacks is **BEST** mitigated by using strong passwords?

A. Man-in-the-middle attack
B. Brute force attack
C. Remote buffer overflow
D. Root kit

B is the correct answer.

Justification:
A. Man-in-the-middle attacks intercept network traffic and must be protected by encryption.
B. Strong passwords mitigate brute force attacks.
C. Buffer overflow attacks may not be protected by passwords.
D. Root kits hook into the operating system's kernel and, therefore, operate underneath any authentication mechanism.

S2-37 Phishing is **BEST** mitigated by which of the following?

A. Security monitoring software
B. Encryption
C. Two-factor authentication
D. User awareness

D is the correct answer.

Justification:
A. Security monitoring software is generally incapable of detecting a phishing attack.
B. Encryption would not mitigate this threat.
C. Two-factor authentication would not mitigate this threat.
D. Phishing is a type of email attack that attempts to convince a user that the originator is genuine, but with the intention of obtaining information for use in social engineering. It can best be mitigated by appropriate user awareness.

S2-38 Risk assessments should be repeated at regular intervals because:

A. business threats are constantly changing.
B. omissions in earlier assessments can be addressed.
C. repetitive assessments allow various methodologies.
D. they help raise awareness on security in the business.

A is the correct answer.

Justification:
A. As business objectives and methods change, the nature and relevance of threats change as well.
B. Omissions in earlier assessments do not, by themselves, justify regular reassessment.
C. Use of various methodologies is not a business reason for repeating risk assessments at regular intervals.
D. Risk assessments may help raise business awareness, but there are better ways of raising security awareness than by performing a risk assessment.

S2-39 Which of the following steps in conducting a risk assessment should be performed **FIRST**?

A. Identify business assets
B. Identify business risk
C. Assess vulnerabilities
D. Evaluate key controls

A is the correct answer.

Justification:
A. Risk assessment first requires that the business assets that need to be protected be identified before identifying the threats.
B. The second step in risk assessment is to establish whether those threats represent business risk by identifying the likelihood and effect of occurrence.
C. Assessing the vulnerabilities that may affect the security of the asset follows identifying business assets and risk.
D. Risk evaluation after analysis is used to determine whether controls address the risk to meet the criteria for acceptability.

S2-40 What is a reasonable expectation to have of a risk management program?

A. It removes all inherent risk.
B. It maintains residual risk at an acceptable level.
C. It implements preventive controls for every threat.
D. It reduces control risk to zero.

B is the correct answer.

Justification:
A. Risk management is not intended to remove every identified risk because this may not be cost-effective.
B. The goal of risk management is to ensure that all residual risk is maintained at a level acceptable to the business.
C. Risk management is not intended to implement controls for every threat because not all threats pose a risk, and this would not be cost-effective.
D. Control risk is the risk that a control may not be effective; it is a component of the program but is unlikely to be reduced to zero.

S2-41 In which phase of the development process should risk assessment be **FIRST** introduced?

A. Programming
B. Specification
C. User testing
D. Feasibility

D is the correct answer.

Justification:
A. Assessment would not be relevant in the programming phase.
B. Risk should also be considered in the specification phase, where the controls are designed, but this would still be based on the assessment carried out in the feasibility study.
C. Assessment would not be relevant in the user testing phase.
D. Risk should be addressed as early as possible in the development cycle. The feasibility study should include risk assessment so that the cost of controls can be estimated before the project proceeds.

S2-42 In conducting an initial technical vulnerability assessment, which of the following choices should receive top priority?

A. Systems impacting legal or regulatory standing
B. Externally facing systems or applications
C. Resources subject to performance contracts
D. Systems covered by business interruption insurance

D is the correct answer.

Justification:
A. Legal and regulatory considerations are evaluated in the same manner as other forms of risk.
B. Externally facing systems or applications are not necessarily high-impact systems. The prioritization of a vulnerability assessment needs to be made on the basis of impact.
C. Although the impact associated with the loss of any resource subject to a performance contract is clearly quantifiable, it may not necessarily be a critical resource. If the loss of a contract system poses a significant impact to the organization, additional measures such as business interruption insurance will be in place.
D. Maintaining business operations is always the priority. If a system is covered by business interruption insurance, it is a clear indication that management deems it to be a critical system.

S2-43 Why would an organization decide not to take any action on a denial-of-service vulnerability found by the risk assessment team?

A. There are sufficient safeguards in place to prevent this risk from happening.
B. The needed countermeasures are too complicated to deploy.
C. The cost of countermeasures outweighs the value of the asset and potential loss.
D. The likelihood of the risk occurring is unknown.

C is the correct answer.

Justification:
A. The safeguards need to match the risk level. You can never be certain of having sufficient safeguards because threats are always evolving.
B. While countermeasures could be too complicated to deploy, this is not the most compelling reason.
C. An organization may decide to live with specific risk because it would cost more to protect the organization than the value of the potential loss.
D. It is unlikely that a global financial institution would not be exposed to such attacks, and the likelihood could not be predicted.

S2-44 Which of the following types of information would the information security manager expect to have the **LOWEST** level of security protection in a publicly traded, multinational enterprise?

A. Strategic business plan
B. Upcoming financial results
C. Customer personal information
D. Previous financial results

D is the correct answer.

Justification:
A. The strategic business plan is private information and should only be accessed by authorized entities.
B. Upcoming financial results are private information and should only be accessed by authorized entities.
C. Customer personal information is private information and should only be accessed by authorized entities.
D. Previous financial results are public; all of the other choices are private information and should only be accessed by authorized entities.

S2-45 What is the **PRIMARY** purpose of using risk analysis within a security program?

A. The risk analysis helps justify the security expenditure.
B. The risk analysis helps prioritize the assets to be protected.
C. The risk analysis helps inform executive management of the residual risk.
D. The risk analysis helps assess exposures and plan remediation.

D is the correct answer.

Justification:
A. Risk analysis indirectly supports the security expenditure, but justifying the security expenditure is not its primary purpose.
B. Helping businesses prioritize the assets to be protected is an indirect benefit of risk analysis but not its primary purpose.
C. Informing executive management of residual risk value is not directly relevant.
D. Risk analysis explores the degree to which an asset needs protecting so this can be managed effectively.

S2-46 Which of the following is the **PRIMARY** prerequisite to implementing data classification within an organization?

A. Defining job roles
B. Performing a risk assessment
C. Identifying data owners
D. Establishing data retention policies

C is the correct answer.

Justification:
A. Defining job roles is not relevant.
B. Performing a risk assessment is important but will require the participation of data owners (who must first be identified).
C. Identifying the data owners is the first step and is essential to implementing data classification.
D. Establishing data retention policies may occur at any time.

S2-47 An online banking institution is concerned that a breach of customer personal information will have a significant financial impact due to the need to notify and compensate customers whose personal information may have been compromised. The institution determines that residual risk will always be too high and decides to:

A. mitigate the impact by purchasing insurance.
B. implement a circuit-level firewall to protect the network.
C. increase the resiliency of security measures in place.
D. implement a real-time intrusion detection system.

A is the correct answer.

Justification:
A. Residual risk is the remaining risk after management has implemented a risk response. Because residual risk will always be too high, the only practical solution is to mitigate the financial impact by purchasing insurance. Purchasing insurance is also known as risk transfer.
B. The organization has determined the residual risk will always be too high and chosen to transfer the risk, so there is no need to attempt further mitigation.
C. The organization has determined the residual risk will always be too high and chosen to transfer the risk, so there is no need to attempt further mitigation.
D. The organization has determined the residual risk will always be too high and chosen to transfer the risk, so there is no need to attempt further mitigation.

S2-48 What mechanism should be used to identify deficiencies that would provide attackers with an opportunity to compromise a computer system?

A. Business impact analysis
B. Security gap analysis
C. System performance metrics
D. Incident response processes

B is the correct answer.

Justification:
A. A business impact analysis does not identify vulnerabilities.
B. Security gap analysis is a process that measures all security controls in place against control objectives, which will identify gaps.
C. System performance metrics may indicate security weaknesses, but that is not their primary purpose.
D. Incident response processes exist for cases in which security weaknesses are exploited.

S2-49 A project manager is developing a developer portal and requests that the security manager assign a public Internet Protocol address so that it can be accessed by in-house staff and by external consultants outside the organization's local area network. What should the security manager do **FIRST**?

A. Understand the business requirements of the developer portal.
B. Perform a vulnerability assessment of the developer portal.
C. Install an intrusion detection system.
D. Obtain a signed nondisclosure agreement from the external consultants before allowing external access to the server.

A is the correct answer.

Justification:
A. The information security manager cannot make an informed decision about the request without first understanding the business requirements of the developer portal.
B. Performing a vulnerability assessment of developer portal is prudent but is subsequent to understanding the requirements.
C. Installing an intrusion detection system may be useful but not as essential as understanding the requirements.
D. Obtaining a signed nondisclosure agreement is a prudent practice but is secondary to understanding requirements.

S2-50 A mission-critical system has been identified as having an administrative system account with attributes that prevent locking and change of privileges and name. Which would be the **BEST** approach to prevent a successful brute force attack of the account?

A. Prevent the system from being accessed remotely.
B Create a strong random password.
C. Ask for a vendor patch.
D. Track usage of the account by audit trails.

B is the correct answer.

Justification:

A. Preventing the system from being accessed remotely is not always an option in mission-critical systems and still leaves local access risk.
B. Creating a strong random password reduces the risk of a successful brute force attack by exponentially increasing the time required.
C. Vendor patches are not always available.
D. Tracking usage is a detective control and will not prevent an attack.

X **S2-51** Attackers who exploit cross-site scripting vulnerabilities take advantage of:

A. a lack of proper input validation controls.
B. weak authentication controls in the web application layer.
C. flawed cryptographic Secure Sockets Layer implementations and short key lengths.
D. implicit web application trust relationships.

A is the correct answer.

Justification:

A. Cross-site scripting attacks inject malformed input.
B. Attackers who exploit weak application authentication controls can gain unauthorized access to applications, but this has little to do with cross-site scripting vulnerabilities.
C. Attackers who exploit flawed cryptographic Secure Sockets Layer implementations and short key lengths can sniff network traffic and crack keys to gain unauthorized access to information. This has little to do with cross-site scripting vulnerabilities.
D. Web application trust relationships do not relate directly to the attack.

S2-52 Which of the following would **BEST** address the risk of data leakage?

A. File backup procedures
B. Database integrity checks
C. Acceptable use policies
D. Incident response procedures

C is the correct answer.

Justification:

A. File backup procedures ensure the availability of information in alignment with data retention requirements but do nothing to prevent leakage.
B. Database integrity checks verify the allocation and structural integrity of all the objects in the specified database but do nothing to prevent leakage.
C. An acceptable use policy establishes an agreement between users and the enterprise and defines for all parties the ranges of use that are approved before gaining access to a network or the Internet.
D. Incident response procedures provide detailed steps that help an organization minimize the impact of an adverse event and do not directly address data leakage.

S2-53 A company recently developed a breakthrough technology. Because this technology could give this company a significant competitive edge, which of the following would **FIRST** govern how this information is to be protected?

A. Access control policy
B. Data classification policy
C. Encryption standards
D. Acceptable use policy

B is the correct answer.

Justification:
A. Without a mandated ranking of degree of protection, it is difficult to determine what access controls should be in place.
B. Data classification policies define the level of protection to be provided for each category of data based on business value.
C. Without a mandated ranking of degree of protection, it is difficult to determine what levels of encryption should be in place.
D. An acceptable use policy is oriented more toward the end user and, therefore, would not specifically address what controls should be in place to adequately protect information.

S2-54 What is the **BEST** technique to determine which security controls to implement with a limited budget?

A. Risk analysis
B. Annual loss expectancy calculations
C. Cost-benefit analysis
D. Impact analysis

C is the correct answer.

Justification:
A. Risk analysis quantifies risk to prioritize risk responses.
B. The annual loss expectancy is the monetary loss that can be expected for an asset due to a risk over a one-year period but does nothing to prioritize controls.
C. Cost-benefit analysis is performed to ensure that the cost of a safeguard does not outweigh its benefit and that the best safeguard is provided for the cost of implementation.
D. An impact analysis is a study to prioritize the criticality of information resources for the enterprise based on costs (or consequences) of adverse events. In an impact analysis, threats to assets are identified and potential business losses determined for different time periods. This assessment is used to justify the extent of safeguards that are required and recovery time frames. This analysis is the basis for establishing the recovery strategy.

S2-55 A company's mail server allows anonymous File Transfer Protocol access, which could be exploited. What process should the information security manager deploy to determine the necessity for remedial action?

A. A penetration test
B. A security baseline review
C. A risk assessment
D. A business impact analysis

C is the correct answer.

Justification:
A. A penetration test may identify the vulnerability but not potential threats or the remedy.
B. A security baseline review may identify the vulnerability but not the remedy.
C. A risk assessment will identify the business impact of the vulnerability being exploited and the remedial options.
D. A business impact analysis will identify the impact of the loss of the mail server and requirements for restoration.

S2-56 Which of the following measures would be **MOST** effective against insider threats to confidential information?

A. Role-based access control
B. Audit trail monitoring
C. Privacy policy
D. Defense in depth

A is the correct answer.

Justification:
A. Role-based access control is a preventive control that provides access according to business needs; therefore, it reduces unnecessary access rights and enforces accountability.
B. Audit trail monitoring is a detective control, which is "after the fact."
C. Privacy policy is not relevant to this risk.
D. Defense in depth primarily focuses on external threats and control layering.

X

S2-57 After a risk assessment study, a bank with global operations decided to continue conducting business in certain regions of the world where identity theft is rampant. The information security manager should encourage the business to:

A. increase its customer awareness efforts in those regions.
B. implement monitoring techniques to detect and react to potential fraud.
C. outsource credit card processing to a third party.
D. make the customer liable for losses if they fail to follow the bank's advice.

B is the correct answer.

Justification:
A. While customer awareness helps mitigate risk, this is insufficient on its own to control fraud risk.
B. Implementing monitoring techniques, which will detect and deal with potential fraud cases, is the most effective way to deal with this risk.
C. If the bank outsources its processing, the bank still retains liability.
D. While it is an unlikely possibility to make the customer liable for losses, the bank needs to be proactive in managing risk.

S2-58 Which of the following is the **BEST** basis for determining the criticality and sensitivity of information assets?

A. A threat assessment
B. A vulnerability assessment
C. A resource dependency assessment
D. An impact assessment

D is the correct answer.

Justification:

A. Threat assessment lists only the threats that the information asset is exposed to; it does not consider the value of the asset and impact of the threat on the value.
B. Vulnerability assessment lists only the vulnerabilities inherent in the information asset that can attract threats. It does not consider the value of the asset and the impact of perceived threats on the value.
C. Resource dependency assessments provide process needs, but not impact.
D. The criticality and sensitivity of information assets depends on the impact of the likelihood of the threats exploiting vulnerabilities in the asset and takes into consideration the value of the assets and the impairment of the value.

S2-59 Which program element should be implemented **FIRST** in asset classification and control?

A. Risk assessment
B. Classification
C. Valuation
D. Risk mitigation

C is the correct answer.

Justification:

A. Risk assessment is performed to identify and quantify threats to information assets that are selected by the first step, valuation.
B. Classification is a step following valuation.
C. Valuation is performed first to identify and understand the value of assets needing protection.
D. Risk mitigation is a step following valuation based on the valuation.

S2-60 Which of the following is the **MOST** important consideration when performing a risk assessment?

A. Management supports risk mitigation efforts.
B. Annual loss expectancies have been calculated for critical assets.
C. Assets have been identified and appropriately valued.
D. Attack motives, means and opportunities are understood.

C is the correct answer.

Justification:

A. Management support is always important, but is not relevant when performing a risk assessment except to the extent that a lack of support may present a risk.
B. The annual loss expectancy calculations can be used in risk analysis, which is subsequent to assets first being identified and properly valued.
C. Identification and valuation of assets provides the essential basis for risk assessment efforts. Without knowing an asset exists and its value to the organization, the risk and impact cannot be determined.
D. Motives, means and opportunities are considered as a part of risk identification but must be considered in the context of identified and valued assets.

S2-61 Why is asset classification important to a successful information security program?

A. It determines the priority and extent of risk mitigation efforts.
B. It determines the amount of insurance needed in case of loss.
C. It determines the appropriate level of protection to the asset.
D. It determines how protection levels compare to peer organizations.

C is the correct answer.

Justification:

A. Classification does not determine the priority and extent of the risk mitigation efforts; prioritization of risk mitigation efforts is generally based on risk analysis or a business impact analysis.
B. Classification does not establish the amount of insurance needed; insurance is often not a viable option.
C. Classification is based on the value of the asset to the organization and helps establish the protection level in proportion to the value of the asset.
D. Classification schemes differ from organization to organization and are often not suitable for benchmarking.

S2-62 What is the **BEST** strategy for risk management?

A. Achieve a balance between risk and organizational goals.
B. Reduce risk to an acceptable level.
C. Ensure that policy development properly considers organizational risk.
D. Ensure that all unmitigated risk is accepted by management.

B is the correct answer.

Justification:

A. Achieving balance between risk and organizational goals is not always practical.
B. The best strategy for risk management is to reduce risk to an acceptable level, as this will take into account the organization's appetite for risk and the fact that it is not possible to eliminate all risk.
C. Policy development must consider organizational risk as well as business objectives but is not a strategy.
D. It may be prudent to ensure that management understands and accepts risk that it is not willing to mitigate, but that is a practice and is not sufficient to be considered a strategy.

S2-63 Which of the following is the **MOST** important factor to be considered in the loss of mobile equipment with unencrypted data?

A. Disclosure of personal information
B. Sufficient coverage of the insurance policy for accidental losses
C. Potential impact of the data loss
D. Replacement cost of the equipment

C is the correct answer.

Justification:

A. Personal information is not defined in the question as the data that were lost.
B. If insurance is available, it is unlikely to compensate for all potential impact.
C. When mobile equipment is lost or stolen, the information contained on the equipment matters most in determining the impact of the loss. The more sensitive the information, the greater the liability. If staff carries mobile equipment for business purposes, an organization must develop a clear policy as to what information should be kept on the equipment and for what purpose.
D. Cost of equipment would be a less important issue.

S2-64 An organization has to comply with recently published industry regulatory requirements—compliance that potentially has high implementation costs. What should the information security manager do **FIRST**?

A. Consult the security committee.
B. Perform a gap analysis.
C. Implement compensating controls.
D. Demand immediate compliance.

B is the correct answer.

Justification:
A. Consulting the steering committee before knowing the extent of the issues would not be the first step.
B. Because they are regulatory requirements, a gap analysis would be the first step to determine the level of compliance already in place.
C. Implementing compensating controls would not be the first step.
D. Demanding immediate compliance without knowing the extent of possible noncompliance would not be a prudent first step.

S2-65 Which of the following would be **MOST** relevant to include in a cost-benefit analysis of a two-factor authentication system?

A. Annual loss expectancy of incidents
B. Frequency of incidents
C. Total cost of ownership
D. Approved budget for the project

C is the correct answer.

Justification:
A. Annual loss expectancy could help measure the benefit but does not address the costs.
B. The potential reduction in the frequency of incidents could also help measure the benefit but does not address cost.
C. The total cost of ownership would be the most relevant piece of information to determine both the total cost and the benefit.
D. The approved budget for the project is not relevant to the cost-benefit analysis.

S2-66 What is a reasonable approach to determine control effectiveness?

A. Determine whether the control is preventive, detective or corrective.
B. Review the control's capability of providing notification of failure.
C. Confirm the control's ability to meet intended objectives.
D. Assess and quantify the control's reliability.

C is the correct answer.

Justification:
A. The type of control is not relevant.
B. Notification of failure is not determinative of control effectiveness.
C. Control effectiveness requires a process to verify that the control process works as intended. Examples such as dual-control or dual-entry bookkeeping provide verification and assurance that the process operated as intended.
D. Reliability is not an indication of control strength; weak controls can be highly reliable, even if they are ineffective controls.

S2-67 Of the following, what does a network vulnerability assessment expect to identify?

A. Zero-day vulnerabilities
B. Malicious software and spyware
C. Security design flaws
D. Misconfiguration and missing updates

D is the correct answer.

Justification:
A. Zero-day vulnerabilities by definition are not previously known and, therefore, are undetectable.
B. Malicious software and spyware are normally addressed through antivirus and antispyware policies.
C. Security design flaws require a deeper level of analysis.
D. A network vulnerability assessment intends to identify known vulnerabilities based on common misconfigurations and missing updates.

S2-68 Which of the following roles is responsible for ensuring that information is classified?

A. Senior management
B. The security manager
C. The data owner
D. The data custodian

C is the correct answer.

Justification:

A. Senior management is ultimately responsible for the organization.
B. The security manager is responsible for applying security protection relative to the level of classification specified by the owner.
C. The data owner is responsible for applying the proper classification to the data.
D. The technology group is delegated the custody of the data by the data owner, but the group does not classify the information.

S2-69 After a risk assessment, it is determined that the cost to mitigate the risk is much greater than the benefit to be derived. The information security manager should recommend to business management that the risk be:

A. transferred.
B. treated.
C. accepted.
D. terminated.

C is the correct answer.

Justification:
A. Transferring the risk is of limited benefit if the cost of that control is more than the potential cost of the risk manifesting.
B. Treating the risk is of limited benefit if the cost of that control is more than the cost of the risk being exploited.
C. When the cost of the control is more than the cost of the risk, the risk should be accepted.
D. If the value of the activity is greater than the potential cost of compromise, then terminating the activity would not be the appropriate advice.

S2-70 The **PRIMARY** reason for initiating a policy exception process is when:

A. operations are too busy to comply.
B. the risk is justified by the benefit.
C. policy compliance would be difficult to enforce.
D. users may initially be inconvenienced.

B is the correct answer.

Justification:
A. Being busy is not a justification for policy exceptions.
B. Exceptions to policy are warranted in circumstances where compliance may be difficult or impossible and the risk of noncompliance is outweighed by the benefits.
C. The fact that compliance cannot be enforced is not a justification for policy exceptions.
D. User inconvenience is not a reason to automatically grant exception to a policy.

S2-71 Which of the following would be the **MOST** relevant factor when defining the information classification policy?

A. Quantity of information
B. Available IT infrastructure
C. Benchmarking
D. Requirements of data owners

D is the correct answer.

Justification:
A. The quantity of information is not a factor in defining the information classification policy.
B. The availability of IT infrastructure would not be a significant factor in determining the policy.
C. Benchmarking would not be a factor in defining the classification policy.
D. When defining the information classification policy, the requirements of the data owners need to be identified.

S2-72 Who should be assigned as data owner for sensitive customer data that is used only by the sales department and stored in a central database?

A. The sales department
B. The database administrator
C. The chief information officer
D. The head of the sales department

D is the correct answer.

Justification:
A. The sales department cannot be the owner of the asset because that removes personal responsibility.
B. The database administrator is a custodian.
C. The chief information officer (CIO) is not an owner of this database because the CIO is less likely to be knowledgeable about the specific needs of sales operations and security concerns.
D. The owner of the information asset should be the individual with the decision-making power in the department deriving the most benefit from the asset. In this case, it is the head of the sales department.

S2-73 What activity should information security management perform **FIRST** when assessing the potential impact of new privacy legislation on the organization?

A. Develop an operational plan for achieving compliance with the legislation.
B. Identify systems and processes that contain privacy components.
C. Restrict the collection of personal information until compliant.
D. Identify privacy legislation in other countries that may contain similar requirements.

B is the correct answer.

Justification:
A. Developing an operational plan for achieving compliance with the legislation is incorrect because it is not the first step.
B. Identifying the relevant systems and processes is the best first step.
C. Restricting the collection of personal information comes later.
D. Identifying privacy legislation in other countries would not add much value.

S2-74 When should risk assessments be performed for optimum effectiveness?

A. At the beginning of security program development
B. On a continuous basis
C. While developing the business case for the security program
D. During the business change management process

B is the correct answer.

Justification:
A. The beginning of a security program is only one time a risk assessment should be performed.
B. Risk assessment needs to be performed on a continuous basis because of organizational and technical changes. Risk assessment must take into account all significant changes in order to be effective.
C. Part of developing the business case is another point where risk assessment should occur.
D. Risk should be assessed during the change management process but is only one point.

S2-75 There is a delay between the time when a security vulnerability is first published, and the time when a patch is delivered. Which of the following should be carried out **FIRST** to mitigate the risk during this time period?

A. Identify the vulnerable systems and apply compensating controls.
B. Minimize the use of vulnerable systems.
C. Communicate the vulnerability to system users.
D. Update the signatures database of the intrusion detection system.

A is the correct answer.

Justification:
A. The best protection is to identify the vulnerable systems and apply compensating controls until a patch is installed.
B. Minimizing the use of vulnerable systems could be a compensating control but would not be the first course of action.
C. Communicating the vulnerability to system users would not be of much benefit.
D. Updating the signatures database of the intrusion detection system (IDS) does not address the timing of when the IDS signature list would be updated to accommodate those vulnerabilities that are not yet publicly known. Therefore, this approach should not always be considered as the first option.

S2-76 Which of the following techniques **MOST** clearly indicates whether specific risk-reduction controls should be implemented?

A. Cost-benefit analysis
B. Penetration testing
C. Frequent risk assessment programs
D. Annual loss expectancy calculation

A is the correct answer.

Justification:

A. In a cost-benefit analysis, the annual cost of safeguards is compared with the expected cost of loss. This can then be used to justify a specific control measure.
B. Penetration testing may indicate the extent of a weakness but, by itself, will not establish the cost-benefit of a control.
C. Frequent risk assessment programs will certainly establish what risk exists but will not determine the cost of controls.
D. Annual loss expectancy is a measure that will contribute to the potential cost associated with the risk but does not address the benefit of a control.

S2-77 Which of the following is the **MAIN** reason for performing risk assessment on a continuous basis?

A. Justification of the security budget must be continually made.
B. New vulnerabilities are discovered every day.
C. The risk environment is constantly changing.
D. Management needs to be continually informed about emerging risk.

C is the correct answer.

Justification:

A. Justification of a budget should never be the main reason for performing a risk assessment.
B. New vulnerabilities should be managed through a patch management process.
C. The risk environment is impacted by factors such as changes in technology and business strategy. These changes introduce new threats and vulnerabilities to the organization. As a result, risk assessment should be performed continuously.
D. Informing management about emerging risk is important, but is not the main driver for determining when a risk assessment should be performed.

S2-78 Which of the following roles is **PRIMARILY** responsible for determining the information classification levels for a given information asset?

A. Manager
B. Custodian
C. User
D. Owner

D is the correct answer.

Justification:

A. Management is responsible for higher-level issues such as providing and approving budget, supporting activities, etc.
B. The information custodian is responsible for day-to-day security tasks such as protecting information, backing up information, etc.
C. Users are the lowest level. They use the data but do not classify the data. The owner classifies the data.
D. Although the information owner may be in a management position and is also considered a user, the information owner role has the responsibility for determining information classification levels.

X **S2-79** The **PRIMARY** reason for classifying information resources according to sensitivity and criticality is to:

A. determine inclusion of the information resource in the information security program.
B. define the appropriate level of access controls.
C. justify the costs of each information resource.
D. determine the overall budget of the information security program.

B is the correct answer.

Justification:

A. The assignment of sensitivity and criticality takes place with the information assets that have already been included in the information security program.
B. The assigned class of sensitivity and criticality of the information resource determines the level of access controls to be put in place.
C. Classification is unrelated to the costs of the information resource.
D. The overall security budget is not directly related to classification.

S2-80 When performing a qualitative risk analysis, which of the following will **BEST** produce reliable results?

A. Estimated productivity losses
B. Possible scenarios with threats and impacts
C. Value of information assets
D. Vulnerability assessment

B is the correct answer.

Justification:

A. Estimated productivity losses are better suited to quantitative analysis but, without threats being considered, would not produce useful results.
B. Listing all reasonable scenarios that could occur, along with threats and impacts, will best frame the range of risk and facilitate a more informed discussion and decision.
C. Value of information assets would be part of a quantitative analysis requiring threat to be considered as well.
D. Vulnerability assessments would be better analyzed as a part of a quantitative analysis when threat is considered.

S2-81 Which of the following is the **BEST** method to ensure the overall effectiveness of a risk management program?

A. User assessments of changes
B. Comparison of the program results with industry standards
C. Assignment of risk within the organization
D. Participation by all members of the organization

D is the correct answer.

Justification:
A. User assessments are most likely focused on their convenience and ease of use rather than effectiveness of the program.
B. Comparing results with industry standards is a meaningless gauge; however, comparing results to program objectives would be very useful.
C. Assigning ownership of risk is a good first step in improving accountability and, therefore, probably effectiveness.
D. Effective risk management requires participation, support and acceptance by all applicable members of the organization, beginning with the executive levels. Personnel must understand their responsibilities and be trained on how to fulfill their roles.

S2-82 The **MOST** effective use of a risk register is to:

A. identify risk and assign roles and responsibilities for mitigation.
B. identify threats and probabilities.
C. facilitate a thorough review of all IT-related risk on a periodic basis.
D. record the annualized financial amount of expected losses due to risk.

C is the correct answer.

Justification:
A. Identifying risk and assigning roles and responsibilities for mitigation are elements of the register.
B. Identifying threats and probabilities are two elements that are defined in the risk matrix, as differentiated from the broader scope of content in, and purpose for, the risk register.
C. A risk register is more than a simple list—it should be used as a tool to ensure comprehensive documentation, periodic review and formal update of all risk elements in the enterprise's IT and related organization.
D. While the annual loss expectancy should be included in the register, this quantification is only a single element in the overall risk analysis program.

S2-83 Which of the following are the essential ingredients of a business impact analysis?

A. Downtime tolerance, resources and criticality
B. Cost of business outages in a year as a factor of the security budget
C. Business continuity testing methodology being deployed
D. Structure of the crisis management team

A is the correct answer.

Justification:

A. **A business impact analysis (BIA) is an exercise that determines the impact of losing the support of any resource to an enterprise, establishes the escalation of that loss over time, identifies the minimum resources needed to recover, and prioritizes the recovery of processes and the supporting system. The main inputs into a BIA are criticality of a business function or process, associated resources and maximum tolerable downtime.**
B. Cost of business outages is associated with business continuity planning but is not related to the BIA.
C. Business continuity testing methodology is associated with business continuity planning but is not related to the BIA.
D. Structure of the crisis management team is associated with business continuity planning but is not related to the BIA.

S2-84 Which of the following is the **MOST** effective way to treat a risk such as a natural disaster that has a low probability and a high impact level?

A. Implement countermeasures.
B. Eliminate the risk.
C. Transfer the risk.
D. Accept the risk.

C is the correct answer.

Justification:

A. Implementing countermeasures may not be possible or the most cost-effective approach to security management.
B. Eliminating the risk may not be possible.
C. **Risk is typically transferred to insurance companies when the probability of an incident is low but the impact is high. Examples include hurricanes, tornados and earthquakes.**
D. Accepting the risk would leave the organization vulnerable to a catastrophic disaster that may cripple or ruin the organization. It would be more cost-effective to pay recurring insurance costs than to be affected by a disaster from which the organization cannot financially recover.

S2-85 What activity needs to be performed for previously accepted risk?

A. Risk should be reassessed periodically because risk changes over time.
B. Accepted risk should be flagged to avoid future reassessment efforts.
C. Risk should be avoided next time to optimize the risk profile.
D. Risk should be removed from the risk log after it is accepted.

A is the correct answer.

Justification:

A. Acceptance of risk should be regularly reviewed to ensure that the rationale for the initial risk acceptance is still valid within the current business context. The rationale for initial risk acceptance may no longer be valid due to change(s), and therefore, risk cannot be accepted permanently.
B. Even risk that has been accepted should be monitored for changing conditions that could alter the original decision.
C. Risk is an inherent part of business, and avoiding it to improve the risk profile would be misleading and dangerous.
D. Even risk that has been accepted should be maintained in the risk log and monitored for changing conditions that could alter the original decision.

S2-86 An information security manager is advised by contacts in law enforcement that there is evidence that the company is being targeted by a skilled gang of hackers known to use a variety of techniques, including social engineering and network penetration. The **FIRST** step that the security manager should take is to:

X

A. perform a comprehensive assessment of the organization's exposure to the hackers' techniques.
B. initiate awareness training to counter social engineering.
C. immediately advise senior management of the elevated risk.
D. increase monitoring activities to provide early detection of intrusion.

C is the correct answer.

Justification:

A. The security manager should assess the risk, but senior management should be immediately advised.
B. It may be prudent to initiate an awareness campaign subsequent to sounding the alarm if awareness training is not current.
C. Information about possible significant new risk from credible sources should be provided to management along with advice on steps that need to be taken to counter the threat.
D. Monitoring activities should also be increased after notifying management.

S2-87 Abnormal server communication from inside the organization to external parties may be monitored to:

A. record the trace of advanced persistent threats.
B. evaluate the process resiliency of server operations.
C. verify the effectiveness of an intrusion detection system.
D. support a nonrepudiation framework in e-commerce.

A is the correct answer.

Justification:

A. The most important feature of target attacks as seen in advanced persistent threats is that malware secretly sends information back to a command and control server. Therefore, monitoring of outbound server communications that do not follow predefined routes will be the best control to detect such security events.

B. Server communications are usually not monitored to evaluate the resiliency of server operations.

C. The effectiveness of an intrusion detection system may not be verified by monitoring outbound server communications.

D. Nonrepudiation may be supported by technology, such as a digital signature. Server communication itself does not support the effectiveness of an e-commerce framework.

S2-88 Which of the following authentication methods prevents authentication replay?

A. Password hash implementation
B. Challenge/response mechanism
C. Wired equivalent privacy encryption usage
D. Hypertext Transfer Protocol basic authentication

B is the correct answer.

Justification:

A. Capturing the authentication handshake and replaying it through the network will not work. Using hashes by itself will not prevent a replay.

B. A challenge/response mechanism prevents replay attacks by sending a different random challenge in each authentication event. The response is linked to that challenge.

C. A wired equivalent privacy key will not prevent sniffing, but it will take the attacker longer to break the WEP key if he/she does not already have it). Therefore, it will not be able to prevent recording and replaying an authentication handshake.

D. Hypertext Transfer Protocol basic authentication is cleartext and has no mechanisms to prevent replay.

S2-89 IT-related risk management activities are **MOST** effective when they are:

A. treated as a distinct process.
B. conducted by the IT department.
C. integrated within business processes.
D. communicated to all employees.

C is the correct answer.

Justification:

A. IT risk is part of the broader risk landscape and must be integrated into overall risk management activities.

B. To ensure an objective, holistic approach, IT risk management must be addressed on an enterprisewide basis, making it separate from the IT department.

C. IT is an enabler of business activities, and to be effective, it must be integrated into business processes.

D. Communication alone does not necessarily correlate with successful execution of the process.

S2-90 What is the **PRIMARY** purpose of segregation of duties?

A. Employee monitoring
B. Reduced supervisory requirements
C. Fraud prevention
D. Enhanced compliance

C is the correct answer.

Justification:

A. Segregation of duties (SoD) is unrelated to monitoring.
B. As a secondary benefit, some reduction in supervision may be possible.
C. SoD is primarily used to prevent fraudulent activities.
D. If SoD is a policy requirement, then a secondary benefit is enhanced compliance. However, the policy exists to reduce fraud.

S2-91 What is the **PRIMARY** basis for the selection and implementation of products to protect the IT infrastructure?

A. Regulatory requirements
B. Technical expert advisories
C. State-of-the-art technology
D. A risk assessment

D is the correct answer.

Justification:

A. Regulatory requirements drive business requirements.
B. An expert advisory may not be aligned with business needs.
C. A risk assessment is the main driver for selecting technologies.
D. A risk assessment helps identify control gaps in the IT infrastructure and prioritize mitigation plans, which will help drive selection of security solutions.

S2-92 What is the **BEST** means to standardize security configurations in similar devices?

A. Policies
B. Procedures
C. Technical guides
D. Baselines

D is the correct answer.

Justification:

A. Policies set high-level direction, not technical details.
B. Procedures are used to provide instructions on accomplishing specific tasks.
C. Technical guides provide support but not necessarily the requirements.
D. Baselines describe the minimum configuration requirements across similar devices, activities or resources.

S2-93 Which of the following provides the **BEST** defense against the introduction of malware in end-user computers via the Internet browser?

A. Input validation checks on structured query language injection
B. Restricting access to social media sites
C. Deleting temporary files
D. Restricting execution of mobile code

D is the correct answer.

Justification:

A. Validation of checks on structured query language injection does not apply to this scenario.
B. Restricting access to social media sites may be helpful but is not the primary source of malware.
C. Deleting temporary files is not applicable to this scenario.
D. Restricting execution of mobile code is the most effective way to avoid introduction of malware into the end user's computers.

S2-94 An enterprise is transferring its IT operations to an offshore location. An information security manager should **PRIMARILY** focus on:

A. reviewing new laws and regulations.
B. updating operational procedures.
C. validating staff qualifications.
D. conducting a risk assessment.

D is the correct answer.

Justification:

A. Reviewing new laws and regulations may or may not be identified as a mitigating measure based on the risk determined by the assessment.
B. Updating operational procedures may or may not be identified as a mitigating measure based on the risk determined by the assessment.
C. Validating staff qualifications may or may not be identified as a mitigating measure based on the risk determined by the assessment.
D. A risk assessment should be conducted to determine new risk introduced by the outsourced processes.

S2-95 A social media application system has a process to scan posted comments in search of inappropriate disclosures. Which of the following choices would circumvent this control?

A. An elaborate font setting
B. Use of a stolen identity
C. An anonymous posting
D. A misspelling in the text

D is the correct answer.

Justification:

A. Depending on the font style, text messages may become illegible; however, character codes stay the same behind the scene. Therefore, scanning may not be affected by font settings.
B. Even when a message is posted using a stolen identity, scanning will be able to catch an inappropriate posting by checking text against a predefined vocabulary table.
C. Absence of the identity of the user who posted an inappropriate message may not be a major issue in conducting the scanning of posted information.
D. Intentional misspellings are hard to detect by fixed rules or keyword search because it is difficult for the system to consider the possible misspellings. The computer may ignore misspelled items. Because humans can understand the context, it is rather easy for humans to sense the true intention hidden behind the misspelling.

S2-96 The use of insurance is an example of which of the following?

A. Risk mitigation
B. Risk acceptance
C. Risk elimination
D. Risk transfer

D is the correct answer.

Justification:

A. The effects from a potential event can be shared by procuring assurance, but the risk is not mitigated.
B. Acceptance of risk is a decision by the enterprise to assume the impact of the effects of an event.
C. Risk is never fully eliminated, unless the activity that causes the risk is stopped or avoided.
D. Insurance is a method of offsetting the financial loss that might be incurred as a result of an adverse event. Some, but not all, of the potential costs are transferred to the insurance company.

S2-97 After residual risk has been determined, the enterprise should **NEXT**:

A. transfer the remaining risk to a third party.
B. acquire insurance against the effects of the residual risk.
C. validate that the residual risk is acceptable.
D. formally document and accept the residual risk.

C is the correct answer.

Justification:

A. Transfer of the risk is a step that might be taken after initial validation occurs.
B. Acquiring insurance is a step taken after initial validation occurs.
C. After residual risk has been determined, the next step should be to validate that the risk is acceptable (or not) and within the enterprise's risk tolerance.
D. Formally documenting and accepting the residual risk is a step taken after initial validation occurs.

S2-98 An information security manager is performing a security review and determines that not all employees comply with the access control policy for the data center. The **FIRST** step to address this issue should be to:

A. assess the risk of noncompliance.
B. initiate security awareness training.
C. prepare a status report for management.
D. increase compliance enforcement.

A is the correct answer.

Justification:

A. Assessing the risk of noncompliance will provide the information needed to determine the most effective remediation requirements.
B. If awareness is adequate, training may not help and increased compliance enforcement may be indicated.
C. A report may be warranted but will not directly address the issue that is normally a part of the information security manager's responsibilities.
D. Increased enforcement is not warranted if the problem is a lack of effective communication about security policy.

X **S2-99** Which of the following factors will **MOST** affect the extent to which controls should be layered?

A. The extent to which controls are procedural
B. The extent to which controls are subject to the same threat
C. The total cost of ownership for existing controls
D. The extent to which controls fail in a closed condition

B is the correct answer.

Justification:
A. Whether controls are procedural or technical will not affect layering requirements.
B. To manage the aggregate risk of total risk, common failure modes in existing controls must be addressed by adding or modifying controls so that they fail under different conditions.
C. The total cost of ownership is unlikely to be reduced by adding additional controls.
D. Controls that fail in a closed condition pose a risk to availability, whereas controls that fail in an open condition may require additional control layers to prevent compromise.

S2-100 Who should generally determine the classification of an information asset?

A. The asset custodian
B. The security manager
C. Senior management
D. The asset owner

D is the correct answer.

Justification:
A. The custodian enforces protection of assets depending on their classification.
B. The security manager develops the structure and standards for classification and may classify the information under their ownership.
C. Senior management generally does not determine classification levels unless they are also the information owner.
D. Classifying an information asset is the responsibility of the asset owner.

S2-101 Which of the following is a preventive measure?

A. A warning banner
B. Audit trails
C. An access control
D. An alarm system

C is the correct answer.

Justification:
A. A warning banner is a deterrent control, which provides a warning that can deter potential compromise.
B. Audit trails are an example of a detective control.
C. Preventive controls inhibit attempts to violate security policies. An example of such a control is an access control.
D. An alarm system is an example of a detective control.

S2-102 Which of the following is the **BEST** resolution when a security standard conflicts with a business objective?

A. Changing the security standard
B. Changing the business objective
C. Performing a risk analysis
D. Authorizing a risk acceptance

C is the correct answer.

Justification:
A. The security standard may be changed once it is determined by analysis that the risk of doing so is acceptable.
B. It is highly improbable that a business objective could be changed to accommodate a security standard.
C. Conflicts between a security standard and a business objective should be resolved based on a risk analysis of the costs and benefits of allowing or disallowing an exception to the standard.
D. Risk acceptance is a process that derives from the risk analysis once the risk is determined to be acceptable.

S2-103 A business unit intends to deploy a new technology in a manner that places it in violation of existing information security standards. What immediate action should an information security manager take?

A. Enforce the existing security standard.
B. Change the standard to permit the deployment.
C. Perform a risk analysis to quantify the risk.
D. Perform research to propose use of a better technology.

C is the correct answer.

Justification:
A. Enforcing existing standards is a good practice; however, standards need to be continuously examined in light of new technologies and the risk they present and business requirements.
B. Standards should not be changed without an appropriate risk assessment.
C. Resolving conflicts of this type should be based on a sound risk analysis of the costs and benefits of allowing or disallowing an exception to the standard. A blanket decision should never be given without conducting such an analysis.
D. It would not be the job of the security manager to research alternative technologies.

S2-104 Logging is an example of which type of defense against systems compromise?

A. Containment
B. Detection
C. Reaction
D. Recovery

B is the correct answer.

Justification:
A. Examples of containment defenses are awareness, training and physical security defenses.
B. Detection defenses include logging as well as monitoring, measuring, auditing, detecting viruses and intrusion.
C. Examples of reaction defenses are incident response, policy and procedure change, and control enhancement.
D. Examples of recovery defenses are backups and restorations, failover and remote sites, and business continuity plans and disaster recovery plans.

S2-105 Temporarily deactivating some monitoring processes, even if supported by an acceptance of operational risk, may not be acceptable to the information security manager if:

A. it implies compliance risk.
B. short-term impact cannot be determined.
C. it violates industry security practices.
D. changes in the roles matrix cannot be detected.

A is the correct answer.

Justification:
A. Monitoring processes are also required to guarantee fulfillment of laws and regulations of the organization; therefore, the information security manager will be obligated to advise compliance with the law.
B. Even if short-term impact cannot be determined, it is a business decision to accept the risk.
C. Industry security practices do not override the business decision to accept the risk.
D. Changes in the roles matrix do not override the business decision to accept the risk.

S2-106 Which of the following is the **MOST** important to keep in mind when assessing the value of information?

A. The potential financial loss
B. The cost of recreating the information
C. The cost of insurance coverage
D. Regulatory requirements

A is the correct answer.

Justification:
A. The potential for financial loss is always a key factor when assessing the value of information.
B. The cost of recreating the information may be a contributor but not the key factor.
C. The cost of insurance coverage may be a contributor but not the key factor.
D. Regulatory requirements may be a contributor but not the key factor.

S2-107 When a proposed system change violates an existing security standard, the conflict would be **BEST** resolved by:

A. calculating the risk.
B. enforcing the security standard.
C. redesigning the system change.
D. implementing mitigating controls.

A is the correct answer.

Justification:
A. Decisions regarding security should always weigh the potential loss from a risk against the benefits derived from the change.
B. It is a management decision to determine if the change in risk is worth the benefit.
C. Redesigning the proposed change might not always be the best option because it might not meet the business needs.
D. Implementing additional controls might be an option, but this would be done after the change in risk is known.

S2-108 The information classification scheme should:

A. consider possible impact of a security breach.
B. classify personal information in electronic form.
C. be performed by the information security manager.
D. be based on a risk assessment.

A is the correct answer.

Justification:

A. **Data classification is determined by the business value of the asset (i.e., the potential impact on the business of the loss, corruption or disclosure of information).**
B. Classification of personal information in electronic form is an incomplete answer because it addresses a subset of organizational data.
C. Information classification is performed by the data owner based on accepted security criteria.
D. The risk to a particular asset is not the basis for classification, rather the potential impact from compromise is the basis.

S2-109 Which of the following is the **BEST** method to provide a new user with their initial password for email system access?

A. Provide a system-generated complex password by interoffice mail with 30 days expiration.
B. Provide a temporary password over the telephone set for immediate expiration.
C. Require no password but force the user to set their own in 10 days.
D. Set initial password equal to the user ID with expiration in 30 days.

B is the correct answer.

Justification:

A. Documenting the password on paper is not the best method even if sent through interoffice mail—if the password is complex and difficult to memorize, the user will likely keep the printed password, and this creates a security concern.
B. **A temporary password that will need to be changed upon first logon is the best method because it is reset immediately and is replaced with the user's choice of password, which will make it easier for the user to remember. If it is given to the wrong person, the legitimate user will likely notify security if still unable to access the system; therefore, the security risk is low.**
C. Setting an account with no initial password is a security concern even if it is just for a few days.
D. Choice D provides the greatest security threat because user IDs are typically known by both users and security staff, thus compromising access for up to 30 days.

S2-110 An operating system noncritical patch to enhance system security cannot be applied because a critical application is not compatible with the change. Which of the following is the **BEST** solution?

A. Rewrite the application to conform to the upgraded operating system.
B. Compensate for not installing the patch with mitigating controls.
C. Alter the patch to allow the application to run in a privileged state.
D. Run the application on a test platform; tune production to allow patch and application.

B is the correct answer.

Justification:
A. Rewriting the application is not a viable option.
B. Because the operating system (OS) patch will adversely impact a critical application, a mitigating control should be identified that will provide an equivalent level of security.
C. Altering the OS patch to allow the application to run in a privileged state is likely to create new security weaknesses.
D. Running a production application on a test platform is not an acceptable alternative because it will mean running a critical production application on a platform not subject to the same level of security controls.

S2-111 Who should **PRIMARILY** provide direction on the impact of new regulatory requirements that may lead to major application system changes?

A. The internal audit department
B. System developers/analysts
C. Key business process owners
D. Corporate legal counsel

C is the correct answer.

Justification:
A. Internal auditors would not be in as good a position to fully understand all the business ramifications.
B. System developers would not be aware of the impact on business operations.
C. Business process owners are in the best position to understand how new regulatory requirements may affect their systems.
D. Legal counsel would not be in a position to understand the ramifications.

S2-112 The IT function has declared that it is not necessary to update the business impact analysis when putting a new application into production because it does not produce modifications in the business processes. The information security manager should:

A. verify the decision with the business units.
B. check the system's risk analysis.
C. recommend update after postimplementation review.
D. request an audit review.

A is the correct answer.

Justification:
A. Verifying the decision with the business units is the correct answer because it is not the IT function's responsibility to decide whether a new application modifies business processes.
B. Checking the system's risk analysis does not consider the change in the applications.
C. Recommending the update after postimplementation review delays the update.
D. Requesting an audit review delays the update.

S2-113 An internal review of a web-based application system reveals that it is possible to gain access to all employees' accounts by changing the employee's ID used for accessing the account on the uniform resource locator. The vulnerability identified is:

A. broken authentication.
B. unvalidated input.
C. cross-site scripting.
D. structured query language injection.

A is the correct answer.

Justification:

A. The authentication process is broken because, although the session is valid, the application should reauthenticate when the input parameters are changed.
B. The review provided valid employee IDs, and valid input was processed. The problem here is the lack of reauthentication when the input parameters are changed.
C. Cross-site scripting is not the problem in this case because the attack is not transferred to any other user's browser to obtain the output.
D. Structured query language (SQL) injection is not a problem because input is provided as a valid employee ID and no SQL queries are injected to provide the output.

S2-114 What is the **MOST** cost-effective method of identifying new vendor vulnerabilities?

A. External vulnerability reporting sources
B. Periodic vulnerability assessments performed by consultants
C. Intrusion prevention software
D. Honeypots located in the demilitarized zone (DMZ)

A is the correct answer.

Justification:

A. External vulnerability sources are the most cost-effective method of identifying these vulnerabilities.
B. The cost involved in periodic vulnerability assessments would be much higher.
C. Intrusion prevention software would not identify new vendor vulnerabilities.
D. Honeypots may or may not identify vulnerabilities and may create their own security risk.

S2-115 Of the following, retention of business records should be **PRIMARILY** based on:

A. periodic vulnerability assessment.
B. business requirements.
C. device storage capacity and longevity.
D. legal requirements.

B is the correct answer.

Justification:

A. Retention of records is unrelated to vulnerability assessments.
B. Business requirements are the primary driver for records retention. This includes any legal or regulatory requirements to the extent determined by business requirements.
C. Device storage capacity and longevity are secondary considerations.
D. Legal requirements as determined by the organization to meet business requirements.

S2-116 Which is the **BEST** way to assess aggregate risk derived from a chain of linked system vulnerabilities?

A. Vulnerability scans
B. Penetration tests
C. Code reviews
D. Security audits

B is the correct answer.

Justification:

A. Security assessments, such as vulnerability scans, can help give an extensive and thorough risk and vulnerability overview but will not be able to test or demonstrate the final consequence of having several vulnerabilities linked together.
B. A penetration test is normally the only security assessment that can link vulnerabilities together by exploiting them sequentially. This gives a good measurement and prioritization of risk. Penetration testing can give risk a new perspective and prioritize based on the end result of a sequence of security problems.
C. Code reviews are very time-consuming and unlikely to occur on different parts of a system at the same time making the discovery of linked system vulnerabilities unlikely.
D. Audits are unlikely to assess aggregate risk from linked system vulnerabilities.

S2-117 What is the **PRIMARY** basis for the selection of controls and countermeasures?

A. Eliminating IT risk
B. Cost-benefit balance
C. Resource management
D. The number of assets protected

B is the correct answer.

Justification:

A. The focus must include procedural, operational and other risk—not just IT risk.
B. The balance between cost and benefits should direct controls selection.
C. Resource management is not directly related to controls.
D. The implementation of controls is based on the impact and risk, not on the number of assets.

S2-118 What is the **PRIMARY** purpose of performing an internal attack and penetration test?

A. Identify weaknesses in network and server security.
B. Identify ways to improve the incident response process.
C. Identify attack vectors on the network perimeter.
D. Identify the optimum response to internal hacker attacks.

A is the correct answer.

Justification:

A. Internal attack and penetration tests are designed to identify weaknesses in network and server security.
B. Internal attack and penetration tests do not focus on incident response.
C. The network perimeter is about external attacks.
D. Possible responses can be a secondary follow on effort after the internal attack and penetration test.

S2-119 An organization has learned of a security breach at another company that uses similar technology. The **FIRST** thing the information security manager should do is:

A. assess the likelihood of incidents from the reported cause.
B. discontinue the use of the vulnerable technology.
C. report to senior management that the organization is not affected.
D. remind staff that no similar security breaches have taken place.

A is the correct answer.

Justification:

A. The security manager should first assess the likelihood of a similar incident occurring, based on available information.
B. Discontinuing the use of the vulnerable technology would not necessarily be practical because it would likely be needed to support the business.
C. Reporting to senior management that the organization is not affected due to controls already in place would be premature until the information security manager can first assess the impact of the incident.
D. Until this has been researched, it is not certain that no similar security breaches have taken place.

S2-120 Which of the following tasks should the information security manager do **FIRST** when business information has to be shared with external entities?

A. Execute a nondisclosure agreement.
B. Review the information classification.
C. Establish a secure communication channel.
D. Enforce encryption of information.

B is the correct answer.

Justification:

A. Execution of a nondisclosure agreement may be needed after the classification of the data to be shared is determined.
B. The information security manager should first determine whether sharing the information poses a risk for the organization based on the information classification.
C. Whether a secure channel is needed is a function of the classification of data to be shared.
D. Encryption requirements will be determined as a function of the classification of data to be shared.

X **S2-121** When introducing public cloud computing technology to the business, which of the following situations would be a **MAJOR** concern?

A. An upward curve in the running cost triggered by the scale expansion
B. A difficulty in identifying the origination of business transactions
C. An unawareness of risk scenarios that need to be included in the risk profile
D. An increased chance to be hit by attacks to exploit vulnerabilities

C is the correct answer.

Justification:

A. In general, ease of scaling is the benefit of a cloud solution. Scaling is flexible with cloud computing technology at a predictable cost.
B. Identification of the origination point of a transaction may be a separate issue from cloud technology. Therefore, it is unnecessary to raise this concern for a cloud computing solution.
C. Cloud computing involves the interaction with a third party, as does any other outsourcing arrangement. Therefore, a cloud computing solution has a chance of introducing new risk that is not currently recognized by the organization's risk profile. It is essential for the review risk profile to cover new risk scenarios.
D. The organization may come under attack regardless of the introduction of a cloud computing solution. If proper security management for cloud computing is in place, the chance of being compromised may be lower.

S2-122 Which of the following is the **MOST** important element of information asset classification?

A. Residual risk
B. Segregation of duties
C. Potential impact
D. Need to know

C is the correct answer.

Justification:

A. Residual risk is unrelated to asset classification.
B. Segregation of duties is a control unrelated to asset classification.
C. Classification levels must be based on the level of impact that would occur as a result of compromise.
D. Need to know is a control indirectly related to asset classification.

S2-123 Which of the following would be the **BEST** indicator of an asset's value to an organization?

A. Risk assessment
B. Security audit
C. Certification
D. Classification

D is the correct answer.

Justification:

A. Assessing the risk to resources will not determine their importance to the business.
B. Security audits may provide an indication of the importance of particular resources but will be more focused on risk, vulnerabilities and compliance.
C. Certification is the process of assessing compliance with a standard.
D. Classification is the process of determining criticality and sensitivity of information resources (i.e., business value).

S2-124 In controlling information leakage, management should **FIRST** establish:

A. a data leak prevention program.
B. user awareness training.
C. an information classification process.
D. a network intrusion detection system.

C is the correct answer.

Justification:
A. Only after data are determined critical to the organization can a data leak prevention program be properly implemented.
B. User awareness training can be helpful but only after data have been classified.
C. Information classification must be conducted first.
D. Network intrusion detection is a technology that can support the data leak prevention program, but it is not a primary consideration.

S2-125 Which of the following processes is **CRITICAL** for deciding prioritization of actions in a business continuity plan?

A. Business impact analysis
B. Risk assessment
C. Vulnerability assessment
D. Business process mapping

A is the correct answer.

Justification:
A. The business impact analysis (BIA) is the critical process for deciding prioritization of restoration of the information system/business processes in case of a security incident.
B. Risk assessment provides information on the likelihood of occurrence of a security incident and assists in the selection of countermeasures, but not in prioritization of restoration.
C. A vulnerability assessment provides information regarding the security weaknesses of the system, supporting the risk analysis process.
D. Business process mapping assists in conducting a BIA, but additional information obtained during a BIA is needed to determine restoration prioritization.

S2-126 After performing an asset classification, the information security manager is **BEST** able to determine the:

A. level of risk to information resources.
B. impact of a compromise.
C. requirements for control strength.
D. annual loss expectancy.

B is the correct answer.

Justification:
A. The value of resources does not provide information on the risk to those resources.
B. Knowledge of an information resource's value provides an understanding of the potential impact of the loss of the resource.
C. Information regarding potential impact is not adequate to determine control strength requirements; risk levels must also be understood.
D. The annual loss expectancy can only be calculated after determining the magnitude of the loss and frequency of occurrence.

S2-127 Which of the following actions is involved when conducting a business impact analysis?

A. Identifying security threats and vulnerabilities
B. Developing notification and activation procedures
C. Listing investigative priorities
D. Listing critical business resources

D is the correct answer.

Justification:

A. Identifying security threats is part of a risk assessment, not a business impact analysis (BIA).
B. Notification and activation procedures are not part of a BIA but should be part of a business continuity plan.
C. Listing investigative priorities is not part of a BIA.
D. Key results of a BIA include listing critical business resources, identifying disruption impacts and allowable outage times and developing recovery priorities.

S2-128 What is the **TYPICAL** output of a risk assessment?

A. A list of appropriate controls for reducing or eliminating risk
B. Documented threats to the organization
C. Evaluation of the consequences to the entity
D. An inventory of risk that may impact the organization

D is the correct answer.

Justification:

A. A list of appropriate controls for reducing risk is subsequent to the assessment.
B. Documented threats are a part of the input for a risk assessment.
C. Evaluation of the consequences is subsequent to the assessment.
D. An inventory of risk is the output of a risk assessment.

S2-129 The assessment of risk is always subjective. To improve accuracy, which of the following is the **MOST** important action to take?

A. Train or calibrate the assessor.
B. Use only standardized approaches.
C. Ensure the impartiality of the assessor.
D. Use multiple methods of analysis.

A is the correct answer.

Justification:

A. Studies show that training or calibrating the assessor improves accuracy and reduces the subjectivity of risk assessments.
B. A standardized approach is less effective in preventing overestimating risk.
C. Assessor impartiality is important but does not compensate for the tendency to overestimate risk.
D. Multiple methods of analysis may help accuracy but training risk assessors is the most effective.

S2-130 Tightly integrated IT systems are **MOST** likely to be affected by:

A. aggregated risk.
B. systemic risk.
C. operational risk.
D. cascading risk.

D is the correct answer.

Justification:

A. Aggregated risk can occur in homogenous systems where one threat vector can compromise many systems whether integrated or not.
B. Systemic risk is unrelated to the degree of integration.
C. Operational risk is also unrelated to the degree of integration.
D. Tightly integrated systems are more susceptible to cascading risk because the failure of one element causes a sequence of failures.

S2-131 Security risk assessments are **MOST** cost-effective to a software development organization when they are performed:

A. before system development begins.
B. at system deployment.
C. before developing a business case.
D. at each stage of the system development life cycle.

D is the correct answer.

Justification:

A. A risk assessment performed before system development will not find vulnerabilities introduced during development.
B. Performing a risk assessment at system deployment is generally not cost-effective and can miss a key risk.
C. If performed prior to business case development, a risk assessment will not discover risk introduced during the system development life cycle (SDLC).
D. Performing risk assessments at each stage of the SDLC is the most cost-effective method because it ensures that vulnerabilities are discovered as soon as possible.

S2-132 Which of the following is the **BEST** quantitative indicator of an organization's current risk appetite?

A. The number of incidents and the subsequent mitigation activities
B. The number, type and layering of deterrent control technologies
C. The extent of risk management requirements in policies and standards
D. The ratio of cost to insurance coverage for business interruption protection

D is the correct answer.

Justification:

A. Incident history can provide only an approximation of the organization's efforts to mitigate further occurrences after consequences have been determined. Incident history may also indicate a lack of risk awareness.
B. Controls deployment can provide a rough qualitative estimation of risk appetite as long as technologies are tested and effectiveness is determined.
C. Requirements set in policies and standards can only serve as a qualitative approximation of risk appetite.
D. The cost of a business interruption can be accurately determined. The comparison of this expense (added to any deductible) with the total cost of premiums paid for a specific amount of insurance can serve as an accurate indicator of how much the organization will spend to protect against a defined loss.

S2-133 What is the **PRIMARY** deficiency in using annual loss expectancy to predict the annual extent of losses?

A. It is based on at least some subjective information.
B. The overall process and computations are time-consuming.
C. Effective use of the approach takes specialized training.
D. The approach is not recognized by international standards.

A is the correct answer.

Justification:
A. When used for information risk, the annual loss expectancy (ALE) is based on at least some subjective information.
B. Information security does not possess sufficient historic data to complete actuarial tables and provide highly refined predictions of the occurrence of events (e.g., accident data for the automotive industry).
C. Time and training requirements are less important factors than the subjectivity that is inherent to ALE when assessing IT risk.
D. Some international standards do recognize ALE, and even if this were not the case, it would not be a primary concern in most instances.

S2-134 Value at risk can be used:

A. as a qualitative approach to evaluating risk.
B. to determine maximum probable loss over a period of time.
C. for risk analysis applicable only to financial organizations.
D. as a useful tool to expedite the assessment process.

B is the correct answer.

Justification:
A. Value at risk (VAR) is an analysis tool, not an assessment tool and is quantitative rather than qualitative.
B. VAR provides a quantitative value of the maximum probable loss in a given time period—typically at 95 or 99 percent certainty.
C. While primarily being used by financial organizations, applicability to information security has been demonstrated.
D. VAR calculations are typically complex and time-consuming.

X **S2-135** Which of the following is the **MOST** important reason to include an effective threat and vulnerability assessment in the change management process?

— A. To reduce the need for periodic full risk assessments.
B. To ensure that information security is aware of changes.
C. To ensure that policies are changed to address new threats.
D. To maintain regulatory compliance.

A is the correct answer.

Justification:
A. By assessing threats and vulnerabilities during the change management process, changes in risk can be determined and a risk assessment can be updated incrementally. This keeps the risk assessment current without the need to complete a full reassessment.
B. Information security should have notification processes in place to ensure awareness of changes that might impact security other than threat and vulnerability assessments.
C. Policies should rarely require adjustment in response to changes in threats or vulnerabilities.
D. While including an effective threat and vulnerability assessment may assist in maintaining compliance, it is not the primary reason for the change management process.

S2-136 Once the objective of performing a security review has been defined, the **NEXT** step for the information security manager is to determine:

A. constraints.
B. approach.
C. scope.
D. results.

C is the correct answer.

Justification:
A. Constraints must be determined to understand the limits of the review, but this is not the next step.
B. Approach must be defined after scope and constraints.
C. Scope is defined after objectives are determined.
D. Results are last after scope, constraints, and approach.

S2-137 The fact that an organization may suffer a significant disruption as the result of a distributed denial-of-service (DDoS) attack is considered:

A. an intrinsic risk.
B. a systemic risk.
C. a residual risk.
D. an operational risk.

D is the correct answer.

Justification:
A. Intrinsic risk is the result of underlying internal and external factors that are not readily subject to controls.
B. Systemic risk refers to the collapse of an entire system as a result of the risk imposed by system interdependencies.
C. Residual risk is the level of risk remaining after controls and countermeasures are implemented, and it may approach intrinsic risk.
D. Operational risk is the risk to an organization as a result of its internal and external operations.

S2-138 Which one of the following factors of a risk assessment typically involves the **GREATEST** amount of speculation?

A. Exposure
B. Impact
C. Vulnerability
D. Likelihood

D is the correct answer.

Justification:
A. Exposure can be determined within a range.
B. Impact can be determined within a range.
C. Vulnerability can be determined within a range.
D. The likelihood of a threat encountering a susceptible vulnerability can only be estimated statistically.

S2-139 The acquisition of new IT systems that are critical to an organization's core business can create significant risk. To effectively manage the risk, the information security manager should **FIRST**:

A. ensure that the IT manager accepts the risk of the technology choices.
B. require the approval of auditors prior to deployment.
C. obtain senior management approval for IT purchases.
D. ensure that appropriate procurement processes are employed.

D is the correct answer.

Justification:
A. Acceptance of identified risk associated with particular technologies is the responsibility of the business process owner, and possibly of senior management, but would happen after the risk was identified during the procurement process.
B. Auditors may identify risk but are not responsible for managing it.
C. Senior management will typically be involved in IT acquisitions only from a budgetary perspective.
D. Appropriate procurement processes will include processes to initially identify the risk that may be introduced by the new system.

S2-140 Under what circumstances is it **MOST** appropriate to reduce control strength?

A. Assessed risk is below acceptable levels.
B. Risk cannot be determined.
C. The control cost is high.
D. The control is not effective.

A is the correct answer.

Justification:
A. It is appropriate to reduce control strength if it exceeds mitigation requirements set by acceptable risk levels.
B. An inability to determine risk is not a justification for reducing control strength.
C. Excessive control cost is not a reason to reduce strength, although it suggests that a redesign of the control is needed.
D. Control effectiveness does not change the control strength requirement.

S2-141 The **MOST** effective approach to ensure the continued effectiveness of information security controls is by:

A. ensuring inherent control strength.
B. ensuring strategic alignment.
C. using effective life cycle management.
D. using effective change management.

C is the correct answer.

Justification:
A. Inherent strength will not ensure that controls do not degrade over time.
B. Maintaining strategic alignment will help identify life cycle stages of controls but by itself will not address control degradation.
C. Managing controls over their life cycle will allow for compensation of decreased effectiveness over time.
D. Change management strongly supports life cycle management but by itself does not address the complete cycle.

S2-142 Inherent control strength is **PRIMARILY** a function of which of the following?

A. Implementation
B. Design
C. Testing
D. Policy

B is the correct answer.

Justification:
A. Improper implementation can affect design control strength; however, even good implementation is not likely to overcome poor design.
B. Inherent control strength is mainly achieved by proper design.
C. Testing is important to determine whether design strength has been achieved but will generally not solve design problems.
D. Policy support for appropriate controls is important but is generally too high level to ensure that a design has inherent control strength.

S2-143 Which of the following is the **MOST** useful indicator of control effectiveness?

A. The extent to which the control provides defense in depth
B. Whether the control fails open or closed
C. How often the control has failed
D. The extent to which control objectives are achieved

D is the correct answer.

Justification:
A. Defense in depth is an important standard concept but is a metric only to the extent that it meets control objectives.
B. Whether the control fails open or closed is only relevant as a metric to the extent identified in defined control objectives.
C. Without knowing the reason a control has failed, how often the control fails is not a good indication of control effectiveness.
D. The extent to which control objectives are achieved is the only true indicator of control effectiveness. It is a measurement with a point of reference.

S2-144 Addressing risk at various life cycle stages is **BEST** supported by:

A. change management.
B. release management.
C. incident management.
D. configuration management.

A is the correct answer.

Justification:
A. Change management is the overall process to assess and control risk introduced by changes. It is involved in the greatest range of the system life cycle.
B. Release management is the specific process to manage risk of production system deployment.
C. Incident management is not directly relevant to life cycle stages.
D. Configuration management is the specific process to manage risk associated with systems configuration, but change management addresses a broader range of risk.

S2-145 The **PRIMARY** reason to consider information security during the first stage of a project life cycle is:

A. the cost of security is higher in later stages.
B. information security may affect project feasibility.
C. information security is essential to project approval.
D. it ensures proper project classification.

B is the correct answer.

Justification:

A. Introducing security at later stages can cause projects to exceed budgets and can also create issues with project schedules and delivery dates, but this is generally avoided if security issues are assessed in feasibility.
B. Project feasibility can be directly impacted by information security requirements and is the primary reason to introduce information security requirements at this stage. The cost of security must be factored into any business case that will support project feasibility, and sometimes the cost of doing something securely exceeds the benefits that the project is anticipated to produce.
C. Project approval is a business decision that may be influenced by information security considerations but is not essential.
D. Considering information security during the first stage will not ensure proper project classification.

S2-146 Which of the following **BEST** supports the principle of security proportionality?

A. Release management
B. Ownership schema
C. Resource dependency analysis
D. Asset classification

D is the correct answer.

Justification:

A. Release management provides no indication that protection is proportionate to the value of the asset.
B. An implemented ownership schema is one step in achieving proportionality, but other steps must also occur.
C. Resource dependency analysis can reveal the level of protection afforded a particular system, but that may be unrelated to the level of protection of other assets.
D. Classification provides the basis for protecting resources in relation to their importance to the organization; more important assets get a proportionally higher level of protection.

S2-147 A permissive controls policy would be reflected in which one of the following implementations?

A. Access is allowed unless explicitly denied.
B. IT systems are configured to fail closed.
C. Individuals can delegate privileges.
D. Control variations are permitted within defined limits.

A is the correct answer.

Justification:

A. A permissive controls policy allows activities that are not explicitly denied.
B. Configuration to fail closed is a restrictive controls policy.
C. Delegation of privileges refers to discretionary access control.
D. Standards permit control variations within defined limits.

S2-148 The chief information security officer (CISO) has recommended several information security controls (such as antivirus) to protect the organization's information systems. Which one of the following risk treatment options is the CISO recommending?

A. Risk transfer
B. Risk mitigation
C. Risk acceptance
D. Risk avoidance

B is the correct answer.

Justification:
A. Risk transfer involves transferring the risk to another entity such as an insurance company.
B. By implementing security controls, the company is trying to decrease risk to an acceptable level, thereby mitigating risk.
C. Risk acceptance involves accepting the risk in the system and doing nothing further.
D. Risk avoidance stops the activity causing the risk.

S2-149 Which of the following is **MOST** effective in preventing disruptions to production systems?

A. Patch management
B. Security baselines
C. Virus detection
D. Change management

D is the correct answer.

Justification:
A. Patch management involves the correction of software vulnerabilities as they are discovered by modifying the software with a "patch," which may or may not prevent production system disruptions.
B. Security baselines provide minimum recommended settings and do not necessarily prevent introduction of control weaknesses.
C. Virus detection is an effective tool but primarily focuses on malicious code from external sources.
D. Change management controls the process of introducing changes to systems. Changes that are not properly reviewed before implementation can disrupt or alter established controls in an otherwise secure, stable environment.

S2-150 The **PRIMARY** objective of a vulnerability assessment is to:

A. reduce risk to the business.
B. ensure compliance with security policies.
C. provide assurance to management.
D. measure efficiency of services provided.

C is the correct answer.

Justification:

A. It is necessary to identify vulnerabilities in order to mitigate them. Actual reduction of risk is accomplished through deployment of controls and is a business decision based on a cost-benefit analysis.
B. A security policy may mandate a vulnerability assessment program, but such a program is not established primarily to comply with policy.
C. A vulnerability assessment identifies vulnerabilities so that they may be considered for mitigation. By giving management a complete picture of the vulnerabilities that exist, a vulnerability assessment program allows management to prioritize those vulnerabilities deemed to pose the greatest risk.
D. Vulnerability assessment is not concerned with efficiency of services.

S2-151 An organization's IT change management process requires that all change requests be approved by the asset owner and the information security manager. The **PRIMARY** objective of getting the information security manager's approval is to ensure that:

A. changes comply with security policy.
B. risk from proposed changes is managed.
C. rollback to a current status has been considered.
D. changes are initiated by business managers.

B is the correct answer.

Justification:

A. A change affecting a security policy is not handled by an IT change process.
B. Changes in the IT infrastructure may have an impact on existing risk. An information security manager must ensure that the proposed changes do not adversely affect the security posture.
C. Rollback to a current state may cause a security risk event and is normally part of change management, but it is not the primary reason that security is involved in the review.
D. The person who initiates a change has no effect on the person who reviews and authorizes an actual change.

S2-152 An information security manager observed a high degree of noncompliance for a specific control. The business manager explained that noncompliance is necessary for operational efficiency. The information security manager should:

A. evaluate the risk due to noncompliance and suggest an alternate control.
B. ignore the issue of operational efficiency and insist on compliance for the control.
C. change the security policies to reduce the amount of noncompliance risk.
D. conduct an awareness session for the business manager to emphasize compliance.

A is the correct answer.

Justification:

A. The information security manager must consider the business requirements of the control and assess the risk of noncompliance.
B. Information security cannot ignore issues related to operational efficiency. The business can decide to accept the risk.
C. Changing the information security policies may not reduce the risk.
D. Conducting an awareness session may be a good idea, but it may not resolve the issue in this situation.

S2-153 The information security policies of an organization require that all confidential information must be encrypted while communicating to external entities. A regulatory agency insisted that a compliance report must be sent without encryption. The information security manager should:

A. extend the information security awareness program to include employees of the regulatory authority.
B. send the report without encryption on the authority of the regulatory agency.
C. initiate an exception process for sending the report without encryption.
D. refuse to send the report without encryption.

C is the correct answer.

Justification:

A. Although this choice may not be possible, the information security manager can discuss and understand the reason for insisting on an unencrypted report and try to convince the regulatory authority.
B. If the information security manager chooses to ignore the regulatory authority's request (which may not be possible in many parts of the world), it is necessary that a comparative risk assessment be conducted.
C. The information security manager should first assess the risk in sending the report to the regulatory authority without encryption. The information security manager can consider alternate communication channels that will address the risk and provide for the exception.
D. The information security policy states that confidential information must be encrypted when sent to external entities. The information security manager's role is to find a way within the policy to complete the task. The best way to do this is to initiate an exception.

S2-154 Which of the following is the **MOST** cost-effective approach to test the security of a legacy application?

A. Identify a similar application and refer to its security weaknesses.
B. Recompile the application using the latest library and review the error codes.
C. Employ reverse engineering techniques to derive functionalities.
D. Conduct a vulnerability assessment to detect application weaknesses.

D is the correct answer.

Justification:

A. Many applications that appear to be functionally similar may be remarkably dissimilar at the code implementation level. Even a newer version of the same software may have been entirely rewritten, and any software developed in-house is necessarily unique to the environment. Referring to the weaknesses of what appears to be a similar application may lead to a false sense of security or a time-consuming list of false positives.
B. Recompiling a legacy application is possible only when source code is available and may not function properly if underlying libraries or coding standards have changed.
C. Reverse engineering a legacy application is likely to cost significantly more than a vulnerability assessment, and deriving the functionalities of the application is not the goal.
D. Identifying vulnerabilities will allow an organization to determine what compensating controls may be needed to continue operating a legacy application where replacement is not an option. Vulnerability assessments are not necessarily comprehensive in all cases, but they are generally effective when planned properly.

S2-155 An information security manager's **MOST** effective efforts to manage the inherent risk related to a third-party service provider will be the result of:

A. limiting organizational exposure.
B. a risk assessment and analysis.
C. strong service level agreements.
D. independent audits of third parties.

A is the correct answer.

Justification:

A. It is likely to be more effective to control the organization's vulnerabilities to third-party risk by limiting organizational exposure than to control the third party's actions.
B. It is essential to know the risk, but this does not manage the risk.
C. Defining contractual responsibilities of third parties is important but will not directly manage risk.
D. Audits may indicate the threats posed by third parties but will not ensure that the risk is managed.

S2-156 The **BEST** process for assessing an existing risk level is a(n):

A. impact analysis.
B. security review.
C. vulnerability assessment.
D. threat analysis.

B is the correct answer.

Justification:

A. An impact analysis is used to determine potential impact in the event of the loss of a resource.
B. A security review is used to determine the current state of security for various program components.
C. While vulnerability assessments help identify and classify weakness in the design, implementation, operation or internal control of a process, they are only one aspect of a security review.
D. A threat analysis is not normally a part of a security review. Threat assessments evaluate the type, scope and nature of events or actions that can result in adverse consequences; identification is made of the threats that exist against organization assets.

S2-157 Which of the following is the **BEST** approach to deal with inadequate funding of the information security program?

A. Eliminate low-priority security services.
B. Require management to accept the increased risk.
C. Use third-party providers for low-risk activities.
D. Reduce monitoring and compliance enforcement activities.

C is the correct answer.

Justification:

A. Prioritizing information security activities is always useful, but eliminating even low-priority security services is a last resort.
B. If budgets are seriously constrained, management is already addressing increases in other risk and is likely to be aware of the issue. A proactive approach to doing more with less will be well received.
C. Outsourcing of some information security activities can cut costs and increase resources for other security activities in a proactive manner, as can automation of some security procedures.
D. Reducing monitoring activities may unnecessarily increase risk when lower-cost options to perform those functions may be available.

S2-158 A cost-benefit analysis is performed on any proposed control to:

A. define budget limitations.
B. demonstrate due diligence to the budget committee.
C. verify that the cost of implementing the control is within the security budget.
D. demonstrate the costs are justified by the reduction in risk.

D is the correct answer.

Justification:

A. A cost-benefit analysis does not define budget constraints; the board of directors or senior management of the organization will do that based on a variety of factors.
B. The purpose of the analysis is not to show that due diligence was performed, but to establish a result that will show the cost of the control and the reduction in risk.
C. A cost-benefit analysis does not help verify that the cost of a control is within the security budget; it may, however, help identify controls that require additional expenses that exceed the established security budget.
D. Senior management can weigh the cost of the risk against the cost of the control and show that the control will reduce that risk by some measure.

S2-159 Which of the following represents the **MAJOR** focus of privacy regulations?

A. Unrestricted data mining
B. Identity theft
C. Human rights protection
D. Identifiable personal data

D is the correct answer.

Justification:

A. Data mining is an accepted tool for *ad hoc* reporting; it could pose a threat to privacy only if it violates regulatory provisions.
B. Identity theft is a potential consequence of privacy violations but not the main focus of many regulations.
C. Human rights protection addresses privacy issues but is not the main focus of regulations.
D. Protection of identifiable personal data is the major focus of privacy regulations such as the Health Insurance Portability and Accountability Act (HIPAA).

S2-160 Addressing risk scenarios at various information system life cycle stages is **PRIMARILY** a function of:

A. change management.
B. release management.
C. incident management.
D. configuration management.

A is the correct answer.

Justification:

A. Change management is the overall process to assess and control risk scenarios introduced by changes.
B. Release management is the process to manage risk scenarios of production system deployment and is a component of change management.
C. Incident management addresses impacts when or after they occur.
D. Configuration management is the specific process to manage risk scenarios associated with systems configuration and is a component of change management.

S2-161 The **PRIMARY** objective of asset classification is to:

A. maximize resource management.
B. comply with IT policy.
C. define information architecture.
D. determine protection level.

D is the correct answer.

Justification:
A. Classification is one of many parts of resource management.
B. The IT policy of an organization is determined based on business policies.
C. Asset classification is an input to information architecture.
D. Classification allows the appropriate protection level to be assigned to the asset.

S2-162 A control for protecting an IT asset, such as a laptop computer, is **BEST** selected if the cost of the control is less than the:

A. cost of the asset.
B. impact on the business if the asset is lost or stolen.
C. available budget.
D. net present value.

B is the correct answer.

Justification:
A. While the control may be more expensive than the cost of the physical asset, such as a laptop computer, the impact to the business may be much higher and thus justify the cost of the control.
B. Controls are selected based on their impact on the business due to the nonavailability of the asset rather than on the cost of the asset or the available budget.
C. Budget availability is a consideration; however, this is not as important as the overall impact to the business if the asset is compromised.
D. Net present value (NPV) calculations are not useful to determine the cost of a control. While a laptop computer might be fully amortized (or even expensed), the impact of the loss of the asset may be much higher than its NPV.

S2-163 Which of the following choices **BEST** reveals the evolving nature of attacks in an online environment?

A. A high-interaction honeypot
B. A rogue access point
C. Industry tracking groups
D. A vulnerability scanner

C is the correct answer.

Justification:
A. A honeypot is used to lure a hacker and learn the methods of attacks. However, an attacker may or may not use known methods of attacks. Also, the honeypot will only reveal attacks directed against the organization, not the overall nature of attacks occurring in the broader online environment.
B. A rogue access point is put in place by an attacker to lure legitimate users to connect to it.
C. Industry tracking groups, such as Infraguard, US Computer Emergency Readiness Team (CERT) and Internet Storm Center, provide insight into what sort of attacks are affecting organizations on a national or global scale.
D. Even if a vulnerability scanner is updated regularly, it will reveal vulnerabilities, not attacks.

S2-164 The information security manager should treat regulatory compliance requirements as:

A. an organizational mandate.
B. a risk management priority.
C. a purely operational issue.
D. just another risk.

D is the correct answer.

Justification:

A. While it is generally preferable to be as compliant as reasonably possible, the extent and level of regulatory compliance is a management decision, not a mandate.
B. All risk should be prioritized, and regulation may not be the highest priority.
C. Regulatory compliance is not just an operational issue but primarily a management issue.
D. Many regulations exist that must be considered. Priority should be given to those with the greatest impact, just as other risk is considered with priority given to feasibility, level of enforcement, possible sanctions and costs of compliance.

S2-165 Management decided that the organization will not achieve compliance with a recently issued set of regulations. Which of the following is the **MOST** likely reason for the decision?

A. The regulations are ambiguous and difficult to interpret.
B. Management has a low level of risk tolerance.
C. The cost of compliance exceeds the cost of possible sanctions.
D. The regulations are inconsistent with the organizational strategy.

C is the correct answer.

Justification:

A. Management should address ambiguous regulations by requesting clarification from the issuer or the legal department.
B. Management decisions on compliance should be based on a cost-benefit analysis.
C. Management may decide it is less expensive to deal with possible sanctions than to attempt to be in compliance.
D. The fact that the regulations are inconsistent with the organizational strategy is not a major factor in deciding not to comply.

S2-166 Asset classification should be **MOSTLY** based on:

A. business value.
B. book value.
C. replacement cost.
D. initial cost.

A is the correct answer.

Justification:

A. Classification should be based on the value of the asset to the business, generally in terms of revenue production or potential impact on loss or disclosure of sensitive information.
B. Book value is not an appropriate basis for classification.
C. Replacement cost is not an appropriate basis for classification.
D. Initial cost is not an appropriate basis for classification.

S2-167 Control baselines are **MOST** directly related to the:

A. organization's risk appetite.
B. external threat landscape.
C. effectiveness of mitigation options.
D. vulnerability assessment.

A is the correct answer.

Justification:

A. Control baselines are designed to mitigate risk and will depend on the organization's risk appetite.
B. The viability and existence of threats will have a direct bearing on control baselines, but only to the extent that they can exploit vulnerabilities and create a risk of potential impact.
C. In some cases the effectiveness may modify the control objectives if it is not feasible to mitigate the risk, but generally that will not change the objectives.
D. Vulnerability assessments are conducted against a control baseline.

S2-168 The **MOST** likely reason that management would choose not to mitigate a risk that exceeds the risk appetite is that it:

A. is the residual risk after controls are applied.
B. is a risk that is expensive to mitigate.
C. falls within the risk tolerance level.
D. is a risk of relatively low frequency.

C is the correct answer.

Justification:

A. The residual risk may or may not be considered appropriate depending on the level of acceptable risk and the tolerance for variation to that level.
B. If mitigation is too expensive, management should consider other treatment options and not simply choose not to address it.
C. Risk tolerance is the acceptable level of variation that management is willing to allow for any particular risk as the enterprise pursues its objectives.
D. Even if a risk occurs infrequently, the information security manager should address the risk if the magnitude is substantial.

S2-169 Which of the following is the **BEST** indicator of the level of acceptable risk in an organization?

A. The proportion of identified risk that has been remediated
B. The ratio of business insurance coverage to its cost
C. The percentage of the IT budget allocated to security
D. The percentage of assets that has been classified

B is the correct answer.

Justification:

A. The proportion of unremediated risk may be an indicator, but there are many other factors unrelated to acceptable risk such as treatment feasibility, availability of controls, etc.
B. The amount of business insurance coverage carried and the cost provide a directly quantifiable indication of the level of risk the organization will accept and at what cost.
C. The percentage of the IT budget allocated to security is an indicator but does not quantify acceptable levels of risk.
D. Classifying assets will indicate which assets are more important than others but does not quantify the acceptability of risk.

S2-170 The aspect of governance that is **MOST** relevant to setting security baselines is:

A. policies.
B. acceptable risk.
C. impacts.
D. standards.

D is the correct answer.

Justification:
A. Policies may require that baselines be defined, but the specifics will be in the standards.
B. Acceptable risk will define the control objectives, which are then expressed in the standards.
C. Potential impacts will help determine acceptable risk expressed in the standards, which collectively set the baseline.
D. Standards taken together define the lowest limits of security, thereby defining the baseline.

S2-171 Which of the following is the **FIRST** action to be taken when the information security manager notes that the controls for a critical application are inadequate?

A. Perform a risk assessment to determine the level of exposure.
B. Classify the risk as acceptable to senior management.
C. Deploy additional countermeasures immediately.
D. Transfer the remaining risk to another organization.

A is the correct answer.

Justification:
A. It is most important to perform a risk assessment to determine the exposure if additional controls are not deployed.
B. The exposure level needs to be redetermined and compared with the residual risk before this decision can be made.
C. Additional countermeasures may be deployed after determining possible losses so as to not overprotect or underprotect the asset.
D. Risk transfer is an action that may be taken after reviewing the results of the risk assessment of the current situation.

S2-172 When assessing the maturity of the risk management process, which of the following findings raises the **GREATEST** concern?

A. Organizational processes are not adequately documented.
B. Multiple frameworks are used to define the desired state.
C. Required security objectives are not well defined.
D. The desired state is not based on the business objectives.

D is the correct answer.

Justification:
A. It is expected that the organization will start work to improve the system beyond this level. It is not wrong for an organization to start off at the base of the maturity model.
B. This method could be unnecessarily expensive if not well planned and may result in conflicts between the frameworks.
C. It is very important that qualitative and quantitative objectives be well defined for a gap analysis to be effective. However, defining a desired state without input from the business strategy invalidates the entire process.
D. Risk management is about the business. Defining a desired state without consideration of business objectives implies that the stated desired outcome may not be effective, even if attained.

S2-173 Which of the following is the **GREATEST** concern for an organization in which there is a widespread use of mobile devices?

A. There is an undue reliance on public networks.
B. Batteries require constant recharges.
C. There is a lack of operating system standardization.
D. Mobile devices can be easily lost or stolen.

D is the correct answer.

Justification:
A. The fact that mobile devices must be connected to public networks creates a security risk that can be exploited in the public space, but appropriate security controls can mitigate the risk.
B. The need to constantly recharge batteries is not a significant security concern.
C. While the lack of operating system standardization is a concern, it is not as great as the loss of devices.
D. Because of their size, mobile devices can be easily lost or stolen and sensitive information disclosed.

S2-174 Due to limited storage media, an IT operations employee has requested permission to overwrite data stored on a magnetic tape. The decision of the authorizing manager will **MOST** likely be influenced by the data:

A. classification policy.
B. retention policy.
C. creation policy.
D. leakage protection.

B is the correct answer.

Justification:
A. The data classification policy addresses who can access or modify data. It is more focused on ensuring that confidential data do not fall into the wrong hands.
B. The data retention policy will specify the time that must lapse before data can be overwritten or deleted.
C. The data creation policy will address conditions that must be satisfied before new data are recognized in an organization and who is authorized for such tasks.
D. Leakage protection ensures confidentiality of corporate data.

S2-175 Which of the following **BEST** assists the information security manager in identifying new threats to information security?

A. Performing more frequent reviews of the organization's risk factors
B. Developing more realistic information security risk scenarios
C. Understanding the flow and classification of information used by the organization
D. A process to monitor postincident review reports prepared by IT staff

C is the correct answer.

Justification:

A. Risk factors determine the business impact or frequency of risk and are not related to the identification of threats.
B. Risk scenarios are not used to identify threats as much as they are used to identify the impact and frequency of threats exploiting vulnerabilities within the information security architecture.
C. Understanding the business objectives of the organization and how data are to be used by the business assists management in assessing whether an information security event should be considered as a new information security threat.
D. The analysis of postincident reviews assists managers in identifying IS threats that have materialized into incidents and does not necessarily assist IT managers in identifying threats that pose a risk to information security.

S2-176 Retention of business records should **PRIMARILY** be based on:

A. business strategy and direction.
B. regulatory and legal requirements.
C. storage capacity and longevity.
D. business case and value analysis.

B is the correct answer.

Justification:

A. Business strategy and direction would address the issue of business record retention among numerous others but would typically not be the primary focus.
B. Retention of business records is primarily driven by legal and regulatory requirements.
C. Storage capacity and longevity are important but secondary issues.
D. Business case and value analysis would be secondary to complying with legal and regulatory requirements.

S2-177 Which of the following is **MOST** important to achieve proportionality in the protection of enterprise information systems?

A. Asset classification
B. Risk assessment
C. Security architecture
D. Configuration management

A is the correct answer.

Justification:

A. Asset classification is based on the criticality and sensitivity of information assets with the goal of providing the appropriate and, therefore, proportional degree of protection.

B. Proper risk assessment requires assets to be classified; asset classification most directly impacts the mitigation efforts an organization will implement.

C. Security architecture will be affected by asset classification and, to some extent, may affect how assets are classified; asset classification most directly impacts the mitigation efforts an organization will implement.

D. Configuration management is likely to be affected by asset classification levels but is not directly related to information security.

S2-178 The **MOST** important component of a privacy policy is:

A. notifications.
B. warranties.
C. liabilities.
D. standards

A is the correct answer.

Justification:

A. Privacy policies must contain notification requirements in the event of unauthorized disclosure and opt-out provisions.

B. Privacy policies do not address warranties, which are generally unrelated to a privacy policy.

C. Privacy policies may address liabilities as a consequence of unauthorized disclosure, but that is not the most important component.

D. Standards regarding privacy would be separate and not a part of the policy.

S2-179 Highly integrated enterprise IT systems pose a challenge to the information security manager when attempting to set security baselines **PRIMARILY** from the perspective of:

A. increased difficulty in problem management.
B. added complexity in incident management.
C. determining the impact of cascading risk.
D. less flexibility in setting service delivery objectives.

C is the correct answer.

Justification:

A. Determining root causes in problem management may be more difficult in highly integrated systems because of the many interconnected functions, but that is not the primary risk concern.
B. Incident management may be affected by the added complexity of highly integrated systems when attempting to quickly isolate and ascertain the source of a problem along a chain of tightly coupled functions; however, this is not the primary issue.
C. Highly integrated systems are more susceptible to cascading risk where the failure or compromise of any one element has the possibility of causing a domino effect of failures.
D. Setting service delivery objectives will be constrained by the extent of the integration because most elements require the same level of functionality. This is due to a lower service level of any component reducing functionality of all dependent elements; but this is not the primary consideration.

S2-180 Which of the following is the **MOST** important prerequisite to undertaking asset classification?

A. Threat analysis
B. Impact assessment
C. Controls evaluation
D. Penetration testing

B is the correct answer.

Justification:

A. Threat analysis only identifies the threats that exist against enterprise assets. It is useful but is not the most important for asset classification.
B. The classification level is an indication of the value or importance of the asset to the enterprise. Impact assessments are needed to determine criticality and sensitivity, which is the basis for the classification level.
C. Controls evaluation is needed after classification levels have been determined to ensure that the asset is protected according to the classification level.
D. Penetration testing is not important to the classification process.

S2-181 Vulnerabilities discovered during an assessment should be:

A. handled as a risk, even though there is no threat.
B. prioritized for remediation solely based on impact.
C. a basis for analyzing the effectiveness of controls.
D. evaluated for threat, impact and cost of mitigation.

D is the correct answer.

Justification:

A. Vulnerabilities may not be exposed to potential threats. Also, there may be no threat or possibly little or no impact even if exploited. While threats are always evolving, without additional information, the appropriate treatment cannot be determined.
B. Vulnerabilities should be prioritized for remediation based on probability of compromise (which is affected by the level of exposure), impact and cost of remediation.
C. Vulnerabilities discovered will to some extent show whether existing controls are in place to address a potential risk but does not indicate the control effectiveness.
D. Vulnerabilities uncovered should be evaluated and prioritized based on whether there is a credible threat, the impact if the vulnerability is exploited and the cost of mitigation. If there is a potential threat but little or no impact if the vulnerability is exploited, there is little risk, and it may not be cost-effective to address it.

X

S2-182 An appropriate risk treatment method is:

A. the method that minimizes risk to the greatest extent.
B. based on the organization's risk appetite.
C. an efficient approach to achieve control objectives.
D. the method that maximizes risk mitigation.

C is the correct answer.

Justification:

A. While minimizing risk is generally preferable, doing so beyond what is acceptable is likely too costly and counterproductive.
B. The risk appetite triggers the risk response; however, it does not define the actual treatment method.
C. Control objectives will have been determined based on acceptable risk and the least costly or most efficient approach to do so will be the most appropriate.
D. Mitigation is just one treatment option and may not be the most appropriate.

S2-183 Which of the following factors **BEST** helps determine the appropriate protection level for an information asset?

A. The cost of acquisition and implementation of the asset
B. Knowledge of vulnerabilities present in the asset
C. The degree of exposure to known threats
D. The criticality of the business function supported by the asset

D is the correct answer.

Justification:

A. The criticality of the asset is determined by the business value of the asset, not just the cost of the asset. The value is determined by the cost of acquisition and implementation of the asset.
B. Knowledge of vulnerabilities helps in determining the protection method; however, it is implemented based on the business value of the asset compared with the cost of protection method.
C. The degree of exposure may require certain treatment options, but the degree and extent of protection is still determined by criticality.
D. Although all the options may help in determining the protection level of the asset, the criticality of the business function supported by the asset is the most important because nonavailability might affect the delivery of services.

X **S2-184** Which of the following vulnerabilities allowing attackers access to the application database is the **MOST** serious?

— A. Validation checks are missing in data input pages.
B. Password rules do not allow sufficient complexity.
C. Application transaction log management is weak.
D. Application and database share a single access ID.

A is the correct answer.

Justification:

A. Attackers are able to exploit the weaknesses that exist in the application layer. For example, they can submit a part of a structured query language (SQL) statement (SQL injection attack) to illegally retrieve application data. Validation control is an effective countermeasure.
B. Noncomplex passwords may make accounts vulnerable to brute force attacks, but these can be countered in other ways besides complexity (e.g., lockout thresholds).
C. There is a chance that confidential information is inadvertently written to the application transaction log; therefore, sufficient care should be given to log management. However, it is uncommon for attackers to use the log server to steal database information.
D. Although developers may embed a single ID in the program to establish a connection from application to database, if the original account is sufficiently secure, then the overall risk is low.

S2-185 The cost of implementing and operating a security control should not exceed the:

A. annual loss expectancy.
B. cost of an incident.
C. asset value.
D. acceptable loss level

C is the correct answer.

Justification:
A. Annual loss expectancy represents the losses that are expected to happen for the entire organization during a single calendar year.
B. A security mechanism may cost more than the cost of a single incident and still be cost-effective.
C. The cost of implementing security controls should not exceed the business value of the asset.
D. The cost of a control may well exceed the acceptable loss level in order to achieve this loss level objective.

S2-186 The **MOST** important factor in planning for the long-term retention of electronically stored business records is to take into account potential changes in:

A. storage capacity and shelf life.
B. regulatory and legal requirements.
C. business strategy and direction.
D. application systems and media.

D is the correct answer.

Justification:
A. Storage capacity and shelf life are important but secondary issues.
B. Legal and regulatory requirements do not generally apply to long-term retention of electronically stored business records.
C. Business strategy and direction do not generally apply to long-term retention of electronically stored business records.
D. Long-term retention of business records may be severely impacted by changes in application systems and media. For example, data stored in nonstandard formats that can only be read and interpreted by previously decommissioned applications may be difficult, if not impossible, to recover.

S2-187 Which of the following actions should the information security manager take **FIRST** on finding that current controls are not sufficient to prevent a serious compromise?

A. Strengthen existing controls.
B. Reassess the risk.
C. Set new control objectives.
D. Modify security baselines.

B is the correct answer.

Justification:

A. Until a clear picture of risk has been developed, the extent of control increases needed cannot be determined.
B. Control decisions are driven by risk. Risk should be carefully reassessed and analyzed to correct potential misjudgment in the original assessment.
C. A control objective is a statement of the desired result or purpose to be achieved by implementing control procedures in a particular process. Changes to control objectives should be made after risk has been reassessed.
D. Security baselines set by appropriate standards are the minimum security requirements for different trust domains across the enterprise. Baselines may need to be strengthened after risk has been reassessed.

S2-188 What is the **MOST** important reason to periodically test controls?

A. To meet regulatory requirements
B. To meet due care requirements
C. To ensure that objectives are met
D. To achieve compliance with standard policy

C is the correct answer.

Justification:

A. Not all organizations are required to periodically test controls.
B. Periodically testing controls does not help meet due care requirements. Due care is what a reasonable person of similar competency would do under similar circumstances.
C. Periodically testing controls ensures that controls continue to meet control objectives.
D. Compliance with policy is not the most important factor for periodically testing controls.

S2-189 Which of the following approaches is **BEST** for addressing regulatory requirements?

A. Treat regulatory compliance as any other risk.
B. Ensure that policies address regulatory requirements.
C. Make regulatory compliance mandatory.
D. Obtain insurance for noncompliance.

A is the correct answer.

Justification:

A. There are many regulatory requirements with varying degrees of enforcement and possible sanctions. These should be assessed and treated as any other risk.
B. Policies addressing compliance with regulatory requirements are not by themselves sufficient to deal with those requirements.
C. Mandatory compliance with all regulatory mandates without determining the risk and potential impact may not be cost-effective.
D. Insurance for regulatory noncompliance may not be available.

S2-190 The output of the risk management process is an input for making:

A. business plans.
B. audit charters.
C. security policy decisions.
D. software design decisions.

C is the correct answer.

Justification:

A. Business plans are an output of management translating strategic aspirations into attainable business goals. Business plans provide background, goal statements and plans for reaching those goals.
B. Audit charters are documents describing the purpose, rights and responsibilities of the audit function. They do not rely on the risk assessment process.
C. The risk management process is about making specific, security-related decisions such as the selection of specific risk responses.
D. Software design decisions are based on stakeholder needs, not the risk management process.

S2-191 What is a **PRIMARY** advantage of performing a risk assessment on a consistent basis?

A. It lowers costs of assessing risk.
B. It provides evidence of attestation.
C. It is a necessary part of third-party audits.
D. It provides trends in the evolving risk profile.

D is the correct answer.

Justification:

A. There may be some minor cost benefits to performing risk assessments on a consistent basis, but that is not the main benefit.
B. An assessment deals with a review of a process, not a person's claim of the process being in place. An attestation is a claim without the supporting evidence.
C. External audits do not require risk assessments, although it is encouraged.
D. Tracking trends in evolving risk is of significant benefit to managing risk and ensuring that appropriate controls are in place.

S2-192 A new regulation for safeguarding information processed by a specific type of transaction has come to the attention of an information security officer. The officer should **FIRST**:

A. meet with stakeholders to decide how to comply.
B. analyze key risk in the compliance process.
C. assess whether existing controls meet the regulation.
D. update the existing security/privacy policy.

C is the correct answer.

Justification:

A. While meeting with stakeholders to decide how to comply is appropriate and important, this action is subsequent to assessing whether existing controls meet the regulation and will depend on whether there is an existing control gap.
B. While analyzing key risk in the compliance process is appropriate and important, this action is subsequent to assessing whether existing controls meet the regulation and will depend on whether existing controls are adequate.
C. If the organization is in compliance through existing controls, the need to perform other work related to the regulation is not a priority.
D. While updating the existing security/privacy policy is appropriate and important, this action is subsequent to assessing whether existing controls meet the regulation and will depend on whether there is an existing control gap.

S2-193 Which of the following choices **BEST** helps determine appropriate levels of information resource protection?

A. A business case
B. A vulnerability assessment
C. Asset classification
D. Asset valuation

C is the correct answer.

Justification:

A. A business case may be useful to support the need for asset classification but does not by itself provide a basis for assignment at the individual resource level.
B. Vulnerability assessment does not take into account criticality or sensitivity, which is the basis for assigning levels of information resource protection.
C. Asset classification based on criticality and sensitivity provides the best basis for assigning levels of information resource protection.
D. Asset valuation is not an adequate basis for determining the needed level of protection. For example, an asset can be very valuable from a cost standpoint but be neither critical to operations nor sensitive if exposed.

S2-194 It is **MOST** important that a privacy statement on a company's e-commerce web site includes:

A. a statement regarding what the company will do with the information it collects.
B. a disclaimer regarding the accuracy of information on its web site.
C. technical information regarding how information is protected.
D. a statement regarding where the information is being hosted.

A is the correct answer.

Justification:

A. Most privacy laws and regulations require disclosure on how information will be used.
B. A disclaimer may be prudent but is not necessary because it does not refer to data privacy.
C. Technical details regarding how information is protected are not mandatory to publish on the web site and would not be desirable.
D. It is not mandatory to say where information is being hosted.

S2-195 From an information security perspective, information that no longer supports the main purpose of the business should be:

A. analyzed under the retention policy.
B. protected under the information classification policy.
C. analyzed under the backup policy.
D. assessed by a business impact analysis.

A is the correct answer.

Justification:

A. Information analyzed under the retention policy will determine whether the organization is required to maintain the data for business, legal or regulatory reasons. Keeping data that are no longer required unnecessarily consumes resources and, in the case of sensitive personal information, can increase the risk of data compromise.
B. Information that is protected under the information classification policy is an attribute that should be considered in the destruction and retention policy.
C. There is little reason to back up information no longer of use to the organization, and it should be considered as a part of the retention policy.
D. A business impact analysis could help determine that this information does not support the main objective of the business but does not indicate the action to take.

X **S2-196** When corporate standards change due to new technology, which of the following choices is **MOST** likely to be impacted?

A. Organizational policies
B. The risk assessment approach
C. Control objectives
D. Systems security baselines

D is the correct answer.

Justification:

A. Properly developed organizational policies are not likely to require any changes when corporate standards change due to new technology.
B. Risk assessment is a process used to identify and evaluate risk and its potential effects. Approaches to assessing risk probably will not need to change when corporate standards change due to new technology.
C. A control objective is a statement of the desired result or purpose to be achieved by implementing control procedures in a particular process. Properly developed control objectives are not likely to require any changes when corporate standards change due to new technology.
D. Because security baselines are set by standards, it is most likely that a change in some standards will necessitate a review and possible changes in baseline security.

X **S2-197** What are the essential elements of risk?

A. Impact and threat
B. Likelihood and consequence
C. Threat and exposure
D. Sensitivity and exposure

B is the correct answer.

Justification:

A. Threat is an element of risk only in combination with vulnerability.
B. Risk is the combination of the probability of an event and its consequence. (International Organization for Standardization/International Electrotechnical Commission [ISO/IEC] 73) The probability of an event is threat exploiting a vulnerability.
C. Threat and exposure are insufficient to determine risk.
D. Sensitivity is a measure of consequence, but does not take into account probability. Exposure moderates risk but is not in itself a component of risk.

S2-198 Monitoring has flagged a security noncompliance. What is the **MOST** appropriate action?

A. Validate the noncompliance.
B. Escalate the noncompliance to management.
C. Update the risk register.
D. Fine-tune the key risk indicator threshold.

A is the correct answer.

Justification:

A. Before any other action is taken, the security manager should ensure that the noncompliance identified by monitoring is not a false positive.
B. The escalation to management should not occur until more is known about the situation, and even then only if it is outside the security manager's scope to address the issue.
C. Updating the risk register is one possible response to a validated noncompliance.
D. Key risk indicator threshold changes would occur only if subsequent investigation found them to be necessary.

S2-199 The facilities department of a large financial organization uses electronic swipe cards to manage physical access. The information security manager requests that facilities provide the manager with read-only access to the physical access data. What is the **MOST** likely purpose?

A. To monitor that personnel are complying with contract provisions
B. To determine who is in the building in case of fire
C. To compare logical and physical access for anomalies
D. To ensure that the physical access control system is operating correctly

C is the correct answer.

Justification:
A. Contract compliance monitoring would usually not be part of an information security manager's role.
B. The physical security and emergency response personnel should be monitoring presence in the building in case of fire.
C. Any differences between physical and logical access may indicate one of several risk scenarios such as personnel not swiping in and tailgating, password sharing, or system compromise and serves as a key risk indicator. Some of the best security metrics come from non–security-related activities.
D. The correct operation of the system is likely the responsibility of IT, although a periodic validation by security is prudent.

S2-200 What is the result of segmenting a highly sensitive database?

A. It reduces threat.
B. It reduces criticality.
C. It reduces sensitivity.
D. It reduces exposure.

D is the correct answer.

Justification:
A. The threat may remain constant, but each segment may represent a different vector against which it must be directed.
B. Criticality of data is not affected by the manner in which it is segmented.
C. Sensitivity of data is not affected by the manner in which it is segmented.
D. Segmenting data reduces the quantity of data exposed as a result of a particular event.

S2-201 Which of the following components is established during the **INITIAL** steps of developing a risk management program?

A. Management acceptance and support
B. Information security policies and standards
C. A management committee to provide oversight for the program
D. The context and purpose of the program

D is the correct answer.

Justification:

A. Although an important component in the development of any managed program, obtaining management acceptance and support ideally occurs well before the development of the program, in the plan and organize phase.
B. Information security policies and standards are a component of the risk management program but do not belong to the initial stages of its development. Information security policies and standards are formed by the decisions made in the planning phase of the program and are developed based on the outcomes and business objectives established by the business.
C. Management and oversight of the risk management program is a monitoring control that is developed to ensure that the program is satisfying the outcomes and business objectives established by the business. This process is designed at the latter stages of development once the purpose of the program and the mechanics of its deployment have been established. This oversight process could be integrated with internal audit activities or other compliance program processes.
D. An initial requirement is to determine the organization's purpose for creating an information security risk management program, determine the desired outcomes and define objectives.

S2-202 As part of system development, how should an organization determine which element of the confidentiality, integrity and availability triad requires the **MOST** protection?

A. It should be based on the threat to each of the elements.
B. Availability is most important.
C. It should be based on the likelihood and impact to each element if compromised.
D. All elements are equally important.

C is the correct answer.

Justification:

A. Even if the threat of compromise is high, the impact may be low; the best basis to determine where to implement the most protection is the risk to the specific element.
B. While it may seem that availability is the most important, if the system is down, there is no access to the data. There are many cases in which the standard business processes can continue, even if the system is down, but stringent controls must be maintained around confidentiality and integrity of information. The level of control should be based on the risk to the specific element.
C. The probability of compromise and the impact on the organization are combined to determine which element requires the greatest protection with emphasis on impact.
D. It is very unlikely that all elements of the confidentiality, integrity or availability triad require equal levels of protection.

S2-203 Which of the following approaches would be **BEST** to address significant system vulnerabilities that were discovered during a network scan?

A. All significant vulnerabilities must be mitigated in a timely fashion.
B. Treatment should be based on threat, impact and cost considerations.
C. Compensating controls must be implemented for major vulnerabilities.
D. Mitigation options should be proposed for management approval.

B is the correct answer.

Justification:
A. Some vulnerabilities may not have significant impact and may not require mitigation.
B. The treatment should consider the degree of exposure and potential impact and the costs of various treatment options.
C. Compensating controls are considered only when there is a viable threat and impact, and only if the primary control is inadequate.
D. Management approval may not be required in all cases.

S2-204 An organization's board of directors has learned of recent legislation requiring organizations within the industry to enact specific safeguards to protect confidential customer information. What actions should the board take next?

A. Direct information security on what they need to do.
B. Research solutions to determine the proper solutions.
C. Require management to report on compliance.
D. Do nothing; information security does not report to the board.

C is the correct answer.

Justification:
A. The board would not direct information security activities; this would be the function of executive management.
B. The board would not undertake research but might direct the executive to see that it is completed.
C. Information security governance is the responsibility of the board of directors and executive management. In this instance, the appropriate action is to ensure that a plan is in place for implementation of needed safeguards and to require updates on that implementation.
D. The board has oversight responsibilities, and doing nothing would not be a prudent course of action.

S2-205 An organization is using a vendor-supplied critical application which has a maximum password length that does not comply with organizational security standards. Which of the following approaches **BEST** helps mitigate the weakness?

A. Shorten the password validity period.
B. Encourage the use of special characters.
C. Strengthen segregation of duties.
D. Introduce compensating controls.

D is the correct answer.

Justification:

A. Periodic change of password is a good control against password theft. However, it would not compensate for the shortcoming in password length.
B. Use of special characters will enhance password complexity. However, it will not fully replace the shortcoming in password length.
C. Segregation of duties will tighten the control against fraud. However, it will not resolve password noncompliance.
D. Vendor systems are sometimes unable to provide a security control that meets the policy of the organization. In such cases, compensating controls should be sought (e.g., password lockout on failed attempts).

S2-206 The information security manager has determined that a risk exceeds risk appetite, yet the manager does not mitigate the risk. What is the **MOST** likely reason that management would consider this course of action appropriate?

A. The risk is the residual risk after controls are applied.
B. The risk is expensive to mitigate.
C. The risk falls within the risk tolerance level.
D. The risk is of relatively low frequency.

C is the correct answer.

Justification:

A. Even if the risk is residual, if it exceeds the risk appetite, then it is acceptable only if it falls within the risk tolerance. The residual risk may or may not be considered appropriate depending on the level of acceptable risk and the tolerance for variation to that level.
B. If mitigation is too expensive compared to the benefit, the information security manager should consider other treatment options. Just knowing the expense is not enough.
C. Risk tolerance is the acceptable level of variation that management is willing to allow for any particular risk as the enterprise pursues its objectives.
D. Low frequency alone does not warrant ignoring a risk.

S2-207 Why should the analysis of risk include consideration of potential impact?

A. Potential impact is a central element of risk.
B. Potential impact is related to asset value.
C. Potential impact affects the extent of mitigation.
D. Potential impact helps determine the exposure.

C is the correct answer.

Justification:
A. Impact is distinct and separate from risk and is not a central element of risk.
B. Impact is related to the loss of the value that the asset provides but is not relevant to the question.
C. The extent of the potential impact in the event of compromise coupled with the likelihood of occurrence will largely determine the extent of mitigation measures.
D. Knowing the impact will not determine the extent to which an asset is exposed to a threat.

S2-208 To be effective, risk management should be applied to:

A. all organizational activities.
B. those elements identified by a risk assessment.
C. any area that exceeds acceptable risk levels.
D. only those areas that have potential impact.

A is the correct answer.

Justification:
A. While not all organizational activities will pose an unacceptable risk, the practice of risk management is still applied to determine which risk requires treatment.
B. Risk assessment is part of the risk management function. Risk assessment does not precede inclusion of the activity in the risk management program.
C. Whether a risk level is acceptable can be determined only when the risk is known.
D. Potential impact can be evaluated only when the risk is known and the value of the asset is determined.

S2-209 What is the **BEST** risk response for risk scenarios where the likelihood of a disruptive event for an asset is very low, but the potential financial impact is very high?

A. Accept the high cost of protection.
B. Implement detective controls.
C. Ensure that asset exposure is low.
D. Transfer the risk to a third party.

D is the correct answer.

Justification:
A. Where there is a high cost of protection paired with a low likelihood, organizations generally find it more cost-effective to transfer risk.
B. A detective control alone does nothing to limit the impact.
C. The fact that the likelihood is low suggests that exposure is already minimal. Additional reductions to exposure would do nothing to limit impact.
D. High-impact, low-likelihood situations are typically most cost-effectively covered by transferring the risk to a third party (e.g., insurance).

S2-210 Faced with numerous risk scenarios, the prioritization of treatment options will be **MOST** effective if based on the:

A. existence of identified threats and vulnerabilities.
B. likelihood of compromise and subsequent impact.
C. results of vulnerability scans and remediation cost.
D. exposure of corporate assets and operational risk.

B is the correct answer.

Justification:

A. Threats and vulnerabilities are the measure of risk, but without knowing potential impact, the most cost-effective treatment options will not be clear.
B. Probability of compromise coupled with the likely impact will be the most important considerations for selecting treatment options.
C. Vulnerabilities and the cost to remediate without considering impact do not provide enough information to make the best treatment selection.
D. Exposure of assets will modify the effective risk by affecting the likelihood that a vulnerability will be exploited but is also insufficient information to choose the best treatment option. Operational risk is only one part of overall risk.

S2-211 Which of the following items determines the acceptable level of residual risk in an organization?

A. Management discretion
B. Regulatory requirements
C. Inherent risk
D. Internal audit findings

A is the correct answer.

Justification:

A. Deciding what level of risk is acceptable to an organization is fundamentally a function of management. At its discretion, organizational management may decide to accept risk. The target risk level for a control is ultimately subject to management discretion.
B. Failure to comply with regulatory requirements has consequences, but those consequences are considered in the context of organizational risk. In some cases, the cost of failure to comply may be lower than the cost of compliance; in this case, management may decide to accept the risk.
C. Inherent risk is the risk that exists before controls are applied.
D. The results of an internal audit are used to determine the actual level of residual risk, but whether this level is acceptable is fundamentally a function of management.

S2-212 A risk management process is **MOST** effective in achieving organizational objectives if:

A. asset owners perform risk assessments.
B. the risk register is updated regularly.
C. the process is overseen by a steering committee.
D. risk activities are embedded in business processes.

D is the correct answer.

Justification:
A. Performing a risk assessment does not ensure mitigation as a part of the business process.
B. Maintaining a risk register may be good for identifying issues, but does not mitigate risk.
C. Centralizing risk management under a steering committee is less effective than integrating it into each business process.
D. The primary objective of the risk management process is that risk is identified, assessed, communicated and addressed. This objective is most effectively achieved by embedding risk management activities in business processes (e.g., change management, incident response, new product design, sales campaign, etc.).

S2-213 Reducing exposure of a critical asset is an effective mitigation measure because it reduces:

A. the impact of a compromise.
B. the likelihood of being exploited.
C. the vulnerability of the asset.
D. the time needed for recovery.

B is the correct answer.

Justification:
A. The impact of a successful exploit will not change.
B. Reducing exposure reduces the likelihood of a vulnerability being exploited.
C. The vulnerabilities of the asset will not change because exposure is reduced.
D. The recovery time is not affected by a reduction in exposure.

S2-214 The classification level of an asset must be **PRIMARILY** based on which of the following choices?

A. Criticality and sensitivity
B. Likelihood and impact
C. Valuation and replacement cost
D. Threat vector and exposure

A is the correct answer.

Justification:
A. The extent to which an asset is critical to business operations or can damage the organization if disclosed is the primary consideration for the level of protection required.
B. Asset classification is driven by criticality and sensitivity, not likelihood of compromise.
C. Probability and frequency are considerations of risk and not the main consideration of asset classification.
D. Threat vector and exposure together do not provide information on impact needed for classification.

S2-215 What is the goal of risk aggregation?

A. To combine homogenous elements to reduce overall risk
B. To influence the organization's risk acceptance methodologies
C. To group individual acceptable risk events for simplified risk reporting
D. To identify significant overall risk from a single threat vector

D is the correct answer.

Justification:
A. Combining homogenous elements does not in itself reduce risk; it may actually increase risk.
B. Aggregation does not affect the methodology used for risk acceptance.
C. Risk reporting is not a primary consideration of risk aggregation.
D. Individual risk with minimal impact may constitute a significant overall risk if each risk can be exploited from the same threat vector. The threat vector is the method used to exploit the target.

X **S2-216** Which of the following choices is **MOST** likely to achieve cost-effective risk mitigation across the organization?

A. A chief risk officer
B. Consistent risk assessments
C. Assurance process integration
D. Defined acceptable risk levels

C is the correct answer.

Justification:
A. A chief risk officer is usually helpful in identifying many types of risk faced by an organization, but remediation is a function of many different organizational units, and unless these activities are integrated, there is the possibility of duplicated efforts or gaps in protection.
B. Risk assessments are helpful in exposing risk but, by themselves, do not serve to mitigate the identified risk.
C. Integrating the risk mitigation of the typical organization's many different risk management and assurance functions will best ensure that there are no gaps in protection efforts and a minimum of duplicated efforts, which is likely to result in the best coverage at the lowest cost.
D. Defining acceptable risk levels can provide guidance to the organization about the required levels of mitigation required but does not prevent duplication of efforts or gaps in protection.

S2-217 When considering the extent of protection requirements, which of the following choices would be the **MOST** important consideration affecting all the others?

A. Exposure
B. Threat
C. Vulnerability
D. Magnitude

A is the correct answer.

Justification:

A. Exposure is the quantified potential for loss that may occur due to an adverse event, calculated as the product of probability and magnitude (impact). Because probability is itself a function of threat and vulnerability, exposure takes into account all three of the other factors and, where known, is the most important consideration.

B. A threat is anything (e.g., object, substance, human) that is capable of acting against an asset in a manner that can result in harm. Threats may cause harm only where they correspond to vulnerability, so the probability of an event can be calculated only when both are known.

C. Vulnerability is a weakness in the design, implementation, operation or internal control that could expose the system to adverse threats from threat events. Vulnerability may lead to harm only when acted on by a corresponding threat, so the probability of an event can be calculated only when both are known.

D. Magnitude (or impact) measures the potential severity of loss from a realized event/scenario. Whether such an event will be realized depends on its probability (likelihood), which requires assessment of both threat and vulnerability.

S2-218 Which of the following should be understood before defining risk management strategies?

A. Risk assessment criteria
B. Organizational objectives and risk appetite
C. IT architecture complexity
D. Enterprise disaster recovery plans

B is the correct answer.

Justification:

A. The assessment criteria are not relevant to defining risk management strategies.

B. The risk management strategy must be designed to achieve organizational objectives as well as provide adequate controls to limit risk to be consistent with the risk appetite.

C. IT architecture complexity may pose a challenge to the risk assessment process but should not affect the risk management strategy directly.

D. Disaster recovery plans are an element of the risk management strategy but are addressed by organizational objectives and risk appetite.

S2-219 Management requests that an information security manager determine which regulations regarding disclosure, reporting and privacy are the most important for the organization to address. The recommendations for addressing these legal and regulatory requirements will be **MOST** useful if based on which of the following choices?

A. The extent of enforcement actions
B. The probability and consequences
C. The sanctions for noncompliance
D. The amount of personal liability

B is the correct answer.

Justification:

A. The extent of enforcement is a measure of probability. Without knowing the scope of consequences, probability cannot be viewed in context.

B. Legal and regulatory requirements should be treated as any other risk to the organization, calculated as the probability of enforcement and the magnitude of possible sanctions (impact or consequences).

C. Sanctions or impact must be considered in the context of the enforcement mechanisms. If sanctions have little probability of being implemented due to lax enforcement, their severity poses less risk to the organization than if they are widely enforced.

D. Except in extreme cases of fraud or other criminal activity, liability for regulatory sanctions generally lies with senior management and the board of directors. It is not a driving factor in the evaluation of regulatory requirements.

X **S2-220** Control objectives are **MOST** closely aligned with:

A. risk tolerance.
B. criticality.
C. risk appetite.
D. sensitivity.

C is the correct answer.

Justification:

A. Risk tolerance is the acceptable level of deviation from acceptable risk and is not directly affected by control objectives.

B. Criticality is the importance to the business and is one of the considerations when control objectives are set in addition to potential impact, exposure, cost and feasibility of possible controls, but criticality plays a lesser role in relationships between risk and control. Criticality is more a need for the business than a control to reduce risk for the environment.

C. Risk appetite is the amount of risk, on a broad level, that an entity is willing to accept in pursuit of its mission. Control objectives are set so that controls can be designed on that basis.

D. Sensitivity is the potential impact of unauthorized disclosure, which will also be one of the considerations in control objectives, but it is not a control itself. Sensitivity creates risk, and that risk is weighed against the controls put in place to reduce that risk, but sensitivity is an identification marker or classification of data or a control and does not define "acceptable risk."

S2-221 An information security manager determines that management of risk is inconsistent across a mature organization, creating a weak link in overall protection. The **MOST** appropriate initial response for the information security manager is to:

A. escalate to the steering committee.
B. review compliance with standards.
C. write more stringent policies.
D. increase enforcement.

B is the correct answer.

Justification:

A. The steering committee may be able to assist in achieving better compliance after it has been established by audit. The steering committee is an executive management-level committee that assists in the delivery of the security strategy, oversees day-to-day management of service delivery and IT projects, and focuses on implementation.
B. A mature organization will have a complete suite of policies and standards, and inconsistent risk treatment is most likely to be inconsistent compliance with standards.
C. Policies need to be reviewed to determine whether they are adequate. The problem may be with enforcement.
D. Enforcement can only be as effective as the policies it supports. Increasing enforcement prior to determining the issues would not be the best initial response.

S2-222 When performing a review of risk treatment options, the **MOST** important benefit to consider is:

A. maximum risk mitigation.
B. savings in control options.
C. alignment with regulatory requirements.
D. achieving control objectives.

D is the correct answer.

Justification:

A. Control objectives are established on the basis of organizational risk appetite, so maximizing mitigation beyond the control objectives means incurring unnecessary cost.
B. Cost is always a consideration, but an option cannot be considered to have "saved" money unless it also meets an objective.
C. Regulatory requirements are considered no differently than any other consideration in the risk assessment process. Control objectives are established on the basis of risk appetite, which may or may not include accepting the risk of not complying with a regulation.
D. Achieving control objectives is the reason for designing controls. No other benefit can offset failure to meet the control objectives.

S2-223 Compliance with legal and regulatory requirements is:

A. a security decision.
B. a business decision.
C. an absolute requirement.
D. conditional and based on cost.

B is the correct answer.

Justification:

A. Information security can advise management on the risk and possible impact of compliance failure, but the business decision must be made by senior management.
B. The extent of compliance with legal and regulatory requirements is a business decision that must be made by senior management.
C. Legal and regulatory requirements are no different than other requirements for purposes of risk assessment and decision making. Each legal or regulatory requirement is evaluated in the context of the risk posed by failure to comply.
D. Cost is only one aspect considered as part of the overall business decision to comply with legal and regulatory requirements.

S2-224 Determining the level of effort needed to meet particular improvement targets in risk management can **BEST** be determined using which of the following tools?

A. A workflow diagram
B. A Gantt chart
C. A gap analysis
D. A return on investment computation

C is the correct answer.

Justification:

A. Workflow diagrams document processes. Having a visual representation of how a risk management process works today versus how it would work in a desired state may be useful as part of proposing or implementing changes, but comparing these two states is not the same as knowing what tasks must be completed to move from the current state to the proposed future state, which is what is needed to determine the level of effort.
B. Gantt charts are used to schedule activities (tasks) needed to complete a project. A fully constructed schedule does include all of the tasks that must be completed and the times that they will take, but building a schedule deals with prioritization and other issues that go beyond what is needed to determine the level of effort.
C. A gap analysis documents the tasks that must be completed to move from the current state to the desired state, and the level of effort may readily be determined. A gap analysis is required for various components of the strategy previously discussed, such as maturity levels, each control objective, and each risk and impact objective.
D. Return on investment, computed in its simplest form by dividing net income by the total investment over the period being considered, is a measure of operating performance and efficiency. It does not measure levels of effort.

S2-225 Which of the following choices would be the **MOST** useful in determining the possible consequences of a major compromise?

A. Risk assessment
B. Asset valuation
C. Penetration testing
D. Architectural review

B is the correct answer.

Justification:

A. A comprehensive risk assessment requires an assessment of probability as well as potential consequences, so it goes beyond what is required.
B. Asset valuation provides a cost representation of what the organization stands to lose in the event of a major compromise.
C. Penetration tests indicate vulnerability rather than the value of what may be affected if that vulnerability is exploited.
D. Architectural review may indicate vulnerability, but like penetration testing, it will not reveal the value of what may be affected if that vulnerability is exploited.

S2-226 Under what circumstances do good information security practices dictate a full reassessment of risk? 

A. After a material control failure
B. When regular assessments show unremediated risk
C. Subsequent to installing an updated operating system
D. After emergency changes have been initiated

A is the correct answer.

Justification:

A. A significant control failure indicates that either the control was poorly designed or the risk was not properly identified and classified.
B. Depending on the nature and extent of unremediated risk, reassessment may be warranted, but, in some cases, the process of change management while addressing the risk will have provided adequate understanding of the risk and adequacy of treatment.
C. Updating an operating system under change management will include an incremental assessment of any new risk, and full reassessment is not likely to be needed.
D. Emergency changes usually require that the change management process be completed subsequently and any specific new risk addressed, making it unlikely that a full risk reassessment is required.

S2-227 The **MOST** important factors to consider when prioritizing control development are:

A. threat and vulnerability.
B. cost and frequency.
C. risk appetite and tolerance.
D. probability and impact.

D is the correct answer.

Justification:

A. Threat and vulnerability are factors in determining probability, but without knowing the magnitude of loss (or impact) associated with a particular event, knowing its probability is an inadequate basis for prioritizing control development.
B. Cost is always a consideration, and resource constraints may lead to certain controls being delayed, but prioritization occurs even among controls of comparable cost.
C. These are considerations when developing control objectives but do not factor into the prioritization of controls.
D. The probability that an adverse event will occur and the consequent impact provide an effective quantitative basis for prioritizing the development of controls.

S2-228 The **MOST** direct way to accurately determine the control baseline in an IT system is to do which of the following activities?

A. Review standards and system compliance.
B. Sample hardware and software configurations.
C. Review system and server logs for anomalies.
D. Perform internal and external penetration tests.

A is the correct answer.

Justification:

A. A control baseline is obtained by reviewing the standards to determine whether the baseline falls within the boundaries set by the standards.
B. Sampling hardware configurations without knowing the control requirements reflected in the standards provides information on the current state but not on how that state relates to the intended state.
C. Anomalies in system logs do not necessarily indicate that baseline security is incorrect nor does an absence of abnormalities mean that the baseline is correct.
D. Penetration tests that reveal vulnerabilities must be evaluated in the context of the control requirements set by the standard.

S2-229 High risk tolerance is useful when:

A. the organization considers high risk acceptable.
B. the uncertainty of risk shown by an assessment is high.
C. the impact from compromise is very low.
D. indicated by a business impact analysis.

B is the correct answer.

Justification:
A. Risk tolerance is the acceptable deviation from acceptable risk and is not related to whether the risk is high or low.
B. High risk tolerance (i.e., a high degree of variability in acceptable risk) addresses the issue of uncertainty in the risk assessment process itself.
C. Risk tolerance is unrelated to impact.
D. The degree of risk tolerance is not indicated by a business impact analysis.

S2-230 An organization has identified a major threat to which it is vulnerable. Which of the following choices is the **BEST** reason why information security management would not be concerned with preventive remediation under these circumstances?

A. The vulnerability is compartmentalized.
B. Incident response procedures are in place.
C. Compensating controls exist if there is any impact.
D. The identified threat has only been found on another continent.

A is the correct answer.

Justification:
A. If the compartmentalization of the vulnerability results in the organization having no exposure, then there is no risk.
B. Prevention is a more prudent approach to dealing with major threats than even the most capable incident response.
C. Compensating controls are a less desirable approach to addressing a major threat than preventive remediation of its corresponding vulnerability.
D. Distance is an inadequate barrier to compromise in the context of information systems.

S2-231 Finding that a lack of adequate compliance with a set of standards poses a significant risk, an information security manager should **FIRST**:

A. review and modify policy to address the risk.
B. create a new set of guidelines to reduce the risk.
C. advise management of the risk and possible consequences.
D. determine whether the standards are consistent with policy.

C is the correct answer.

Justification:
A. The extent of risk mitigation is a business decision, so any action taken to address or reduce risk must be based on input from the business.
B. The extent of risk mitigation is based on policy which is defined with input from the business.
C. If a lack of compliance with standards creates a significant risk, the information security manager should assess possible consequences and advise appropriate managers to determine whether it is acceptable risk.
D. It is generally useful to determine whether standards reflect the intent of policy, but the policies main purpose is to address the risk that might not be included in those standards.

S2-232 Laws and regulations should be addressed by the information security manager:

A. to the extent that they impact the enterprise.
B. by implementing international standards.
C. by developing policies that address the requirements.
D. to ensure that guidelines meet the requirements.

A is the correct answer.

Justification:

A. Legal and regulatory requirements should be assessed based on the extent and nature of enforcement, the probability of enforcement action and sanctions, and the impact of noncompliance or partial compliance balanced against the costs of compliance.
B. International standards may not address the legal requirements in question.
C. Policies should not address particular regulations because regulations are subject to change. Policies should only address the need to assess regulatory requirements and deal with them appropriately based on risk and impact.
D. Guidelines would normally not address regulations, although standards may address regulations based on management's determination of the appropriate level of compliance.

X **S2-233** Which of the following is the **GREATEST** security concern when an incident log is stored on the production database server?

A. Log information may be lost when the database server crashes.
B. The database administrator may tamper with the log information.
C. The capacity to handle large transactions may be compromised.
D. Sensitive information may inadvertently be written to the log file.

B is the correct answer.

Justification:

A. Production data and log data are most likely backed up at the same time. Therefore, it is possible to restore both production data and log data based on the same recovery point.
B. There is a chance that fraud can be committed because the administrator can manipulate the database server. The administrator may alter database transactions and then erase the log. It is best that the log be managed in a separate environment from the production database to serve as a useful detective control.
C. Transaction processing capacity may or may not degrade as the result of log activation. The stem does not provide enough information to affirm process degradation.
D. There may be a case when sensitive information is inadvertently written to a log. This comes from a design failure in the application and/or database system rather than from the fact that the production database and log are on the same server.

S2-234 Quantifying the level of acceptable risk can **BEST** be indicated by which of the following choices?

A. Surveying business process owners and senior managers
B. Determining the percentage of the IT budget allocated to security
C. Determining the ratio of business interruption insurance to its cost
D. Determining the number and severity of incidents impacting the organization

C is the correct answer.

Justification:
A. Surveying management may provide a typically widely varying perspective on acceptable risk.
B. The amount spent on security is an indicator, but does not quantify acceptable levels of risk.
C. The amount of business interruption insurance carried and the cost provides a directly quantifiable level of risk that the organization will accept and at what cost.
D. The history of incidents will show what risk was not addressed and elicit comments about acceptability but will not indicate what the organization is willing to spend on mitigation.

S2-235 At what point in the risk management process is residual risk determined?

A. When evaluating the results of the application of new or existing controls or countermeasures
B. When identifying and classifying information resources or assets that need protection
C. When assessing threats and the consequences of a compromise
D. After the elements of risk have been established, when combining them to form an overall view of risk

A is the correct answer.

Justification:
A. The objective of information risk management is to bring the information security residual risk to an acceptable level, so residual risk is evaluated first on the basis of existing controls and again after any new controls are designed or implemented.
B. Identification and classification of information resources or assets that need protection is the first step of risk management and is followed by assessment of threats and vulnerabilities to determine probability. Probability is an input to calculating initial risk, so there is no basis for calculating residual risk at this stage.
C. Knowledge of the threat environment and consequences of a compromise is inadequate to determine residual risk because it does not take into account vulnerability and exposures.
D. The overall view of risk reflects an initial risk level, which has not yet been reduced by application of controls. After elements of risk are combined to form an overall view of risk, the next step is to identify existing controls or design new controls to bring risk to an acceptable level.

S2-236 Which of the following choices would be the **BEST** measure of the effectiveness of a risk assessment?

A. The time, frequency and cost of assessing risk
B. The scope and severity of new risk discovered
C. The collective potential impact of defined risk
D. The percentage of incidents from unknown risk

D is the correct answer.

Justification:
A. The time and cost of performing a risk assessment is not an indicator of its effectiveness in discovering new risk.
B. The scope and severity of new risk discovered is a useful indicator, but not as good a measure of effectiveness as the risk that is not uncovered and leads to a security incident.
C. The potential impact of defined risk is a secondary measure that may be useful in determining the extent of remedial actions to consider.
D. Incidents from unidentified risk that lead to an incident are the best indicators of how well the risk assessment served to discover risk, thereby indicating effectiveness.

X **S2-237** If a defined threat needs to be addressed and a preventive control is not feasible, the next **BEST** option is to do which of the following activities?

A. Use a deterrent control.
B. Reduce exposure.
C. Use a compensating control.
D. Reassess the risk.

B is the correct answer.

Justification:
A. Using a deterrent control will have only a limited effect on the possibility of compromise.
— **B. Reducing exposure reduces the probability that a risk can be exploited.**
C. Using a compensating control will serve to limit impact, but do nothing to prevent exploitation.
D. Reassessing risk may provide a clearer picture of the risk, but does nothing to reduce exploitation.

S2-238 An information security manager has two identical servers in the network subject to a viable threat, but decides to harden only one of them. The **MOST** likely reason for this choice is that the second server:

A. handles only unimportant information.
B. will be unable to perform required tasks.
C. is placed such that it has no exposure.
D. has constant monitoring that precludes attack.

C is the correct answer.

Justification:
A. Unimportant information may require less protection, but it is unlikely that it should be totally unprotected because it may provide an avenue into the rest of the network.
B. It is unlikely that hardening a server will render it incapable of performing required tasks.
C. If the second server has no exposure, there is no probability that a compromise can occur.
D. Monitoring may indicate when an attack occurs but will not preclude an attack.

S2-239 When conducting a risk assessment, which of the following elements is the **MOST** important?

A. Consequences
B. Threat
C. Vulnerability
D. Probability

A is the correct answer.

Justification:

A. Unless the exploitation of vulnerability by a threat has consequences, there is no risk to the organization.
B. A threat poses no risk absent corresponding vulnerability.
C. Vulnerability poses no risk absent a corresponding threat.
D. Probability is a function of threat and vulnerability, but even a guaranteed event poses no risk to the organization unless there are consequences.

S2-240 Which of the following choices is **MOST** strongly supported by effective management of information assets?

A. An information/data dictionary
B. A data classification program
C. An information-based security culture
D. A business-oriented risk policy

D is the correct answer.

Justification:

A. An information/data dictionary is a useful management tool but is only one aspect of holistic information asset management.
B. A data classification program helps to prioritize asset protection based on business value, but management of information assets goes beyond asset protection.
C. The security culture of an organization does not drive the effectiveness or efficiency of information assets.
D. A risk policy that is oriented to business needs promotes the achievement of organizational objectives. The holistic risk-based approach to the management of information assets includes and addresses a broad range of factors such as data linkages, privacy, business orientation and risk relevance, which in turn help the assets to be managed in an effective and efficient manner.

S2-241 Which of the following considerations is the **MOST** important one in the use of a vulnerability scanning tool?

A. Multiple functions
B. Regular updates
C. Graphical user interface
D. Real-time virus deletion

B is the correct answer.

Justification:

A. Multiple functionalities cannot replace the importance of a scanner being kept current to the latest vulnerabilities.
B. A vulnerability scanner is as good as the last time it was updated.
C. The graphical user interface addresses ease of use rather than the effectiveness of the scanner.
D. A vulnerability scanner does not need to have the ability to delete viruses.

S2-242 Which of the following items is the **BEST** basis for determining the value of intangible assets?

A. Contribution to revenue generation
B. A business impact analysis
C. Threat assessment and analysis
D. Replacement costs

A is the correct answer.

Justification:

A. The value of any business asset is generally based on its contribution to generating revenues for the organization, both now and in the future.
B. A business impact analysis (BIA) is a process to determine the impact of losing the support of any resource. The BIA study will establish the escalation of that loss over time. It is predicated on the fact that senior management, when provided reliable data to document the potential impact of a lost resource, can make the appropriate decision. It may not take into account the long-term impact to revenue of losing intangible assets.
C. Threat analysis is an evaluation of the type, scope and nature of events or actions that can result in adverse consequences; identification of the threats that exist against enterprise assets. The threat analysis usually defines the level of threat and the likelihood of it materializing. Threat assessment is not concerned with asset value but with the probability of compromise.
D. The replacement cost of intangible assets such as trade secrets typically cannot be calculated because replacement is impossible.

S2-243 Which of the following activities is the **FIRST** step toward implementing a bring your own device (BYOD) program?

A. Allow or deny access to devices based on their approval status.
B. Conduct a stringent assessment process prior to approving devices.
C. Implement a plan-do-check-act approach.
D. Review and approve applications in the organization's application store.

B is the correct answer.

Justification:

A. For device access to be determined on the basis of approval status, an assessment process must be in place to grant approval.
B. A stringent assessment process is critical to comply with corporate and regulatory requirements around policies, encryption, detection of jailbreaking or rooted devices, etc.
C. Implementing a plan-do-check-act approach is part of the monitoring and enforcement process, but it is not a prerequisite.
D. Having a review and approval process for applications in the organization's application store applies only to devices granted approval to access the network.

S2-244 How does knowledge of risk appetite help to increase security control effectiveness?

A. It shows senior management that you understand their needs.
B. It provides a basis for redistributing resources to mitigate risk above the risk appetite.
C. It requires continuous monitoring because the entire risk environment is constantly changing.
D. It facilitates communication with management about the importance of security.

B is the correct answer.

Justification:

A. Telling management what their risk appetite is will likely get a "So what?" response. It is meaningless outside the context of control effectiveness.
B. Understanding risk appetite in key security control areas helps redirect resources from risk at or below acceptable levels to risk above the appetite. The result is improved control effectiveness at no additional cost.
C. This answer does not address the value of understanding risk appetite. The risk environment and control effectiveness do change, but continuous monitoring applies more to rapidly changing controls, and to areas of greatest risk. Risk appetite changes are usually more stable.
D. Knowledge of risk appetite does help to facilitate communication with management but is only one small element of effective communication with senior management.

X **S2-245** Which of the following activities **MUST** a financial-services organization do with regard to a web-based service that is gaining popularity among its customers?

A. Perform annual vulnerability mitigation.
B. Maintain third-party liability insurance.
C. Conduct periodic business impact analyses.
D. Architect a real-time failover capability.

C is the correct answer.

Justification:

A. Vulnerability management is an important part of managing any system, but mitigation decisions are made on the basis of risk and are not isolated to an annual activity.
B. The decision of whether to carry liability insurance is a business decision made on the basis of quantified risk.
C. A service that is gaining popularity will increase in value to the organization as it grows, leading to corresponding growth in the magnitude of potential loss should the service be interrupted. Periodic business impact analyses (BIAs) quantify this magnitude and ensure that adequate recovery capabilities can be put in place.
D. Real-time failover capabilities may be warranted, but the decision to design and deploy such capabilities is a business decision based in large part on an accurate BIA quantifying the magnitude of potential loss should the service be interrupted.

X **S2-246** If a security incident is not the result of the failure of a control, then it is **MOST** likely the result of which of the following choices?

A. An incomplete risk analysis
B. The absence of a control
C. A zero-day attack
D. A user error

B is the correct answer.

Justification:

A. An incomplete risk analysis may have the effect of a suitable control not being implemented, but it is not the reason that a compromise occurs.
B. A security incident is inevitably the result of a control failure or the lack of a suitable control.
C. A zero-day attack is difficult to predict, but it will only be successful if a control fails or does not exist.
D. A user error will also only result in a security incident because of control failure or the absence of a control.

S2-247 The effectiveness of managing business risk is **BEST** measured by the number of:

A. significant IT-related incidents that were not identified during risk assessment.
B. security assessments compliant with organizational standards and guidelines.
C. vulnerabilities identified by risk assessment and not properly mitigated.
D. security incidents causing significant financial loss or business disruption.

D is the correct answer.

Justification:

A. Identification of incidents is only one part of effective risk management. If impact is not limited to acceptable levels, the program is not effective. Merely identifying incidents through a risk assessment is insufficient to limit impact.
B. While compliance is important, it is only one aspect of risk management. If impact is not limited to acceptable levels, the program is not effective. Demonstrating that a program is compliant is not a measure of the effectiveness of limiting impact.
C. Identifying unmitigated vulnerabilities is insufficient without knowledge of potential threats, impacts and control measures to determine the potential effectiveness of the risk management program.
D. The goal of risk management is to limit impact and minimize business disruptions. Each instance of a security incident that causes significant financial loss or business disruption is an indication of inadequate risk management.

S2-248 Assuming all options are technically feasible, which of the following would be the **MOST** effective approach for the information security manager to address excessive exposure of a critical customer-facing server?

A. Develop an incident response plan
B. Reduce the attack surface
C. Initiate compartmentalization
D. Implement compensating controls

B is the correct answer.

Justification:

A. Even the most effective incident response plan is unlikely to reduce exposure as effectively as reducing the attack surface.
B. The attack surface determines the extent of exposure. Reducing the attack surface by limiting entry points, ports and protocols and taking other precautions reduces the exposure.
C. Compartmentalization may limit the degree to which impact sustained by one customer results in increased vulnerability or impact for another customer, but the per-customer exposure would not be affected.
D. Compensating controls are appropriate in cases where existing controls are incapable of reducing risk to acceptable levels, but reducing the attack surface will be more effective under circumstances where it is technically feasible.

S2-249 Objectives for preventive controls should be developed **PRIMARILY** on the basis of:

A. risk levels aligned with the enterprise risk appetite.
B. technical requirements directed by industry standards.
C. threat levels as established by monitoring tools.
D. uptime targets specified in service level agreements.

A is the correct answer.

Justification:

A. **Controls are designed and implemented to produce levels of risk aligned with the enterprise risk appetite.**
B. Industry standards offer managers and engineers direction on how desired objectives might be achieved, but organizations adopt them only when doing so aligns with business objectives and the enterprise risk appetite.
C. Monitored threat levels do not provide a comprehensive basis for the design and implementation of preventive controls.
D. The need to meet uptime targets specified in service level agreements is only one of many considerations taken into account when developing preventive controls.

S2-250 The **PRIMARY** purpose of risk evaluation is to:

A. provide a basis on which to select risk responses.
B. ensure that controls are deployed to mitigate risk.
C. provide a means of targeting assessment activities.
D. ensure that risk responses align with control objectives.

A is the correct answer.

Justification:

A. **Risk evaluation provides management with the extent that the risk meets the acceptability criteria and options for response. Response to risk may come in the form of acceptance, transfer (sharing), mitigation or avoidance.**
B. Mitigation is only one possible response to risk.
C. Risk evaluation is the final stage of an assessment activity.
D. Control objectives align with the risk management strategy, which determines risk response.

S2-251 An organization is considering a reciprocal arrangement with a similar organization as a recovery option. Which of the following is the **GREATEST** risk associated with a reciprocal arrangement?

A. Variations between the risk and impact assessments
B. Frequency of testing of the recovery and continuity plans
C. Dissimilarities in infrastructure and capacity
D. Differences in security policies and procedures

C is the correct answer.

Justification:

A. Analyses are predictive, so differences between the organizations will not affect adequacy in the event of recovery.
B. Organizations must collaborate on frequency of testing to ensure that each meets its needs. However, such agreements are generally established when arranging reciprocity and do not constitute ongoing risk.
C. If organizations have dissimilar infrastructure or lack capacity, it may be difficult to implement recovery.
D. Differences in security policies and procedures are generally addressed when establishing reciprocity and can be managed over time through monitoring and reporting.

S2-252 Which of the following is the **MOST** supportable basis for prioritizing risk for treatment?

A. Cost and asset value
B. Frequency and impact
C. Frequency and scope
D. Cost and effort

B is the correct answer.

Justification:

A. Cost to remediate is a major factor only relative to the value of the assets to which remediation applies (i.e., is remediation appropriate for this asset versus another risk treatment option?). It is ineffective as a means of prioritization across different assets, because it does not take into account their business value.
B. The balance between impact and frequency captures the adjusted probability of loss to the organization associated with each risk. Therefore, this provides an immediate and relevant basis for prioritization of treatment, with risks that are high-impact and high-frequency ranking the highest on the list.
C. Breadth of scope is not necessarily equivalent to impact. Prioritizing a risk that affects a broad range of relatively unimportant systems over a risk that impacts a single critical system would not be beneficial to the organization.
D. Effort is a subset of overall cost representing time and expertise. Unto itself, cost is not a suitable basis for prioritization.

S2-253 When the security risk assessment result was reviewed, it was found that the rationale for risk rating varies by department. Which of the following will **BEST** improve this situation?

A. Apply common risk measurement criteria to each department
B. Introduce risk appetite and risk tolerance at the policy level
C. Place increased focus on quantitative risk assessment
D. Implement routine peer review of the risk assessment results

A is the correct answer.

Justification:

A. If departments are reaching different risk ratings for the same outcomes, common risk measurement criteria that can be used across the organization are needed.
B. Risk appetite and risk tolerance inform the acceptance of risk but do not affect the risk ratings.
C. Quantitative risk assessments produces numeric results, but subjectivity in inputs may continue to yield varying risk ratings among departments unless common criteria are applied.
D. Peer review of risk assessments between departments may be hampered by differing expertise among staff members in different job functions. Also, the results of risk assessments generally should not be shared more broadly than is necessary to meet business goals.

S2-254 Outsourcing combined with indemnification:

A. reduces legal responsibility but leaves financial risk relatively unchanged.
B. is more cost-effective as a means of risk transfer than purchasing insurance.
C. eliminates the reputational risk present when operations remain in-house.
D. reduces financial risk but leaves legal responsibility generally unchanged.

D is the correct answer.

Justification:

A. Although indemnification clauses are intended to deflect liability, the legal consequences associated with compromises in information security cannot be fully transferred.
B. The cost-effectiveness of various forms of risk transfer depends on many factors, such as the scope of operations, limits of liability, specialized knowledge that may be required for implementation and criteria for indemnification.
C. Clients deal directly with the organization, not its supply chain. Outsourcing generally has no effect on reputational risk, which remains associated with the organization's own brand regardless of outsourcing arrangements or indemnification clauses.
D. Indemnification clauses can transfer operational risk and financial impacts associated with that risk; however, legal responsibility for the consequences of compromise generally remains with the original entity.

S2-255 Risk management needs to be approached as a regular, ongoing program or activity primarily because:

A. people make mistakes.
B. technology becomes obsolete.
C. the environment changes.
D. standards are updated or replaced.

C is the correct answer.

Justification:

A. People do make mistakes, but mistakes built into a risk management program might well be repeated any number of times in an ongoing program. Therefore, this is not a rationale for risk management as a regular program or activity.
B. Technology is subject to obsolescence over time, but periodic assessment would likely be adequate if this were the primary rationale for risk management.
C. Controls usually degrade over time and are subject to failure, and the threat landscape changes constantly. Therefore, it is important that risk management be performed as an ongoing program or activity in order to capture the implications of these changes and ensure that the organization continues to make risk-treatment decisions consistent with its objectives and risk appetite.
D. Standards do change, and an organization that has identified conforming to a particular standard may be obligated to adjust its technology and processes to remain compliant. However, risk management addresses broader concerns than adherence to standards.

S2-256 The **ULTIMATE** purpose of risk response is to:

A. reduce cost.
B. lower vulnerability.
C. minimize threat.
D. control impact.

D is the correct answer.

Justification:

A. Reducing cost in the short term is rarely the purpose of risk response. Reducing the overall impact of loss associated with risk is only one approach that an organization may take; a level of risk that is already acceptable should generally be accepted regardless of whether it might be further reduced.
B. Lowering vulnerability is only one approach that an organization may take to response to risk.
C. Risk response rarely seeks to reduce threat in the aggregate and is generally unable to minimize it.
D. Organizations respond to risk in ways that control impact by keeping it within acceptable (or tolerable) levels.

S2-257 Why might an organization rationally choose to mitigate a risk that is estimated to be at a level higher than its stated risk appetite but within its stated risk tolerance?

A. The board of directors may insist that all risk be mitigated if it exceeds the appetite.
B. Senior executives may prefer to transfer risk rather than formally accepting it.
C. There may be pressure from key stakeholders to avoid risk that exceeds the appetite.
D. Senior management may have concern that the stated impact is underestimated.

D is the correct answer.

Justification:

A. The board of directors determines the risk appetite and tolerance, so there would be no tolerance in excess of the appetite if the board took this position.
B. The purpose of determining levels of risk appetite and tolerance is to have clear thresholds of when risk can be accepted without mitigation or transfer.
C. Risk avoidance is the best choice for responding to a risk only when it exceeds both the appetite and the tolerance despite all efforts at mitigation or transfer.
D. Risk that exceeds organizational appetite but lies within tolerable levels is not risk that the organization wants to accept. When there is concern that the impact has been underestimated, senior management may prefer to mitigate the risk to acceptable levels rather than unintentionally accept risk whose impact ends up exceeding the tolerance.

S2-258 Controls are effective when:

A. residual risk is at a level acceptable to the organization.
B. continuous monitoring programs are in place.
C. inherent risk is within the organizational risk tolerance.
D. key performance indicators have been identified.

A is the correct answer.

Justification:

A. The purpose of controls is to bring residual risk to acceptable levels. When controls have this result, they are effective by definition.
B. Continuous monitoring provides a means of monitoring the effectiveness of controls, but the existence of a monitoring program does not make controls effective.
C. Inherent risk does not take controls into account.
D. Identifying key performance indicators provides a means by which to gauge performance but does not make controls effective.

S2-259 An organization has implemented several risk mitigation strategies to reduce an identified risk. The risk control measures are sufficient when:

A. the risk acceptance level is less than or equal to the total risk level.
B. the residual risk is less than or equal to the risk acceptance level.
C. risk avoidance is justified by cost-benefit analysis.
D. risk mitigation is equal to annual loss expectancy.

B is the correct answer.

Justification:
A. The risk acceptance level is the level of risk the organization is willing to accept. This does not measure the effectiveness of the controls.
B. Risk controls are adequate once the residual risk is less than or equal to acceptable risk.
C. Risk avoidance is the ceasing the activity associated with the risk, not the implementation of controls.
D. Annual loss expectancy justifies the amount that can be spent on risk mitigation but does not indicate whether the controls are adequate.

S2-260 High risk volatility would be a basis for the information security manager to:

A. base mitigation measures solely on assessed impact.
B. raise the assessed risk level and increase remediation priority.
C. disregard volatility as irrelevant to assessed risk level.
D. perform another risk assessment to validate results.

B is the correct answer.

Justification:
A. Mitigation should be based on likelihood, potential impact and cost benefit.
B. High risk volatility means that the risk is higher during one period and lower in another. The appropriate response is to assess risk at its highest level and due to unpredictability, raise the priority of treatment.
C. Volatility must be considered in terms of maximum risk potential.
D. A second risk assessment will not be as useful as a volatility assessment and unnecessary.

S2-261 Treatment of risk should be prioritized **PRIMARILY** based on:

A. vulnerability and impact.
B. exposure and frequency.
C. frequency and impact.
D. mitigation cost and effectiveness.

C is the correct answer.

Justification:
A. Vulnerability without threat is not risk (no probability of compromise) and, therefore, not a good basis for prioritization.
B. Exposure is not by itself a risk and so is not a basis for prioritization.
C. Frequency and magnitude are the primary basis for prioritizing treatment with high frequency (or likelihood) and magnitude (or impact) being at the top of the list.
D. Mitigation cost and effectiveness will be a consideration after frequency and magnitude have been determined.

S2-262 Which of the following is the **BEST** method to determine classification of data?

A. Assessment of impact associated with compromise of data by the data owner
B. Compliance requirements defined in the information security policy
C. Requirements based on the protection level implemented for different datasets
D. Assessment of risk of data loss by the information security manager

A is the correct answer.

Justification:

A. The classification of data is based upon the potential impact from loss or corruption.
B. Compliance requirements are used as an input to risk assessment by considering risk associated with noncompliance.
C. The protection level is determined based on the classification of data and not the other way around.
D. Classification is not based upon risk; it is based upon impact (criticality or sensitivity or business value). The data owner determines the classification level.

S2-263 An organization is considering the purchase of a new technology that will facilitate better customer interaction and would be integrated into the existing customer relationship management system. Which of the following is the **PRIMARY** risk the information security manager consider related to this purchase?

A. The potential that the new technology will not deliver the promised functionality to support the business
B. The availability of ongoing support for the technology and whether existing staff can provide the support
C. The possibility of the new technology affecting the security or operation of other systems
D. The downtime required to re-configure the existing system to implement and integrate the new technology

C is the correct answer.

Justification:

A. The risk that the new technology will not support business needs is primarily a business manager responsibility not the information security manager.
B. The availability of support is a concern, but this is primarily a responsibility of the IT operations manager.
C. The greatest security risk is that the new technology may bypass existing security or impair the operation of existing systems. The security manager should examine the new system for these issues.
D. The downtime required to implement the new technology is primarily a business and IT department factor.

S2-264 Periodically analyzing the gap between controls and the control objectives is necessary to:

A. prevent an increase in audit findings.
B. address changes in exposure.
C. avoid a substantial increase in cost.
D. maintain alignment with regulatory requirements.

B is the correct answer.

Justification:

A. Although gap analysis may identify shortcomings before they are noted by an audit, correcting deficiencies is distinct from gap analysis.
B. Changes in exposure, business objectives or regulations may occur at any time and have implications for what controls are needed as part of an overall risk management program.
C. Gap analysis does not necessarily avoid increases in cost.
D. Maintaining alignment with regulatory requirements is a component of the stated gap analysis. It is just one of the possible exposures to consider.

S2-265 Which of the following **BEST** describes the outcome of effective risk management?

A. Allows an organization to obtain a continuous overview of vulnerabilities
B. Measures the feasibility of systems compromise and evaluates any related consequences
C. Determines the gap between controls and controls objectives
D. Reduces the incidence of significant adverse impact on an organization

D is the correct answer.

Justification:

A. Vulnerability management is a component of risk management. However, a risk management program that does not reduce significant adverse impact is not effective.
B. Penetration testing, which is a technique for vulnerability assessment, measures the feasibility of systems compromise and evaluates any related consequences. However, unless significant adverse impact to the organization is reduced, a risk management program is ineffective.
C. Gap analysis determines the gap between controls and controls objectives. However, unless identified gaps are addressed in ways that result in reduced impact to the organization, the risk management program is ineffective.
D. Effective risk management serves to reduce the incidence of significant adverse impacts on an organization either by addressing threats, mitigating exposure and/or by reducing vulnerability or impact.

S2-266 Which of the following would be the **FIRST** step in effectively integrating risk management into business processes?

A. Workflow analysis
B. Business impact analysis
C. Threat and vulnerability assessment
D. Analysis of the governance structure

A is the correct answer.

Justification:

A. Analyzing the workflow will be essential to understanding process vulnerabilities and where risk may exist in integrating risk management into business processes.
B. A business impact analysis will be important once the workflow and processes are understood in order to understand unit inputs, outputs and dependencies and the potential consequences of compromise.
C. Threat and vulnerability assessments are properly conducted after the relationship between risk management and business processes has been determined through workflow analysis.
D. The governance structure may be one of the vulnerabilities that pose a potential risk but should be analyzed after the workflow analysis. Ideally, the governance structure should reflect the workflow.

S2-267 Which of the following poses the **GREATEST** challenge to an organization seeking to prioritize risk management activities?

A. An incomplete catalog of information assets
B. A threat assessment that is not comprehensive
C. A vulnerability assessment that is outdated
D. An inaccurate valuation of information assets

D is the correct answer.

Justification:

A. Organizations are only able to prioritize items they know to exist. An incomplete catalog of information assets introduces the possibility that prioritization is overlooking assets that may have substantial value unintentionally resulting in the implicit acceptance of risk that may exceed the risk appetite and tolerance. However, inaccurate valuation of known assets has a greater negative impact on prioritization than the possibility of certain high-value assets not being properly taken into account.
B. Evaluating the threat environment is the most challenging aspect of risk assessment, and it is nearly always the case that a threat assessment excludes one or more threats. As a result, any prioritization effort must assume that the threat assessment is not comprehensive.
C. It is common for a vulnerability assessment to be outdated at the start of each cycle of a risk management program prior to the start of risk management activities, but the influence of outdated vulnerability information is less of a concern than inaccurate valuation of assets.
D. Although prioritization on the basis of risk requires knowledge of threat, vulnerability and potential consequence, it is this last factor expressed in terms of value that is most influential when prioritizing risk management activities. If assets are valued incorrectly, otherwise justifiable decisions of how to prioritize activities may be incorrect.

S2-268 Which of the following is the **MOST** important consideration when choosing between automated fire suppression systems?

A. Probability of fire
B. Cost of maintenance
C. Damage to resources
D. Ownership of the new system

C is the correct answer.

Justification:
A. Probability is part of the justification for adopting an automated fire suppression system, but which system is most appropriate depends on other factors.
B. The cost of maintenance is an important consideration, but because damage is likely to be much more costly than maintenance, it is a later consideration.
C. Fire suppression systems may be harmful to resources; therefore, automated systems that release gas or water automatically have their own pros and cons. Gas-based systems are harmful to human life, whereas water-based systems may damage IT resources. Hence, the selection and implementation must consider these aspects.
D. Ownership of assets, including the new system to be acquired, is required to determine the protection levels of resources. However, it will be based on the organization's roles and responsibility definitions. In any case, resource protection will take priority in considering the choice of solutions.

S2-269 Policies regarding the use of bring your own device (BYOD) should include:

A. the need to return the device when leaving the organization.
B. the requirement to protect sensitive data on the device.
C. limitations on which applications can be installed on the device.
D. the ability for security to seize the device as part of an investigation.

B is the correct answer.

Justification:
A. Because it is a personal device, it is unlikely that the organization can require it to be returned.
B. The organization must proactively ensure that data on personal devices are protected.
C. The organization may require the use of a virtual environment on the personal device to provide isolation, but the organization cannot control the personal applications loaded onto the device.
D. In the event of an investigation, the device may be seized by law enforcement, but it is not expected that security will have the authority to seize the device. Varying standards of privacy and other forms of legal protection around the world make it difficult to apply common standards with regards to private seizure of personal devices even in cases where an internal investigation may be warranted.

S2-270 Which of the following is the **PRIMARY** driver for initial implementation of a risk-based information security program?

A. Prioritization
B. Motivation
C. Optimization
D. Standardization

A is the correct answer.

Justification:

A. Because organizations rarely have adequate resources to address all concerns, a risk-based information security program is typically implemented to provide a basis for efficient allocation of limited resources.

B. Motivation is useful in getting the job done, but is not necessarily a result of implementing a risk-based information security program.

C. Optimization is a long-term benefit associated with a mature risk-based program. It does not present itself during initial implementation.

D. Standardization is a technique that offers numerous benefits and may support risk management activities. It is not the result of a focus on risk.

S2-271 For which of the following types of controls is notification of a verified network intrusion an indication that the control is working properly?

A. Preventative
B. Corrective
C. Detective
D. Deterrent

C is the correct answer.

Justification:

A. Preventative controls, such as authentication mechanisms and encryption, are intended to stop intrusions, so a verified intrusion indicates that preventative controls were ineffective.

B. Corrective controls, such as backups and failover capabilities, are intended to offset the impact caused by successful attacks directed against information systems. Intrusions may not have impact at the time of their detection, so an intrusion does not unto itself offer any indications regarding the workings of corrective controls.

C. Detective controls, such as intrusion detection systems, are designed to alert staff to intrusions when they occur. Notification of a verified network intrusion is an indication that the control is working properly.

D. Deterrent controls, such as warning banners, are intended to reduce the threat level by creating disincentives for threat events. A verified network intrusion indicates that the deterrent was inadequate for the responsible threat actor.

DOMAIN 3—INFORMATION SECURITY PROGRAM DEVELOPMENT AND MANAGEMENT (27%)

S3-1 Who can **BEST** advocate the development of and ensure the success of an information security program?

X

A. Internal auditor
B. Chief operating officer
C. Steering committee
D. IT management

C is the correct answer.

Justification:
A. An internal auditor is a good advocate but is secondary to the influence of senior management.
B. The chief operating officer will be a member of the steering committee.
C. Senior management represented in the security steering committee is in the best position to advocate the establishment of, and continued support for, an information security program.
D. IT management has a lesser degree of influence and would also be part of the steering committee.

S3-2 Which of the following **BEST** ensures that information transmitted over the Internet will remain confidential?

A. A virtual private network
B. Firewalls and routers
C. Biometric authentication
D. Two-factor authentication

A is the correct answer.

Justification:
A. Encryption of data in a virtual private network ensures that transmitted information is not readable, even if intercepted.
B. Firewalls and routers protect access to data resources inside the network and do not protect traffic in the public network.
C. Biometric authentication alone would not prevent a message from being intercepted and read.
D. Two-factor authentication alone would not prevent a message from being intercepted and read.

S3-3 What does the effectiveness of virus detection software **MOST** depend on?

A. Packet filtering
B. Intrusion detection
C. Software upgrades
D. Definition files

D is the correct answer.

Justification:
A. Packet filtering does not focus on virus detection.
B. Intrusion detection does not address virus detection.
C. Software upgrades are related to the periodic updating of the program code, which would not be as critical.
D. The effectiveness of virus detection software depends on virus signatures, which are stored in virus definition files.

S3-4 Which of the following is the **MOST** cost-effective type of access control?

A. Centralized
B. Role-based
C. Decentralized
D. Discretionary

B is the correct answer.

Justification:
A. Centralized access control is not a type of access control but a form of administration.
B. Role-based access control allows users to be grouped into job-related categories, which significantly eases the required administrative overhead because in most organizations there are fewer roles than employees, and roles change far less frequently.
C. Decentralized access control is not a typed of access control but an administration approach.
D. Discretionary access control would require a greater degree of administrative overhead because it is based on each individual rather than groups of individuals.

X **S3-5** Who should be responsible for enforcing access rights to application data?

A. Data owners
B. Business process owners
C. The security steering committee
D. Security administrators

D is the correct answer.

Justification:
A. Data owners are responsible for approving access rights.
B. Business process owners are sometimes the data owners as well and would not be responsible for enforcement.
C. The security steering committee would not be responsible for enforcement.
D. As custodians, security administrators are responsible for enforcing access rights to data.

X **S3-6** When designing an intrusion detection system, the information security manager should recommend that it be placed:

A. outside the firewall.
B. on the firewall server.
C. on a screened subnet.
D. on the external router.

C is the correct answer.

Justification:
A. Placing the intrusion detection system (IDS) on the Internet side of the firewall is not advised because the system will generate alerts on all malicious traffic—even though 99 percent will be stopped by the firewall and never reach the internal network.
B. Because firewalls should be installed on hardened servers with minimal services enabled, it would be inappropriate to install the IDS on the same physical device.
C. An IDS should be placed on a screened subnet, which is a demilitarized zone.
D. Placing the IDS on the external server, if such a thing were feasible, is not advised because the system will generate alerts on all malicious traffic—even though 99 percent will be stopped by the firewall and never reach the internal network.

S3-7 The **BEST** reason for an organization to implement two discrete firewalls connected directly to the Internet and to the same demilitarized zone would be to:

A. provide in-depth defense.
B. separate test and production.
C. permit traffic load balancing.
D. prevent a denial-of-service attack.

C is the correct answer.

Justification:
A. Two firewalls in parallel provide two concurrent paths for compromise and, therefore, do not provide defense in depth. If they were connected in series one behind the other, they would provide defense in depth.
B. As both entry points connect to the Internet and to the same demilitarized zone, such an arrangement is not practical for separating test from production.
C. Having two entry points, each guarded by a separate firewall, is desirable to permit traffic load balancing.
D. Firewalls are not effective at preventing denial-of-service attacks.

S3-8 When designing information security standards for an enterprise, the information security manager should require that an extranet server be placed:

A. outside the firewall.
B. on the firewall server.
C. on a screened subnet.
D. on the external router.

C is the correct answer.

Justification:
A. Placing the extranet server on the Internet side of the firewall would leave it defenseless.
B. Because firewalls should be installed on hardened servers with minimal services enabled, it would be inappropriate to store the extranet on the same physical device.
C. An extranet server should be placed on a screened subnet, which is a demilitarized zone.
D. Placing the extranet server on the external router, although not be possible, would leave it defenseless.

S3-9 Which of the following is the **BEST** metric for evaluating the effectiveness of security awareness training?

A. The number of password resets
B. The number of reported incidents
C. The number of incidents resolved
D. The number of access rule violations

B is the correct answer.

Justification:
A. Password resets may or may not have anything to do with awareness levels.
B. Reported incidents will provide an indicator of the awareness level of staff. An increase in reported incidents could indicate that the staff is paying more attention to security.
C. The number of incidents resolved may not correlate to staff awareness.
D. Access rule violations may or may not have anything to do with awareness levels.

S3-10 What is the **MOST** important contractual element when contracting with an outsourcer to provide security administration?

A. The right-to-terminate clause
B. Limitations of liability
C. The service level agreement
D. The financial penalties clause

C is the correct answer.

Justification:

A. The service level agreement (SLA) provides metrics to which outsourcing firms can be held accountable and will always include the right-to-terminate clause.
B. Limitations of liability will also be included in the SLA.
C. The SLA includes the other options in addition to a number of other conditions, representations and warranties as well as right to inspect, provisions for audits, requirements on termination, etc.
D. Financial penalties clauses are a standard part of SLAs.

X **S3-11** Which of the following is the **BEST** metric for evaluating the effectiveness of an intrusion detection mechanism?

A. Number of attacks detected
B. Number of successful attacks
C. Ratio of false positives to false negatives
D. Ratio of successful to unsuccessful attacks

C is the correct answer.

Justification:

A. The number of attacks detected does not indicate how many attacks were not detected, and therefore, it is no indication of effectiveness.
B. The number of successful attacks does not indicate how many were detected.
C. The ratio of false positives to false negatives will indicate the effectiveness of the intrusion detection system.
D. Without knowing whether attacks were detected or not, the ratio of successful attacks to unsuccessful attacks indicates nothing about the effectiveness of the IDS.

S3-12 Which of the following is **MOST** effective in preventing weaknesses from being introduced into existing production systems?

A. Patch management
B. Change management
C. Security baselines
D. Virus detection

B is the correct answer.

Justification:

A. Patch management involves the correction of software weaknesses and would necessarily follow change management procedures.
B. Change management controls the process of introducing changes to systems. This is often the point at which a weakness will be introduced.
C. Security baselines provide minimum recommended settings and do not prevent introduction of control weaknesses.
D. Virus detection is an effective tool but primarily focuses on malicious code from external sources. It is unrelated to the introduction of vulnerabilities.

S3-13 Which of the following is **MOST** effective in preventing security weaknesses in operating systems?

A. Patch management
B. Change management
C. Security baselines
D. Configuration management

A is the correct answer.

Justification:
A. Patch management corrects discovered weaknesses by applying a correction (a patch) to the original program code.
B. Change management controls the process of introducing changes to systems.
C. Security baselines provide minimum recommended settings.
D. Configuration management controls the updates to the production environment.

S3-14 Which of the following is the **MOST** effective solution for preventing internal users from modifying sensitive and classified information?

A. Baseline security standards
B. System access violation logs
C. Role-based access controls
D. Exit routines

C is the correct answer.

Justification:
A. Baseline security standards will provide for general access controls but not for specific authorizations.
B. Violation logs are detective and do not prevent unauthorized access.
C. Role-based access controls help ensure that users only have access to files and systems appropriate for their job role.
D. Exit routines are dependent upon appropriate role-based access.

S3-15 Which of the following is generally used to ensure that information transmitted over the Internet is authentic and actually transmitted by the named sender?

A. Biometric authentication
B. Embedded steganographic
C. Two-factor authentication
D. Embedded digital signature

D is the correct answer.

Justification:
A. Authentication does not ensure the authenticity of the data, just the identity of the sender.
B. Steganography is a form of encryption that may ensure integrity but not identity.
C. Authentication does not ensure the authenticity of the data, just the identity of the sender.
D. Digital signature ensures both the identity and the integrity of the data.

S3-16 Which of the following is the **MOST** appropriate frequency for updating antivirus signature files for antivirus software on production servers?

A. Daily
B. Weekly
C. Concurrently with operating system patch updates
D. During scheduled change control updates

A is the correct answer.

Justification:

A. **New viruses are being introduced almost daily. The effectiveness of virus detection software depends on frequent updates to its virus signatures, which are stored on antivirus signature files so updates may be carried out several times during the day. At a minimum, daily updating should occur.**
B. Weekly updates may potentially allow new viruses to infect the system.
C. Operating system updates are too infrequent for virus updates.
D. Change control updates are sporadic and not the basis for virus updates.

S3-17 Which of the following devices should be placed within a demilitarized zone?

A. Network switch
B. Web server
C. Database server
D. File/print server

B is the correct answer.

Justification:

A. Switches may bridge a demilitarized zone (DMZ) to another network but do not technically reside within the DMZ network segment.
B. **A web server should normally be placed within a DMZ to shield the internal network.**
C. Database servers may contain confidential or valuable data and should always be placed on the internal network, never on a DMZ that is subject to compromise.
D. File/print servers may contain confidential or valuable data and should always be placed on the internal network, never on a DMZ that is subject to compromise.

S3-18 Where should a firewall be placed?

A. On the web server
B. On the intrusion detection system server
C. On the screened subnet
D. On the domain boundary

D is the correct answer.

Justification:

A. Placing it on a web server, which is a demilitarized zone (DMZ), does not provide any protection.
B. Because firewalls should be installed on hardened servers with minimal services enabled, it is inappropriate to have the firewall and the intrusion detection system on the same physical device.
C. Placing it on a screened subnet, which is a DMZ, does not provide any protection.
D. **A firewall should be placed on a (security) domain boundary.**

S3-19 Where should an intranet server generally be placed?

A. On the internal network
B. On the firewall server
C. On the external router
D. On the primary domain controller

A is the correct answer.

Justification:

A. **An intranet server should be placed on the internal network. An intranet server should stay in the internal network because external people do not need to access it. This reduces the risk of unauthorized access.**
B. Because firewalls should be installed on hardened servers with minimal services enabled, it is inappropriate to store the intranet server on the same physical device as the firewall.
C. Placing the intranet server on an external router leaves it exposed.
D. Primary domain controllers should not share the same physical device as the intranet server.

S3-20 How can access control to a sensitive intranet application by mobile users **BEST** be implemented?

A. Through data encryption
B. Through digital signatures
C. Through strong passwords
D. Through two-factor authentication

D is the correct answer.

Justification:

A. Data encryption does not provide access control.
B. Digital signatures provide assurance of the identity of the sender, not access control.
C. Strong passwords provide an intermediate strength of access controls but not as strong as two-factor authentication.
D. **Two-factor authentication, through the use of strong passwords combined with security tokens, provides the highest level of security.**

S3-21 A control policy is **MOST** likely to address which of the following implementation requirements?

A. Specific metrics
B. Operational capabilities
C. Training requirements
D. Failure modes

D is the correct answer.

Justification:

A. A control policy may specify a requirement for monitoring or metrics but will not define specific metrics.
B. Operational capabilities will likely be defined in specific requirements or a design document rather than in the control policy.
C. There may be a general requirement for training but not control-specific training, which will be dependent on the particular control.
D. **A control policy will state the required failure modes in terms of whether a control fails open or fails closed, which has implications for safety, confidentiality and availability.**

S3-22 Which of the following ensures that newly identified security weaknesses in an operating system are mitigated in a timely fashion?

A. Patch management
B. Change management
C. Security baselines
D. Acquisition management

A is the correct answer.

Justification:

A. Patch management involves the correction of software weaknesses and helps ensure that newly identified exploits are mitigated in a timely fashion.
B. Change management controls the process of introducing changes to systems.
C. Security baselines provide minimum required settings.
D. Acquisition management controls the purchasing process.

S3-23 What is the **MAIN** advantage of implementing automated password synchronization?

A. It reduces overall administrative workload.
B. It increases security between multi-tier systems.
C. It allows passwords to be changed less frequently.
D. It reduces the need for two-factor authentication.

A is the correct answer.

Justification:

A. Automated password synchronization reduces the overall administrative workload of resetting passwords.
B. Automated password synchronization does not increase security between multi-tier systems.
C. Automated password synchronization does not allow passwords to be changed less frequently.
D. Automated password synchronization does not reduce the need for two-factor authentication.

X **S3-24** Which of the following tools is **MOST** appropriate to assess whether information security governance objectives are being met?

A. SWOT (strengths, weaknesses, opportunities, threats) analysis
B. Waterfall chart
C. Gap analysis
D. Balanced scorecard

D is the correct answer.

Justification:

A. A SWOT (strengths, weaknesses, opportunities, threats) analysis addresses strengths, weaknesses, opportunities and threats. Although useful, a SWOT analysis is not as effective a tool as a balanced scorecard (BSC).
B. A waterfall chart is used to understand the flow of one process into another.
C. A gap analysis, while useful for identifying the difference between the current state and the desired future state, is not the most appropriate tool.
D. A BSC is most effective for evaluating the degree to which information security objectives are being met.

S3-25 Which of the following is **MOST** effective in preventing the introduction of a code modification that may reduce the security of a critical business application?

A. Patch management
B. Change management
C. Security metrics
D. Version control

B is the correct answer.

Justification:

A. Patch management corrects discovered weaknesses by applying a correction to the original program code.
B. Change management controls the process of introducing changes to systems. Failure to have good change management may introduce new weaknesses into otherwise secure systems.
C. Security metrics provide a means for measuring effectiveness.
D. Version control is a subset of change management.

S3-26 Which of the following is **MOST** important to the success of an information security program?

A. Security awareness training
B. Achievable goals and objectives
C. Senior management sponsorship
D. Adequate start-up budget and staffing

C is the correct answer.

Justification:

A. Security awareness training, although important, is secondary.
B. Achievable goals and objectives is an important factor but will not ensure success if senior management support is not present.
C. Sufficient senior management support is the most important factor for the success of an information security program.
D. Having adequate budget and staffing is an important factor and unlikely without senior management support and by themselves will not ensure success without senior management support.

S3-27 Which of the following is the **MOST** effective solution for preventing individuals external to the organization from modifying sensitive information on a corporate database?

A. Screened subnets
B. Information classification policies and procedures
C. Role-based access controls
D. Intrusion detection system

A is the correct answer.

Justification:

A. Screened subnets are demilitarized zones and are oriented toward preventing attacks on an internal network by external users.
B. The policies and procedures to classify information will ultimately result in better protection, but they will not prevent actual modification.
C. Role-based access controls would help ensure that users only had access to files and systems appropriate for their job role.
D. Intrusion detection systems are useful to detect invalid attempts, but they will not prevent attempts.

S3-28 Which of the following technologies is utilized to ensure that an individual connecting to a corporate internal network over the Internet is not an intruder masquerading as an authorized user?

A. Intrusion detection system
B. IP address packet filtering
C. Two-factor authentication
D. Embedded digital signature

C is the correct answer.

Justification:
A. An intrusion detection system can be used to detect an external attack but would not help in authenticating a user attempting to connect.
B. IP address packet filtering would protect against spoofing an internal address but would not provide strong authentication.
C. Two-factor authentication provides an additional security mechanism over and above that provided by passwords alone. This is frequently used by mobile users needing to establish connectivity to a corporate network.
D. Digital signatures ensure that transmitted information can be attributed to the named sender.

S3-29 What is an appropriate frequency for updating operating system patches on production servers?

A. During scheduled rollouts of new applications
B. According to a fixed security patch management schedule
C. Concurrently with quarterly hardware maintenance
D. Whenever important security patches are released

D is the correct answer.

Justification:
A. Patches should not be delayed to coincide with other scheduled rollouts.
B. Patches should not be delayed to coincide with other scheduled maintenance.
C. Due to the possibility of creating a system outage, patches should not be deployed during critical periods of application activity such as month-end or quarter-end closing.
D. Patches should be applied whenever important security updates are released after being tested to ensure compatibility.

S3-30 Which of the following **BEST** accomplishes secure customer use of an e-commerce application?

A. Data encryption
B. Digital signatures
C. Strong passwords
D. Two-factor authentication

A is the correct answer.

Justification:
A. Encryption would be the preferred method of ensuring confidentiality in customer communications with an e-commerce application.
B. A digital signature is not a practical solution because there is typically no client-side certificate and integrity of the communication cannot be ensured.
C. Strong passwords, by themselves, would not be sufficient because the data could still be intercepted.
D. Two-factor authentication would be impractical and provide no assurance that data have not been modified through a man-in-the-middle attack.

S3-31 What is the **BEST** defense against a structured query language injection attack?

A. Regularly updated signature files
B. A properly configured firewall
C. An intrusion detection system
D. Strict controls on input fields

D is the correct answer.

Justification:
A. Regularly updated signature files are unrelated to an structured query language (SQL) attack and would fail to prevent it.
B. A properly configured firewall would fail to prevent such an attack.
C. An intrusion detection system would fail to prevent such an attack.
D. SQL injection involves the typing of programming command statements within a data entry field on a web page, usually with the intent of fooling the application into thinking that a valid password has been entered in the password entry field. The best defense against such an attack is to have strict edits on what can be typed into a data input field so that programming commands will be rejected. Code reviews should also be conducted to ensure that such edits are in place and that there are no inherent weaknesses in the way the code is written; software is available to test for such weaknesses.

S3-32 Which of the following is the **MOST** important consideration when implementing an intrusion detection system?

A. Tuning
B. Patching
C. Encryption
D. Packet filtering

A is the correct answer.

Justification:
A. If an intrusion detection system is not properly tuned it will generate an unacceptable number of false positives and/or fail to sound an alarm when an actual attack is underway.
B. Patching is more related to operating system hardening.
C. Encryption would not be as relevant as tuning.
D. Packet filtering would not be as relevant as tuning.

S3-33 Which of the following is the **MOST** important consideration when securing customer credit card data acquired by a point-of-sale cash register?

A. Authentication
B. Hardening
C. Encryption
D. Nonrepudiation

C is the correct answer.

Justification:
A. Authentication of the point-of-sale terminal will not prevent unauthorized reading of the data.
B. Hardening will protect the point-of-sale but will not prevent unauthorized reading of the data.
C. Cardholder data should be encrypted using strong encryption techniques.
D. Nonrepudiation is not relevant to credit card data protection.

X

S3-34 Organizations implement ethics training **PRIMARILY** to provide guidance to individuals engaged in:

A. monitoring user activities.
B. implementing security controls.
C. managing risk tolerance.
D. assigning access.

A is the correct answer.

Justification:

A. Monitoring user activities may result in access to sensitive corporate and personal information. The organization should implement training that provides guidance on appropriate legal behavior to reduce corporate liability and increase user awareness and understanding of data privacy and ethical behavior.
B. While ethics training is good practice for all employees, those that implement security controls are not necessarily privy to sensitive data.
C. Employees who manage risk tolerance may have access to high-level corporate information but not necessarily sensitive or private information. While ethics training is good practice, it is not required to manage risk tolerance for an organization.
D. Employees who manage network access do not necessarily need ethics training.

S3-35 Which of the following is the **MOST** important item to consider when evaluating products to monitor security across the enterprise?

A. Ease of installation
B. Product documentation
C. Available support
D. System overhead

D is the correct answer.

Justification:

A. Ease of installation, while important, would be secondary.
B. Product documentation, while important, would be secondary.
C. Available support, while important, would be secondary.
D. Monitoring products can impose a significant impact on system overhead for servers and networks.

S3-36 Which of the following is the **MOST** important guideline when using software to scan for security exposures within a corporate network?

A. Never use open source tools.
B. Focus only on production servers.
C. Follow a linear process for attacks.
D. Do not interrupt production processes.

D is the correct answer.

Justification:

A. Open source tools are an excellent resource for performing scans.
B. Scans should focus on both the test and production environments because, if compromised, the test environment could be used as a platform from which to attack production servers.
C. The process of scanning for exposures is more of a spiral process than a linear process.
D. The first rule of scanning for security exposures is to not break anything. This includes the interruption of any running production processes.

S3-37 Which of the following **BEST** ensures that modifications made to in-house developed business applications do not introduce new security exposures?

A. Stress testing
B. Patch management
C. Change management
D. Security baselines

C is the correct answer.

Justification:
A. Stress testing ensures that there are no scalability problems.
B. Patch management involves the correction of software weaknesses and helps ensure that newly identified exploits are mitigated in a timely fashion.
C. Change management controls the process of introducing changes to systems to ensure that unintended changes are not introduced; within change management, regression testing is specifically designed to prevent the introduction of new security exposures when making modifications.
D. Security baselines provide minimum required security settings.

S3-38 What is the advantage of virtual private network tunneling for remote users?

A. It helps ensure that communications are secure.
B. It increases security between multi-tier systems.
C. It allows passwords to be changed less frequently.
D. It eliminates the need for secondary authentication.

A is the correct answer.

Justification:
A. Virtual private network (VPN) tunneling for remote users provides an encrypted link that helps ensure secure communications.
B. VPN tunneling does not affect security within the internal network.
C. VPN tunneling does not affect password change frequency.
D. VPN tunneling does not eliminate the need for secondary authentication.

S3-39 Which of the following is **MOST** effective for securing wireless networks as a point of entry into a corporate network?

A. Boundary router
B. Strong encryption
C. Internet-facing firewall
D. Intrusion detection system

B is the correct answer.

Justification:
A. Boundary routers would do little to secure wireless networks.
B. Strong encryption is the most effective means of protecting wireless networks.
C. Internet facing firewall would offer no protection from a local attack on a wireless network.
D. Compromise of weak encryption would not be detected by an intrusion detection system.

S3-40 Which of the following is **MOST** effective in protecting against the attack technique known as phishing?

A. Firewall blocking rules
B. Up-to-date signature files
C. Security awareness training
D. Intrusion detection system monitoring

C is the correct answer.

Justification:
A. Firewall rules are unsuccessful at blocking this kind of attack.
B. Signature files are unrelated to this kind of attack.
C. Phishing relies on social engineering techniques. Providing good security awareness training will best reduce the likelihood of such an attack being successful.
D. Intrusion detection system monitoring is unsuccessful at blocking this kind of attack.

S3-41 Which of the following should automatically occur **FIRST** when a newly installed system for synchronizing passwords across multiple systems and platforms abnormally terminates without warning?

A. The firewall should block all inbound traffic during the outage.
B. All systems should block new logins until the problem is corrected.
C. Access control should fall back to nonsynchronized mode.
D. System logs should record all user activity for later analysis.

C is the correct answer.

Justification:
A. Blocking traffic would be overly restrictive to the conduct of business.
B. Blocking new logins would be overly restrictive to the conduct of business.
C. The best mechanism is for the system to fall back to the original process of logging on individually to each system.
D. Recording all user activity would add little value.

X **S3-42** Which of the following is the **MOST** important risk associated with middleware in a client-server environment?

A. Server patching may be prevented.
B. System backups may be incomplete.
C. Data integrity may be affected.
D. End-user sessions may be hijacked.

C is the correct answer.

Justification:
A. Sever patching is not affected by the presence of middleware.
B. System backups are not affected.
C. The major risk associated with middleware in a client-server environment is that data integrity may be adversely affected if middleware were to fail or become corrupted.
D. Hijacked end-user sessions can occur but can be detected by implementing security checks in the middleware.

S3-43 An outsourced service provider must handle sensitive customer information. Which of the following is **MOST** important for an information security manager to know?

A. Security in storage and transmission of sensitive data
B. Provider's level of compliance with industry standards
C. Security technologies in place at the facility
D. Results of the latest independent security review

A is the correct answer.

Justification:

A. Knowledge of how the outsourcer protects the storage and transmission of sensitive information will allow an information security manager to understand how sensitive data will be protected.
B. The provider's level of compliance with industry standards may or may not be important.
C. Security technologies are not the only components to protect the sensitive customer information.
D. An independent security review may not include analysis on how sensitive customer information would be protected.

S3-44 Which of the following security mechanisms is **MOST** effective in protecting classified data that have been encrypted to prevent disclosure and transmission outside the organization's network?

A. Configuration of firewalls
B. Strength of encryption algorithms
C. Authentication within application
D. Safeguards over keys

D is the correct answer.

Justification:

A. Firewalls can be perfectly configured, but if the keys make it to the other side, they will not prevent the document from being decrypted.
B. Even easy encryption algorithms require adequate resources to break, whereas encryption keys can be easily used.
C. The application "front door" controls may be bypassed by accessing data directly.
D. Key management is the weakest link in encryption. If keys are in the wrong hands, documents will be able to be read regardless of where they are on the network.

S3-45 In the process of deploying a new email system, an information security manager would like to ensure the confidentiality of messages while in transit. Which of the following is the **MOST** appropriate method to ensure data confidentiality in a new email system implementation?

A. Encryption
B. Strong authentication
C. Digital signature
D. Hashing algorithm

A is the correct answer.

Justification:

A. **To preserve confidentiality of a message while in transit, encryption should be implemented.**
B. Strong authentication ensures the identity of the participants but does not secure the message in transit.
C. Digital signatures only authenticate the sender, the receiver and the integrity of the message but do not prevent interception.
D. A hashing algorithm ensures integrity.

S3-46 The **MOST** important reason that statistical anomaly-based intrusion detection systems (stat IDSs) are less commonly used than signature-based IDSs, is that stat IDSs:

A. create more overhead than signature-based IDSs.
B. cause false positives from minor changes to system variables.
C. generate false alarms from varying user or system actions.
D. cannot detect new types of attacks.

C is the correct answer.

Justification:

A. Due to the nature of statistical anomaly-based intrusion detection system (stat IDS) operations (i.e., they must constantly attempt to match patterns of activity to the baseline parameters), a stat IDS requires much more overhead and processing than signature-based versions. However, this is not the most important reason.
B. Due to the nature of a stat IDS—based on statistics and comparing data with baseline parameters—this type of IDS may not detect minor changes to system variables and may generate many false positives. However, this is not the most important reason.
C. **A stat IDS collects data from normal traffic and establishes a baseline. It then periodically samples the network activity based on statistical methods and compares samples to the baseline. When the activity is outside the baseline parameter (clipping level), the IDS notifies the administrator. The baseline variables can include a host's memory or central processing unit usage, network packet types and packet quantities. If actions of the users or the systems on the network vary widely with periods of low activity and periods of frantic packet exchange, a stat IDS may not be suitable, as the dramatic swing from one level to another almost certainly will generate false alarms. This weakness will have the largest impact on the operation of the IT systems.**
D. Because the stat IDS can monitor multiple system variables, it can detect new types of variables by tracing for abnormal activity of any kind.

S3-47 What is the **MOST** important success factor to design an effective IT security awareness program?

A. Customization of content to target audience
B. Representation of senior management
C. Training of staff across all hierarchical levels
D. Replacing technical jargon with concrete examples

A is the correct answer.

Justification:

A. Awareness training can only be effective if it is customized to the expectations and needs of attendees. Needs will be quite different depending on the target audience and will vary between business managers, end users and IT staff; program content and the level of detail communicated will, therefore, be different.

B. Representation of senior management is important; however, the customization of content is the most important factor.

C. Training of staff across all hierarchical levels is important; however, the customization of content is the most important factor.

D. Replacing technical jargon with concrete examples is a good practice; however, the customization of content is the most important factor.

S3-48 Which of the following practices completely prevents a man-in-the-middle attack between two hosts?

A. Use security tokens for authentication
B. Connect through an IP Security v6 virtual private network
C. Use Hypertext Transfer Protocol Secure with a server-side certificate
D. Enforce static media access control addresses

B is the correct answer.

Justification:

A. Using token-based authentication does not prevent a man-in-the-middle attack; however, it may help eliminate reusability of stolen cleartext credentials.

B. IP Security v6 effectively prevents man-in-the-middle attacks by including source and destination Internet Protocols within the encrypted portion of the packet. The protocol is resilient to man-in-the-middle attacks.

C. A Hypertext Transfer Protocol Secure session can be intercepted through domain name system (DNS) or Address Resolution Protocol (ARP) poisoning.

D. ARP poisoning—a specific kind of man-in-the-middle attack—may be prevented by setting static media access control addresses. Nevertheless, DNS and NetBIOS resolution can still be attacked to deviate traffic.

X **S3-49** What is the **MOST** common protocol to ensure confidentiality of transmissions in a business-to-customer financial web application?

A. Secure Sockets Layer
B. Secure Shell
C. IP Security
D. Secure/Multipurpose Internet Mail Extensions

A is the correct answer.

Justification:

A. Secure Sockets Layer is a cryptographic protocol that provides secure communications, providing end point authentication and communications privacy over the Internet. In typical use, all data transmitted between the customer and the business are, therefore, encrypted by the business's web server and remain confidential.

B. Secure Shell (SSH) File Transfer Protocol is a network protocol that provides file transfer and manipulation functionality over any reliable data stream. It is typically used with the SSH-2 protocol to provide secure file transfer.

C. IP Security (IPSec) is a standardized framework for securing Internet Protocol (IP) communications by encrypting and/or authenticating each IP packet in a data stream. There are two modes of IPSec operation: transport mode and tunnel mode.

D. Secure/Multipurpose Internet Mail Extensions (S/MIME) is a standard for public key encryption and signing of email encapsulated in MIME; it is not a web transaction protocol.

X **S3-50** A certificate authority is required for a public key infrastructure:

A. in cases where confidentiality is an issue.
B. when challenge/response authentication is used.
C. except where users attest to each other's identity.
D. in role-based access control deployments.

C is the correct answer.

Justification:

A. The requirement of confidentiality is not relevant to the certificate authority (CA) other than to provide an authenticated user's public key.

B. Challenge/response authentication is not a process used in a public key infrastructure (PKI).

C. The role of the CA is not needed in implementations such as Pretty Good Privacy, where the authenticity of the users' public keys are attested to by others in a "circle of trust."

D. If the role-based access control is PKI-based, either a CA is required or other trusted parties will have to attest to the validity of users.

S3-51 When a user employs a client-side digital certificate to authenticate to a web server through Secure Sockets Layer, confidentiality is **MOST** vulnerable to which of the following?

A. Internet Protocol spoofing
B. Man-in-the-middle attack
C. Repudiation
D. Trojan

D is the correct answer.

Justification:

A. Internet Protocol spoofing will not work because the IP is not used as an authentication mechanism.
B. Man-in-the-middle attacks are not possible if using Secure Sockets Layer with client-side certificates.
C. Repudiation is unlikely because client-side certificates authenticate the user.
D. A Trojan is a program that can give the attacker full control over the infected computer, thus allowing the attacker to hijack, copy or alter information after authentication by the user.

S3-52 Which of the following is the **MOST** relevant metric to include in an information security quarterly report to the executive committee?

A. Security compliant servers trend report
B. Percentage of security compliant servers
C. Number of security patches applied
D. Security patches applied trend report

A is the correct answer.

Justification:

A. The overall trend of security compliant servers provides a metric of the effectiveness of the IT security program.
B. The percentage of compliant servers will be a relevant indicator of the risk exposure of the infrastructure. However, the percentage is less relevant than the overall trend.
C. The number of patches applied would be less relevant, as this would depend on the number of vulnerabilities identified and patches provided by vendors.
D. The security patches applied trend report is a metric indicating the degree of improvement in patching but provides a less complete picture of the effectiveness of the security program.

S3-53 Why is it important to develop an information security baseline? The security baseline helps define:

A. critical information resources needing protection.
B. a security policy for the entire organization.
C. the minimum acceptable security to be implemented.
D. required physical and logical access controls.

C is the correct answer.

Justification:

A. Before determining the security baseline, an information security manager must identify criticality levels of the organization's information resources.
B. The security policy helps define the security baseline.
C. Developing an information security baseline helps to define the minimum acceptable security that will be implemented to protect the information resources in accordance with the respective criticality/classification levels.
D. The security baseline defines the control objectives but not the specific controls required.

X **S3-54** Which of the following choices is the **WEAKEST** link in the authorized user registration process?

A. The certificate authority's private key
B. The registration authority's private key
C. The relying party's private key
D. A secured communication private key

B is the correct answer.

Justification:

A. The certificate authority's (CA's) private key is heavily secured both electronically and physically and is extremely difficult to access by anyone.
B. The registration authority's (RA's) private key is in the possession of the RA, often stored on a smart card or laptop, and is typically protected by a password and, therefore, is potentially accessible. If the RA's private key is compromised, it can be used to register anyone for a certificate using any identity, compromising the entire public key infrastructure for that CA.
C. The relying party's private key, if compromised, only puts that party at risk.
D. The private key used for secure communication will only pose a risk to the parties communicating.

X **S3-55** Which of the following controls is **MOST** effective in providing reasonable assurance of physical access compliance to an unmanned server room controlled with biometric devices?

A. Regular review of access control lists
B. Security guard escort of visitors
C. Visitor registry log at the door
D. A biometric coupled with a personal identification number

A is the correct answer.

Justification:

A. A review of access control lists is a detective control that will enable an information security manager to ensure that authorized persons are entering in compliance with corporate policy.
B. Visitors accompanied by a guard will also provide assurance but may not be cost-effective.
C. A visitor registry is the next cost-effective control but not as secure.
D. A biometric coupled with a personal identification number will strengthen the access control; however, compliance assurance logs will still have to be reviewed to ensure only authorized access.

S3-56 To **BEST** improve the alignment of the information security objectives in an organization, the chief information security officer should:

A. revise the information security program.
B. evaluate a business balanced scorecard.
C. conduct regular user awareness sessions.
D. perform penetration tests.

B is the correct answer.

Justification:

A. Revising the information security program may be a solution, but it is not the best solution to improve alignment of the information security objectives.
B. The business balanced scorecard (BSC) can track the effectiveness of how an organization executes it information security strategy and determine areas of improvement.
C. User awareness is just one of the areas the organization must track through the business BSC.
D. Performing penetration tests does not affect alignment with information security objectives.

S3-57 Assuming that the value of information assets is known, which of the following gives the information security manager the **MOST** objective basis for determining that the information security program is delivering value?

A. Number of controls
B. Cost of achieving control objectives
C. Effectiveness of controls
D. Test results of controls

B is the correct answer.

Justification:

A. Number of controls has no correlation with the value of assets unless the effectiveness of the controls and their cost are also evaluated.
B. Comparison of cost of achievement of control objectives and corresponding value of assets sought to be protected would provide a sound basis for the information security manager to measure value delivery.
C. Effectiveness of controls has no correlation with the value of assets unless their costs are also evaluated.
D. Test results of controls may determine their effectiveness but has no correlation with the value of assets.

S3-58 Question removed.

S3-59 An enterprise is implementing an information security program. During which phase of the implementation should metrics be established to assess the effectiveness of the program over time?

A. Testing
B. Initiation
C. Design
D. Development

C is the correct answer.

Justification:
A. The testing phase is too late because the system has already been developed and is in production testing.
B. In the initiation phase, the basic security objective of the project is acknowledged.
C. In the design phase, security checkpoints are defined and a test plan is developed.
D. Development is the coding phase and is too late to consider test plans.

S3-60 The **MOST** effective way to ensure that outsourced service providers comply with the organization's information security policy would be:

A. service level monitoring.
B. penetration testing.
C. periodically auditing.
D. security awareness training.

C is the correct answer.

Justification:
A. Service level monitoring can only pinpoint operational issues in the organization's operational environment.
B. Penetration testing can identify security vulnerability but cannot ensure information policy compliance.
C. Regular audit exercise can spot any gap in the information security compliance.
D. Training can increase users' awareness on the information security policy but does not ensure compliance.

S3-61 Which of the following should the information security manager implement to protect a network against unauthorized external connections to corporate systems?

A. Strong authentication
B. Internet Protocol antispoofing filtering
C. Network encryption protocol
D. Access lists of trusted devices

A is the correct answer.

Justification:
A. Strong authentication will provide adequate assurance on the identity of the users.
B. Internet Protocol antispoofing is aimed at the device rather than the user.
C. Encryption protocol ensures data confidentiality and authenticity.
D. Access lists of trusted devices are easily exploited by spoofed identity of the clients.

S3-62 What is the **PRIMARY** driver for obtaining external resources to execute the information security program?

A. External resources can contribute cost-effective expertise not available internally.
B. External resources can be made responsible for meeting the security program requirements.
C. External resources can replace the dependence on internal resources.
D. External resources can deliver more effectively on account of their knowledge.

A is the correct answer.

Justification:
A. External resources that can contribute cost-effective expertise that are not available internally represent the primary driver for the information security manager to make use of external resources.
B. The information security manager will continue to be responsible for meeting the security program requirements despite using the services of external resources.
C. The external resources should never completely replace the role of internal resources from a strategic perspective.
D. The external resources cannot have a better knowledge of the business of the information security manager's organization than do the internal resources.

S3-63 It is essential to determine the forces that drive the business need for the information security program. Determining drivers is critical to:

A. establish the basis for the development of metrics.
B. establish the basis for security controls.
C. report risk results to senior management.
D. develop security awareness training modules.

A is the correct answer.

Justification:
A. Determining the drivers of a program establishes objectives and is essential to developing relevant metrics for the organization.
B. Determining drivers may establish objectives of a program, but the controls are determined by risk and impact.
C. Risk reporting goes beyond the specific drivers and will encompass all organizational risk.
D. Drivers may indirectly provide subject matter for training, but security awareness goes beyond just the drivers.

S3-64 Which of the following controls would **BEST** prevent accidental system shutdown from the console or operations area?

A. Redundant power supplies
B. Protective switch covers
C. Shutdown alarms
D. Biometric readers

B is the correct answer.

Justification:

A. Redundant power supplies would not prevent an individual from powering down a device.
B. Protective switch covers would reduce the possibility of an individual accidentally pressing the power button on a device, thereby turning off the device.
C. Shutdown alarms would be after the fact.
D. Biometric readers would be used to control access to the systems.

S3-65 Which of the following is the **MOST** important reason that information security objectives should be defined?

A. Tool for measuring effectiveness
B. General understanding of goals
C. Consistency with applicable standards
D. Management sign-off and support initiatives

A is the correct answer.

Justification:

A. The creation of objectives can be used in part as a source of measurement of the effectiveness of information security management by the extent those objectives have been achieved, which feeds into the overall state of governance.
B. General understanding of goals is useful but is not the primary reasons for having clearly defined objectives.
C. The standards should be consistent with the objectives, not the other way around.
D. Gaining management sign-off and support is important but by itself will not provide the structure for security governance.

S3-66 What is the **BEST** policy for securing data on mobile universal serial bus (USB) drives?

A. Authentication
B. Encryption
C. Prohibit employees from copying data to USB devices
D. Limit the use of USB devices

B is the correct answer.

Justification:

A. Authentication protects access to the data but does not protect the data once the authentication is compromised.
B. Encryption provides the most effective protection of data on mobile devices.
C. Prohibiting employees from copying data to universal serial bus (USB) devices does not prevent copying data and offers minimal protection.
D. Limiting the use of USB devices does not secure the data on them.

S3-67 What should an information security manager focus on when speaking to an organization's human resources department about information security?

A. An adequate budget for the security program
B. Recruitment of technical IT employees
C. Periodic risk assessments
D. Security awareness training for employees

D is the correct answer.

Justification:

A. Budget considerations are more of an accounting function.
B. Recruiting IT-savvy staff may bring in new employees with better awareness of information security, but that is not a replacement for the training requirements of the other employees.
C. Periodic risk assessments may or may not involve the human resources department function.
D. An information security manager has to impress upon the human resources department the need for security awareness training for all employees. The human resources department would become involved once they are convinced of the need of security awareness training.

S3-68 What is the **MOST** important reason for conducting security awareness programs throughout an organization?

A. Reducing the human risk
B. Maintaining evidence of training records to ensure compliance
C. Informing business units about the security strategy
D. Training personnel in security incident response

A is the correct answer.

Justification:

A. People are the weakest link in security implementation, and awareness would reduce this risk.
B. Maintaining evidence of training is useful but far from the most important reason for conducting awareness training.
C. Informing business units about the security strategy is best done through steering committee meetings or other forums.
D. Security awareness training is not generally for security incident response.

S3-69 What is the **MOST** effective way to ensure network users are aware of their responsibilities to comply with an organization's security requirements?

A. Logon banners displayed at every logon
B. Periodic security-related email messages
C. An Intranet web site for information security
D. Circulating the information security policy

A is the correct answer.

Justification:

A. **Logon banners would appear every time the user logs on, and the user would be required to read and agree to the same before using the resources. Also, as the message is conveyed in writing and appears consistently, it can be easily enforceable in any organization.**
B. Security-related email messages are frequently considered as spam by network users and do not, by themselves, ensure that the user agrees to comply with security requirements.
C. The existence of an Intranet web site does not force users to access it and read the information.
D. Circulating the information security policy alone does not confirm that an individual user has read, understood and agreed to comply with its requirements unless it is associated with formal acknowledgment, such as a user's signature of acceptance.

X

S3-70 What is a desirable sensitivity setting for a biometric access control system that protects a high-security data center?

A. A high false reject rate
B. A high false acceptance rate
C. Lower than the crossover error rate
D. Exactly to the crossover error rate

A is the correct answer.

Justification:

A. **Biometric access control systems are not infallible. When tuning the solution, one has to adjust the sensitivity level to give preference either to false reject rate (FRR) (type I error rate) where the system will be more prone to err denying access to a valid user or erring and allowing access to an invalid user. The preferable setting will be in the FRR region of sensitivity.**
B. A high false acceptance rate (FAR) will marginalize security by allowing too much unauthorized access. In systems where the possibility of false rejects is a problem, it may be necessary to reduce sensitivity and thereby increase the number of false accepts.
C. As the sensitivity of the biometric system is adjusted, the FRR and FAR change inversely. At one point, the two values intersect and are equal. This condition creates the crossover error rate, which is a measure of the system accuracy. Lower than the crossover error rate will create too high a FAR for a high-security data center.
D. The crossover rate is sometimes referred to as equal error rate. In a very sensitive system, it may be desirable to minimize the number of false accepts—the number of unauthorized persons allowed access. To do this, the system is tuned to be more sensitive with a lower FAR, which causes the FRR—the number of authorized persons disallowed access—to increase.

S3-71 An organization has adopted a practice of regular staff rotation to minimize the risk of fraud and encourage cross-training. Which type of authorization policy would **BEST** address this practice?

A. Multilevel
B. Role-based
C. Discretionary
D. Mandatory

B is the correct answer.

Justification:
A. Multilevel policies are based on classifications and clearances.
B. A role-based policy will associate data access with the role performed by an individual, thus restricting access to data required to perform the individual's tasks.
C. Discretionary policies leave access decisions to be made by the information resource managers.
D. Mandatory access control requires a clearance equal to or greater than the classification level of the asset. It generally also includes the need to know.

S3-72 Which of the following is the **MOST** important reason for an information security review of contracts?

A. To help ensure the parties to the agreement can perform
B. To help ensure confidential data are not included in the agreement
C. To help ensure appropriate controls are included
D. To help ensure the right to audit is a requirement

C is the correct answer.

Justification:
A. The ability of the parties to perform is normally the responsibility of legal and the business operation involved.
B. Confidential information may be in the agreement by necessity, and while the information security manager can advise and provide approaches to protect the information, the responsibility rests with the business and legal department.
C. Agreements with external parties can expose an organization to information security risk that must be assessed and appropriately mitigated with appropriate controls.
D. Audit rights may be one of many possible controls to include in a third-party agreement but is not necessarily a contract requirement, depending on the nature of the agreement.

S3-73 For virtual private network access to the corporate network, the information security manager is requiring strong authentication. Which of the following is the strongest method to ensure that logging onto the network is secure?

A. Biometrics
B. Symmetric encryption keys
C. Secure Sockets Layer-based authentication
D. Two-factor authentication

D is the correct answer.

Justification:

A. While biometrics provides unique authentication, it is not strong by itself, unless a personal identification number (PIN) or some other authentication factor is used with it. Biometric authentication by itself is also subject to replay attacks.
B. A symmetric encryption method that uses the same secret key to encrypt and decrypt data is not a typical authentication mechanism for end users. This private key could still be compromised.
C. Secure Sockets Layer (SSL) is the standard security technology for establishing an encrypted link between a web server and a browser. If SSL is used with a client certificate and a password, it would be a two-factor authentication.
D. Two-factor authentication requires more than one type of user authentication, typically something you know and something you have, such as a PIN and smart card.

S3-74 An organization's information security manager is planning the structure of the information security steering committee. Which of the following groups should the manager invite?

A. External audit and network penetration testers
B. Board of directors and the organization's regulators
C. External trade union representatives and key security vendors
D. Leadership from IT, human resources and the sales department

D is the correct answer.

Justification:

A. External audit may assess and advise on the program, and testers may be used by the program; however, they are not appropriate steering committee members.
B. The steering committee needs to have practitioner-level executive representation. It may report to the board, but board members would not generally be part of the steering committee, except for its executive sponsor. Regulators would not participate in this committee.
C. External trade union representatives and key security vendors are entities that may need to be consulted as part of program activities, but they would not be members of the steering committee.
D. Leaders from IT, human resources and sales are some of the key individuals who must support an information security program.

S3-75 Which of the following is the **MOST** effective security measure to protect data held on mobile computing devices?

A. Biometric access control
B. Encryption of stored data
C. Power-on passwords
D. Protection of data being transmitted

B is the correct answer.

Justification:
A. Biometric access control limits access but does not protect stored data once access has been breached.
B. Encryption of stored data will help ensure that the actual data cannot be recovered without the encryption key.
C. Power-on passwords do not protect data effectively.
D. Protecting data stored on mobile computing devices does not relate to protecting data in transmission.

S3-76 Which of the following is **MOST** useful in managing increasingly complex security deployments?

X

A. A standards-based approach
B. A security architecture
C. Policy development
D. Senior management support

B is the correct answer.

Justification:
A. Standards may provide metrics for deployment but would not provide significant management tools.
B. Deploying complex security initiatives and integrating a range of diverse projects and activities would be more easily managed with the overview and relationships provided by a security architecture.
C. Policies would guide direction but would not provide significant management tools.
D. Management support is always helpful and may assist in providing resources, but it would be of little direct benefit in managing complex security deployments.

S3-77 An organization is implementing intrusion protection in their demilitarized zone (DMZ). Which of the following steps is necessary to make sure that the intrusion prevention system (IPS) can view all traffic in the DMZ?

A. Ensure that intrusion prevention is placed in front of the firewall.
B. Ensure that all devices that are connected can easily see the IPS in the network.
C. Ensure that all encrypted traffic is decrypted prior to being processed by the IPS.
D. Ensure that traffic to all devices is mirrored to the IPS.

C is the correct answer.

Justification:

A. An intrusion prevention system (IPS) placed in front of the firewall will almost certainly continuously detect potential attacks, creating endless false positives and directing the firewall to block many sites needlessly. Most of actual attacks would be intercepted by the firewall in any case.
B. All connected devices do not need to see the IPS.
C. For IPS to detect attacks, the data cannot be encrypted; therefore, all encryption should be terminated to allow all traffic to be viewed by the IPS. The encryption should be terminated at a hardware Secure Sockets Layer accelerator or virtual private network server to allow all traffic to be monitored.
D. Traffic to all devices is not mirrored to the IPS.

S3-78 Which of the following guarantees that data in a file have not changed?

A. Inspecting the modified date of the file
B. Encrypting the file with symmetric encryption
C. Using stringent access control to prevent unauthorized access
D. Creating a hash of the file, then comparing the file hashes

D is the correct answer.

Justification:

A. The modified date can be modified to reflect any date.
B. Encrypting the file will make it difficult to modify but does not ensure it has not been corrupted.
C. Access control cannot ensure that file data has not been changed.
D. A hashing algorithm can be used to mathematically ensure that data have not been changed by hashing a file and comparing the hashes after a suspected change.

S3-79 Which of the following mechanisms is the **MOST** secure way to implement a secure wireless network?

A. Filter media access control addresses.
B. Use a Wi-Fi Protected Access protocol.
C. Use a Wired Equivalent Privacy key.
D. Use web-based authentication.

B is the correct answer.

Justification:

A. Media access control (MAC) address filtering by itself is not a good security mechanism because allowed MAC addresses can be easily sniffed and then spoofed to get into the network.
B. Wi-Fi Protected Access (WPA2) protocol is currently one of the most secure authentication and encryption protocols for mainstream wireless products.
C. Wired Equivalent Privacy (WEP) is no longer a secure encryption mechanism for wireless communications. The WEP key can be easily broken within minutes using widely available software. Once the WEP key is obtained, all communications of every other wireless client are exposed.
D. A web-based authentication mechanism can be used to prevent unauthorized user access to a network, but it will not solve the wireless network's main security issues, such as preventing network sniffing.

S3-80 Which of the following devices could potentially stop a structured query language injection attack?

A. An intrusion prevention system
B. An intrusion detection system
C. A host-based intrusion detection system
D. A host-based firewall

A is the correct answer.

Justification:

A. Structured query language (SQL) injection attacks occur at the application layer. Most intrusion prevention systems will detect at least basic sets of SQL injection and will be able to stop them.
B. Intrusion detection systems will detect but not prevent.
C. Host-based intrusion detection systems will be unaware of SQL injection problems.
D. A host-based firewall, be it on the web server or the database server, will allow the connection because firewalls do not check packets at an application layer.

S3-81 Which of the following **BEST** ensures nonrepudiation?

A. Strong passwords
B. A digital hash
C. Symmetric encryption
D. Digital signatures

D is the correct answer.

Justification:

A. Strong passwords only ensure authentication to the system and cannot be used for nonrepudiation involving two or more parties.
B. A digital hash in itself helps in ensuring integrity of the contents, but not nonrepudiation.
C. Symmetric encryption would not help in nonrepudiation because the keys are always shared between parties.
D. Digital signatures use a private and public key pair, authenticating both parties. The integrity of the contents exchanged is controlled through the hashing mechanism that is signed by the private key of the exchanging party.

S3-82 Which of the following is the **MOST** important step before implementing a security policy?

A. Communicating to employees
B. Training IT staff
C. Identifying relevant technologies for automation
D. Obtaining sign-off from stakeholders

D is the correct answer.

Justification:

A. Only after sign-off is obtained can communicating to employees begin.
B. Only after sign-off is obtained can training IT staff begin.
C. Only after sign-off is obtained can identifying relevant technologies for automation begin.
D. Sign-off must be obtained from all stakeholders because that would signify formal acceptance of all the policy objectives and expectations of the business along with all residual risk.

S3-83 An organization's security awareness program should focus on which of the following?

A. Establishing metrics for network backups
B. Installing training software which simulates security incidents
C. Communicating what employees should or should not do in the context of their job responsibilities
D. Access levels within the organization for applications and the Internet

C is the correct answer.

Justification:

A. Metrics for network backups is not an awareness issue.
B. Training software simulating security incidents is suitable for incident response teams but not for general awareness training.
C. An organization's security awareness program should focus on employee behavior and the consequences of both compliance and noncompliance with the security policy.
D. Access levels are specific issues, not generally the content of awareness training.

S3-84 The **BEST** defense against successful phishing attacks is:

A. application hardening.
B. spam filters.
C. an intrusion detection system.
D. end-user awareness.

D is the correct answer.

Justification:
A. Application hardening has no effect on phishing attacks.
B. Spam filters may catch some unsophisticated phishing attacks.
C. An intrusion detection system will not detect phishing attacks.
D. Phishing attacks are social engineering attacks and are best defended by end-user awareness training.

S3-85 Which of the following would be the **GREATEST** challenge when developing a standard awareness training program for a global organization?

A. Technical input requirements for IT security staff
B. Evaluating training program effectiveness
C. A diverse culture and varied technical abilities of end users
D. Availability of users either on weekends or after office hours

C is the correct answer.

Justification:
A. IT security staff will require technical inputs, and having a separate training program would not be considered a challenge.
B. Evaluating training program effectiveness is not a problem when developing a standard training program. In fact, the evaluation of training program effectiveness will be easier for a standard training program delivered across the organization.
C. A diverse culture and differences in the levels of IT knowledge and IT exposure pose the most difficulties when developing a standard training program because the learning needs of employees vary.
D. Availability of users on weekends or beyond office hours has no impact on the development of a standard training program.

S3-86 When outsourcing, to ensure that third-party service providers comply with an organization security policy, which of the following should occur?

A. A predefined meeting schedule
B. A periodic security audit
C. Inclusion in the contract of a list of individuals to be called in the event of an incident (call tree)
D. Inclusion in the contract of a confidentiality clause

B is the correct answer.

Justification:
A. A predefined meeting schedule is a contributor to, but does not ensure, compliance.
B. A periodic security audit is a formal and documented way to determine compliance level.
C. A call tree is useful for dealing with incidents but does nothing to ensure compliance.
D. Inclusion of a confidentiality clause does not ensure compliance.

S3-87 Which of the following security controls addresses availability?

A. Least privilege
B. Public key infrastructure
C. Role-based access
D. Contingency planning

D is the correct answer.

Justification:
A. Least privilege is an access control that is concerned with confidentiality.
B. Public key infrastructure is concerned with confidentiality and integrity.
C. Role-based access limits access but does not directly address availability.
D. Contingency planning ensures that the system and data are available in the event of a problem.

S3-88 Which of the following control measures **BEST** addresses integrity?

A. Nonrepudiation
B. Time stamps
C. Biometric scanning
D. Encryption

A is the correct answer.

Justification:
A. Nonrepudiation is a control technique that addresses the integrity of information by ensuring that the originator of a message or transaction cannot repudiate (deny or reject) the message, so the message or transaction can be considered authorized, authentic and valid.
B. Using time stamps is a control that addresses only one component of message integrity.
C. Biometric scanning is a control that addresses access.
D. Encryption is a control that addresses confidentiality and may be an element of a data integrity scheme, but this is not sufficient to achieve the same level of integrity as the set of measures used to ensure nonrepudiation.

S3-89 What is the **BEST** way to ensure that security settings on each platform are in compliance with information security policies and procedures?

A. Perform penetration testing.
B. Establish security baselines.
C. Implement vendor default settings.
D. Link policies to an independent standard.

B is the correct answer.

Justification:
A. Penetration testing will not be as effective and can only be performed periodically.
B. Security baselines will provide the best assurance that each platform meets minimum security criteria.
C. Vendor default settings will not necessarily meet the criteria set by the security policies.
D. Linking policies to an independent standard will not provide assurance that the platforms meet these levels of security.

S3-90 A web-based business application is being migrated from test to production. Which of the following is the **MOST** important management sign-off for this migration?

A. User
B. Network
C. Operations
D. Database

A is the correct answer.

Justification:

A. **As owners of the system, user management sign-off is the most important. If a system does not meet the needs of the business, then it has not met its primary objective.**
B. The needs of the network are secondary to the needs of the business.
C. The needs of operations are secondary to the needs of the business.
D. The needs of database management are secondary to the needs of the business.

S3-91 What is the **BEST** way to ensure that information security policies are followed?

X

A. Distribute printed copies to all employees.
B. Perform periodic reviews for compliance.
C. Include escalating penalties for noncompliance.
D. Establish an anonymous hotline to report policy abuses.

B is the correct answer.

Justification:

A. Distributing printed copies will not motivate individuals as much as the consequences of being found in noncompliance.
B. **The best way to ensure that information security policies are followed is to periodically review levels of compliance.**
C. Escalating penalties will first require a compliance review.
D. Establishing an abuse hotline will not motivate individuals as much as the consequences of being found in noncompliance.

S3-92 Who is in the **BEST** position to determine the level of information security needed for a specific business application?

A. The system developer
B. The information security manager
C. The system custodian
D. The data owner

D is the correct answer.

Justification:

A. The system developer will have specific knowledge on limited areas but will not have full knowledge of the business issues that affect the level of security required.
B. The security manager's responsibility is to ensure that the level of protection required by the data owner is provided.
C. The custodian provides the level of protection required by the owner.
D. **Data owners are the most knowledgeable of the security needs of the business application for which they are responsible.**

X

S3-93 Which of the following will **MOST** likely reduce the chances of an unauthorized individual gaining access to computing resources by pretending to be an authorized individual needing to have his/her password reset?

A. Performing reviews of password resets
B. Conducting security awareness programs
C. Increasing the frequency of password changes
D. Implementing automatic password syntax checking

B is the correct answer.

Justification:
A. Performing reviews of password resets may be desirable, but will not be effective in reducing the likelihood of a social engineering attack.
B. Social engineering can be mitigated best through periodic security awareness training for staff members who may be the target of such an attempt.
C. Changing the frequency of password changes may be desirable, but will not reduce the likelihood of a social engineering attack.
D. Strengthening passwords is desirable, but will not reduce the likelihood of a social engineering attack.

S3-94 Which of the following is the **MOST** likely to change an organization's culture to one that is more security conscious?

A. Adequate security policies and procedures
B. Periodic compliance reviews
C. Security steering committees
D. Security awareness campaigns

D is the correct answer.

Justification:
A. Adequate policies and procedures will have little effect on changing security culture.
B. Compliance reviews can have a minor impact on an organization's security culture.
C. Steering committees that have high-level management representation can affect the security culture.
D. Of these options, security awareness campaigns are likely to be the most effective at improving security consciousness.

S3-95 What is the **BEST** way to ensure that an external service provider complies with organizational security policies?

A. Explicitly include the service provider in the security policies.
B. Receive acknowledgment in writing stating the provider has read all policies.
C. Cross-reference to policies in the service level agreement.
D. Perform periodic reviews of the service provider.

D is the correct answer.

Justification:
A. References in policies will not be as effective because they will not trigger the detection of noncompliance.
B. Assurance that the provider has read the policies does nothing to ensure compliance.
C. Written documents by themselves provide little assurance without confirming oversight.
D. Periodic reviews will be the most effective way of ensuring compliance from the external service provider.

S3-96 What practice should **FIRST** be applied to an emergency security patch that has been received via email? The patch should be:

A. loaded onto an isolated test machine.
B. decompiled to check for malicious code.
C. validated to ensure its authenticity.
D. copied onto write-once media to prevent tampering.

C is the correct answer.

Justification:
A. The patch should be validated to ensure authenticity before any other action is taken.
B. Decompiling the patch is not practical and unlikely to expose malicious code.
C. It is essential to first validate that the patch is authentic.
D. Copying to write-once media is not generally a useful step.

S3-97 In a well-controlled environment, which of the following activities is **MOST** likely to lead to the introduction of weaknesses in security software?

A. Applying patches
B. Changing access rules
C. Upgrading hardware
D. Backing up files

B is the correct answer.

Justification:
A. Security software will generally have a well-controlled process for applying patches.
B. The greatest risk occurs when access rules are changed because they are susceptible to being opened up too much, which can result in the creation of a security exposure.
C. Upgrading hardware will not affect software vulnerabilities.
D. Backup processes provide little opportunity to introduce vulnerabilities.

S3-98 Which of the following is the **BEST** indicator that security awareness training has been effective?

A. Employees sign to acknowledge the security policy.
B. More incidents are being reported.
C. A majority of employees have completed training.
D. No incidents have been reported in three months.

B is the correct answer.

Justification:
A. Acknowledging the security policy is not an indication of awareness.
B. More incidents being reported could be an indicator that the staff is paying more attention to security.
C. The number of individuals trained is not an indication of the effectiveness of awareness training.
D. No recent security incidents reported does not reflect awareness levels but may prompt further research to confirm.

X S3-99 Which of the following metrics would be the **MOST** useful in measuring how well information security is monitoring violation logs?

A. Penetration attempts investigated
B. Violation log reports produced
C. Violation log entries
D. Frequency of corrective actions taken

A is the correct answer.

Justification:

A. The most useful metric is one that measures the degree to which complete follow-through has taken place.
B. The quantity of reports is not indicative of whether investigative action was taken.
C. The entries on reports are not indicative of whether action was taken.
D. The frequency of corrective actions is only relevant in relation to the number of penetration attempts investigated.

X S3-100 The **MOST** important consideration when determining how a control policy is implemented is:

A. the risk of compromise.
B. life safety.
C. the mean time between failures.
D. the nature of a threat.

B is the correct answer.

Justification:

A. The risk of compromise is a major consideration in the level of protection required, but not at the expense of safety. Only in very rare circumstances does risk of compromise outweigh life safety, and even then it is the risk to a larger population that justifies a fail secure configuration.
B. Safety of personnel is always the first consideration. For example, even if a data center has highly confidential data, failure of physical access controls should not fail closed and prevent emergency exit. Only in very rare circumstances does risk of compromise outweigh life safety, and even then it is the risk to a larger population that justifies a fail secure configuration.
C. The mean time between failure is a consideration for technical or mechanical controls and must be considered from a safety perspective.
D. The nature of a threat is a consideration for the type and strength of controls.

S3-101 What is the **MOST** important action prior to having a third party perform an attack and penetration test against an organization?

A. Ensure that the third party provides a demonstration on a test system.
B. Ensure that goals and objectives are clearly defined.
C. Ensure that technical staff has been briefed on what to expect.
D. Ensure that special backups of production servers are taken.

B is the correct answer.

Justification:

A. A demonstration of the test system will reduce the spontaneity of the test.
B. The most important action is to clearly define the goals and objectives of the test.
C. Technical staff should not be briefed as this will reduce the spontaneity of the test.
D. Assuming that adequate backup procedures are in place, special backups should not be necessary.

S3-102 What is the **BEST** action to undertake when a departmental system continues to be out of compliance with an information security policy's password strength requirement?

A. Submit the issue to the steering committee.
B. Conduct a risk assessment to quantify the risk.
C. Isolate the system from the rest of the network.
D. Request a risk acceptance waiver from senior management.

B is the correct answer.

Justification:
A. The issue should not be escalated before understanding the risk of noncompliance.
B. A risk assessment is warranted to determine whether a risk acceptance should be granted and to demonstrate to the department the danger of deviating from the established policy.
C. Isolating the system would not support the needs of the business.
D. Any waiver should be granted only after performing a risk assessment.

S3-103 Determining the nature and extent of activities required in developing an information security program often requires assessing the existing program components. The **BEST** way to accomplish this is to perform a(n):

A. security review.
B. impact assessment.
C. vulnerability assessment.
D. threat analysis.

A is the correct answer.

Justification:
A. A security review is used to determine the current state of security for various program components.
B. An impact assessment is used to determine potential impact in the event of the loss of a resource.
C. Vulnerability is only one specific aspect that can be considered in a security review.
D. A threat analysis would not normally be a part of a security review.

S3-104 Which of the following environments represents the **GREATEST** risk to organizational security? 

A. Locally managed file server
B. Enterprise data warehouse
C. Load-balanced web server cluster
D. Centrally managed data switch

A is the correct answer.

Justification:
A. A locally managed file server will be the least likely to conform to organizational security policies because it is generally subject to less oversight and monitoring.
B. Data warehouses are subject to close scrutiny, good change control practices and monitoring.
C. Web server clusters are located in data centers or warehouses and subject to good management.
D. Centrally managed switches are also part of a data center or warehouse.

S3-105 A multinational organization operating in fifteen countries is considering implementing an information security program. Which factor will **MOST** influence the design of the information security program?

A. Representation by regional business leaders
B. Composition of the board
C. Cultures of the different countries
D. IT security skills

C is the correct answer.

Justification:
A. Representation by regional business leaders may not have a major influence unless it concerns cultural issues.
B. Composition of the board may not have a significant impact compared to cultural issues.
C. Culture has a significant impact on how information security will be implemented.
D. IT security skills are not as key or high impact in designing a multinational information security program as would be cultural issues.

S3-106 Which of the following is the **BEST** method for ensuring that temporary employees do not receive excessive access rights?

A. Mandatory access controls
B. Discretionary access controls
C. Lattice-based access controls
D. Role-based access controls

D is the correct answer.

Justification:
A. Mandatory access controls require users to have a clearance at or above the level of asset classification but providing clearances for temporary employees is time-consuming and expensive.
B. Discretionary access control allows delegation based on the individual but requires administrative action to grant and remove access.
C. Lattice based access control is a mandatory access model based on the interaction between any combination of "objects" (such as resources, computers and applications) and "subjects."
D. Role-based access controls will grant temporary employee access based on the job function to be performed. This provides a better means of ensuring that the access is not more or less than what is required, and removing access requires less effort.

S3-107 With regard to the implementation of security awareness programs in an organization, it is **MOST** relevant to understand that which one of the following aspects can change?

A. The security culture
B. The information technology
C. The compliance requirements
D. The threats and vulnerabilities

D is the correct answer.

Justification:
A. The security culture changes over time in part because of an effective security awareness training program. It is not necessary that the workforce be told that the culture will change.
B. Changes in technology are only one part of security awareness.
C. Changes in compliance requirements are not a primary driver of security awareness training.
D. People tend to think that security awareness training can be completed once and it is good forever. It is important for everyone, including management and the general workforce, to understand that threats and vulnerabilities change constantly, and that regular refresher training is an important part of security awareness.

S3-108 Which of the following do security policies need to be **MOST** closely aligned with?

A. Industry good practices
B. Organizational needs
C. Generally accepted standards
D. Local laws and regulations

B is the correct answer.

Justification:
A. Good practices are generally a substitute for a clear understanding of what exactly is needed in a specific organization and may be too much or too little.
B. Policies must support the needs of the organization.
C. Generally accepted standards do not exist; they are always tailored to the requirements of the organization.
D. Local law and regulation compliance may be identified in policies but would only be a small part of overall policies that must support the needs of the organization.

S3-109 What is the **BEST** way to determine if an anomaly-based intrusion detection system (IDS) is properly installed?

A. Simulate an attack and review IDS performance.
B. Use a honeypot to check for unusual activity.
C. Audit the configuration of the IDS.
D. Benchmark the IDS against a peer site.

A is the correct answer.

Justification:

A. Simulating an attack on the network demonstrates whether the intrusion detection system (IDS) is properly tuned.
B. A honeypot would be a poor test to see if the IDS is working properly because attacking it is discretionary and not representative of all attacks.
C. Reviewing the configuration may or may not reveal weaknesses because an anomaly-based system uses trends to identify potential attacks.
D. Benchmarking against a peer site would generally not be practical or useful.

S3-110 When is the **BEST** time to perform a penetration test?

A. After an attempted penetration has occurred
B. After an audit has reported weaknesses in security controls
C. After various infrastructure changes are made
D. After a high turnover in systems staff

C is the correct answer.

Justification:

A. Conducting a test after an attempted penetration is not as productive because an organization should not wait until it is attacked to test its defenses.
B. Any exposure identified by an audit should be corrected before it would be appropriate to test.
C. Changes in the systems infrastructure are most likely to inadvertently introduce new exposures.
D. A turnover in administrative staff does not warrant a penetration test, although it may warrant a review of password change practices and configuration management.

X **S3-111** Which of the following project activities is the **MAIN** activity in developing an information security program?

A. Security organization development
B. Conceptual and logical architecture designs
C. Development of risk management objectives
D. Control design and deployment

D is the correct answer.

Justification:

A. The security organization is developed to meet the needs of the security program and may evolve over time, based on evolving requirements.
B. Conceptual and logical architecture designs should have been completed as a part of strategy and road map development.
C. Risk management objectives are a part of strategy development.
D. The majority of program development activities will involve designing, testing and deploying controls that achieve the risk management objectives.

S3-112 Which of the following is the **BEST** way to detect an intruder who successfully penetrates a network before significant damage is inflicted?

A. Perform periodic penetration testing
B. Establish minimum security baselines
C. Implement vendor default settings
D. Install a honeypot on the network

D is the correct answer.

Justification:
A. Penetration testing will not detect an intruder.
B. Security baselines set minimum security levels but are not related to detecting intruders.
C. Implementing vendor default settings do not detect intruders and is not the best idea.
D. Honeypots attract hackers away from sensitive systems and files. Because honeypots are closely monitored, the intrusion is more likely to be detected before significant damage is inflicted.

S3-113 Which of the following presents the **GREATEST** threat to the security of an enterprise resource planning (ERP) system?

A. User *ad hoc* reporting is not logged.
B. Network traffic is through a single switch.
C. Operating system security patches have not been applied.
D. Database security defaults to ERP settings.

C is the correct answer.

Justification:
A. Although the lack of logging for user *ad hoc* reporting is not necessarily good, it does not represent as serious a security weakness as the failure to install security patches.
B. Routing network traffic through a single switch is not unusual.
C. The fact that operating system security patches have not been applied is a serious weakness.
D. Database security defaulting to the enterprise resource planning system's settings is not as significant.

S3-114 In a social engineering scenario, which of the following will **MOST** likely reduce the likelihood of an unauthorized individual gaining access to computing resources?

A. Implementing on-screen masking of passwords
B. Conducting periodic security awareness programs
C. Increasing the frequency of password changes
D. Requiring that passwords be kept strictly confidential

B is the correct answer.

Justification:
A. Implementing on-screen masking of passwords is desirable but will not be effective in reducing the likelihood of a successful social engineering attack.
B. Social engineering can best be mitigated through periodic security awareness training for users who may be the target of such an attempt.
C. Increasing the frequency of password changes is desirable but will not be effective in reducing the likelihood of a successful social engineering attack.
D. Requiring that passwords be kept secret in security policies is a good control but is not as effective as periodic security awareness programs that will alert users of the dangers posed by social engineering.

S3-115 Which of the following will **BEST** ensure that management takes ownership of the decision-making process for information security?

A. Security policies and procedures
B. Annual self-assessment by management
C. Security steering committees
D. Security awareness campaigns

C is the correct answer.

Justification:

A. Security policies and procedures are good but do not necessarily result in the taking of ownership by management.
B. Self-assessment exercises do not necessarily indicate management has taken ownership of the security decision-making process.
C. Security steering committees provide a forum for management to express its opinion and take ownership in the decision-making process.
D. Awareness campaigns are no indication that management has taken ownership of the security decision-making process.

S3-116 Which of the following is the **MOST** appropriate individual to implement and maintain the level of information security needed for a specific business application?

A. System analyst
B. Quality control manager
C. Process owner
D. Information security manager

C is the correct answer.

Justification:

A. The system analyst does not possess the necessary knowledge or authority to implement and maintain the appropriate level of business security.
B. Quality control managers do not implement security.
C. Process owners implement information protection controls as determined by the business' needs. Process owners have the most knowledge about security requirements for the business application for which they are responsible.
D. The information security manager will implement the information security framework and develop standards and controls, but the level of security required by a specific business application is determined by the process owner.

S3-117 Which of the following activities is **MOST** likely to increase the difficulty of totally eradicating malicious code that is not immediately detected?

A. Applying patches
B. Changing access rules
C. Upgrading hardware
D. Backing up files

D is the correct answer.

Justification:
A. Applying patches does not significantly increase the level of difficulty.
B. Changing access rules has no effect on eradication of malicious code.
C. Upgrading hardware does not significantly increase the level of difficulty.
D. If malicious code is not immediately detected, it will most likely be backed up as a part of the normal tape backup process. When later discovered, the code may be eradicated from the device but still remain undetected on a backup tape. Any subsequent restores using that tape may reintroduce the malicious code.

S3-118 When should security awareness training be provided to new employees?

A. On an as-needed basis
B. During system user training
C. Before they have access to data
D. Along with department staff

C is the correct answer.

Justification:
A. Providing training on an as-needed basis implies that security awareness training is delivered subsequent to the granting of system access, which may place security as a secondary step.
B. Providing awareness training during system user training implies that security awareness training is delivered subsequent to the granting of system access, which may place security as a secondary step.
C. Security awareness training should occur before access is granted to ensure the new employee understands that security is part of the system and business process.
D. Providing training along with department staff implies that security awareness training is delivered subsequent to the granting of system access.

S3-119 What is the **BEST** method to verify that all security patches applied to servers were properly documented?

A. Trace change control requests to operating system (OS) patch logs.
B. Trace OS patch logs to OS vendor's update documentation.
C. Trace OS patch logs to change control requests.
D. Review change control documentation for key servers.

C is the correct answer.

Justification:

A. Tracing from the documentation to the patch log will not indicate if some patches were applied without being documented.
B. Comparing patches applied to those recommended by the OS vendor's web site does not confirm that these security patches were properly approved and documented.
C. To ensure that all patches applied went through the change control process, it is necessary to use the operating system (OS) patch logs as a starting point and then check to see if change control documents are on file for each of these changes.
D. Reviewing change control documents for key servers does not confirm that security patches were properly approved and documented.

S3-120 What is the **PRIMARY** objective of security awareness?

A. Ensure that security policies are understood.
B. Influence employee behavior.
C. Ensure legal and regulatory compliance.
D. Notify of actions for noncompliance.

B is the correct answer.

Justification:

A. Ensuring that policies are read and understood is important but secondary.
B. It is most important that security-conscious behavior be encouraged among employees through training that influences expected responses to security incidents.
C. Meeting legal and regulatory requirements is important but secondary.
D. Giving employees fair warning of potential disciplinary action is important but secondary.

S3-121 Which of the following will **BEST** protect against malicious activity by a former employee?

A. Preemployment screening
B. Close monitoring of users
C. Periodic awareness training
D. Effective termination procedures

D is the correct answer.

Justification:

A. Preemployment screening is important but not as effective in preventing this type of situation.
B. Monitoring is important but not as effective in preventing this type of situation.
C. Security awareness training is important but not as effective in preventing this type of situation.
D. When an employee leaves an organization, the former employee may attempt to use their credentials to perform unauthorized or malicious activity. Accordingly, it is important to ensure timely revocation of all access at the time an individual is terminated.

S3-122 Which of the following represents a **PRIMARY** area of interest when conducting a penetration test?

A. Data mining
B. Network mapping
C. Intrusion detection system
D. Customer data

B is the correct answer.

Justification:
A. Data mining is associated with *ad hoc* reporting and is a potential target after the network is penetrated.
B. Network mapping is the process of determining the topology of the network one wishes to penetrate. This is one of the first steps toward determining points of attack in a network.
C. The intrusion detection mechanism in place is not an area of focus because one of the objectives is to determine how effectively it protects the network or how easy it is to circumvent.
D. Customer data, together with data mining, is a potential target after the network is penetrated.

S3-123 The return on investment of information security can **BEST** be evaluated through which of the following? 

A. Support of business objectives
B. Security metrics
C. Security deliverables
D. Process improvement models

A is the correct answer.

Justification:
A. One way to determine the return on security investment is to illustrate how information security supports the achievement of business objectives.
B. Security metrics measure improvement and effectiveness within the security practice but do not necessarily tie to business objectives.
C. Listing deliverables does not necessarily tie into business objectives.
D. Creating process improvement models does not necessarily tie directly into business objectives.

S3-124 What activity **BEST** helps ensure that contract personnel do not obtain unauthorized access to sensitive information?

A. Set accounts to pre-expire.
B. Avoid granting system administration roles.
C. Ensure they successfully pass background checks.
D. Ensure their access is approved by the data owner.

B is the correct answer.

Justification:
A. Setting an expiration date is a positive element but will not prevent contract personnel from obtaining access to sensitive information.
B. Contract personnel should not be given job duties that provide them with power user or other administrative roles that they could then use to grant themselves access to sensitive files.
C. Requiring background checks is a positive element but will not prevent contract personnel from obtaining access to sensitive information.
D. Having the data owner approve access is a marginally effective approach to limiting access to sensitive information.

S3-125 What is the **MOST** important success factor in launching a corporate information security awareness program?

A. Adequate budgetary support
B. Centralized program management
C. Top-down approach
D. Experience of the awareness trainers

C is the correct answer.

Justification:

A. Funding is not a primary concern.
B. Centralized management does not provide sufficient support.
C. Senior management support will provide enough resources and will focus attention to the program; training should start at the top levels to gain support and sponsorship.
D. Trainer experience, while important, is not the primary success factor.

S3-126 Which of the following is the **MOST** appropriate method to protect the delivery of a password that opens a confidential file?

A. Delivery path tracing
B. Reverse lookup translation
C. Out-of-band channels
D. Digital signatures

C is the correct answer.

Justification:

A. Delivery path tracing shows the route taken but does not confirm the identity of the sender.
B. Reverse lookup translation involves converting an Internet Protocol address to a username.
C. It is risky to send the password to a file by the same method as the file was sent. An out-of-band channel such as the telephone reduces the risk of interception.
D. Digital signatures prove the identity of the sender of a message and ensure integrity.

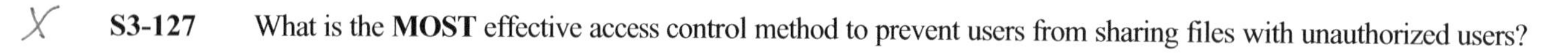

S3-127 What is the **MOST** effective access control method to prevent users from sharing files with unauthorized users?

A. Mandatory
B. Discretionary
C. Walled garden
D. Role-based

A is the correct answer.

Justification:

A. Mandatory access controls restrict access to files based on the security classification of the file. This prevents users from sharing files with unauthorized users.
B. Discretionary access controls are not as effective as mandatory access controls in preventing file sharing.
C. A walled garden is an environment that controls a user's access to web content and services. In effect, the walled garden directs the user's navigation within particular areas and does not necessarily prevent sharing of other material.
D. Role-based access controls grant access according to the role assigned to a user; they do not prevent file sharing with unauthorized users.

S3-128 Which of the following is an inherent weakness of signature-based intrusion detection systems?

A. A higher number of false positives
B. New attack methods will be missed
C. Long duration probing will be missed
D. Attack profiles can be easily spoofed

B is the correct answer.

Justification:
A. False positives are usually lower in signature-based intrusion detection systems (IDSs).
B. Signature-based IDSs do not detect new attack methods for which signatures have not yet been developed.
C. Long duration probing is more likely to fool anomaly-based systems (boiling frog technique).
D. Spoofing is not relevant in this case.

S3-129 Which of the following is the **BEST** way to ensure that a corporate network is adequately secured against external attack?

A. Use an intrusion detection system.
B. Establish minimum security baselines.
C. Implement vendor recommended settings.
D. Perform periodic penetration testing.

D is the correct answer.

Justification:
A. An intrusion detection system may detect an attempted attack, but it will not confirm whether the perimeter is secure.
B. Minimum security baselines are beneficial, but they will not provide the level of assurance that is provided by penetration testing.
C. Vendor recommended settings may be used to harden systems but provide little assurance that other vulnerabilities do not exist, which may be exposed by penetration testing.
D. Penetration testing is the best way to assure that perimeter security is adequate.

S3-130 Which of the following presents the **GREATEST** exposure to internal attack on a network?

A. User passwords are not automatically expired.
B. All network traffic goes through a single switch.
C. User passwords are encoded but not encrypted.
D. All users reside on a single internal subnet.

C is the correct answer.

Justification:
A. Not automatically expiring user passwords does create an exposure but not as great as having unencrypted passwords.
B. Using a single switch does not present a significant exposure.
C. When passwords are sent over the internal network in an encoded format, they can easily be converted to cleartext. All passwords should be encrypted to provide adequate security.
D. Using a subnet does not present a significant exposure.

S3-131 Which of the following provides the linkage to ensure that procedures are correctly aligned with information security policy requirements?

A. Standards
B. Guidelines
C. Security metrics
D. Gap analysis

A is the correct answer.

Justification:

A. Standards set the allowable boundaries for procedures to ensure they comply with the intent of policies.
B. Guidelines are a description of a particular way of accomplishing something that is less prescriptive than a procedure.
C. Security metrics will detect but not necessarily ensure alignment between policies and procedures.
D. Gap analysis is used to determine what is required to move from an existing state to a desired state but is not useful in determining the alignment of procedures and policy.

S3-132 Which of the following are the **MOST** important individuals to include as members of an information security steering committee?

A. Direct reports to the chief information officer
B. IT management and key business process owners
C. Cross-section of end users and IT professionals
D. Internal audit and corporate legal departments

B is the correct answer.

Justification:

A. Direct reports to the chief information officer do not include business process owners, and their input is necessary.
B. Security steering committees provide a forum for management to express its opinion and take some ownership in the decision-making process. It is imperative that business process owners be included in this process.
C. End users and IT professionals would not be part of the steering committee.
D. Internal audit would not be on the steering committee, although legal representation might.

S3-133 What responsibility do data owners normally have?

A. Applying emergency changes to application data
B. Administering security over database records
C. Migrating application code changes to production
D. Determining the level of application security required

D is the correct answer.

Justification:

A. Making emergency changes to data is an infrastructure task performed by custodians of the data.
B. Administering database security is an infrastructure task performed by custodians of the data.
C. Migrating code to production is an infrastructure task performed by custodians of the data.
D. Data owners approve access to data and determine the degree of protection that should be applied (data classification).

S3-134 Which of the following is the **MOST** appropriate individual to ensure that new exposures have not been introduced into an existing application during the change management process?

A. System analyst
B. System user
C. Operations manager
D. Data security officer

B is the correct answer.

Justification:
A. The system analyst would not be as closely involved in testing code changes.
B. System users, specifically the user acceptance testers, would be in the best position to note whether new exposures are introduced during the change management process.
C. The operations manager would not be involved in testing code changes.
D. The data security officer would not be involved in testing code changes.

S3-135 What is the **BEST** way to ensure users comply with organizational security requirements for password complexity?

A. Include password construction requirements in the security standards.
B. Require each user to acknowledge the password requirements.
C. Implement strict penalties for user noncompliance.
D. Enable system-enforced password configuration.

D is the correct answer.

Justification:
A. Standards provide some deterrence but are not as effective as automated controls.
B. Requiring use acknowledgement will help but not to the extent of automatic system enforcement.
C. Penalties for noncompliance may be fairly effective but will not provide the level of assurance provided by automated system enforcement.
D. Automated system enforced password construction provides the highest level of assurance of compliance.

S3-136 Which of the following choices is the **MOST** significant single point of failure in a public key infrastructure?

A. A certificate authority's (CA) public key
B. A relying party's private key
C. A CA's private key
D. A relying party's public key

C is the correct answer.

Justification:
A. The certificate authority's (CA) public key is published and poses no risk.
B. If destroyed, lost or compromised, the private key of any one relying party affects only that party.
C. The CA's private key is the single point of failure for the entire public key infrastructure (PKI) because it is unpublished and the system cannot function if the key is destroyed, lost or compromised.
D. The public key is published and poses no risk.

S3-137 What is the **PRIMARY** reason for using metrics to evaluate information security?

A. To identify security weaknesses
B. To justify budgetary expenditures
C. To enable steady improvement
D. To raise awareness on security issues

C is the correct answer.

Justification:
A. Metrics may not identify vulnerabilities.
B. Metrics can be used to justify budgetary expenditures, but that is not their primary purpose.
C. A primary purpose for metrics is to facilitate and track continuous improvement in security posture.
D. Metrics may serve to raise awareness of security issues, but that would be for the purpose of improving security.

S3-138 What is the **BEST** method to confirm that all firewall rules and router configuration settings are adequate?

A. Periodic review of network configuration
B. Review of intrusion detection system logs for evidence of attacks
C. Periodically perform penetration tests
D. Daily review of server logs for evidence of hacker activity

C is the correct answer.

Justification:
A. Due to the complexity of firewall rules and router tables, plus the sheer size of intrusion detection systems (IDSs) and server logs, a physical review would be complex, time-consuming and probably insufficient.
B. Reviewing IDS logs for evidence of attacks would not indicate whether the settings are adequate.
C. The best approach for confirming the adequacy of these configuration settings is to periodically perform attack and penetration tests.
D. Evidence of hacker activity has little to do with configuration adequacy.

S3-139 Which of the following is **MOST** important for measuring the effectiveness of a security awareness program?

A. Reduced number of security violation reports
B. A quantitative evaluation to ensure user comprehension
C. Increased interest in focus groups on security issues
D. Increased number of security violation reports

B is the correct answer.

Justification:
A. A reduction in the number of violation reports may not be indicative of a high level of security awareness.
B. To truly judge the effectiveness of security awareness training, some means of measurable testing is necessary to confirm user comprehension.
C. Focus groups may or may not provide meaningful feedback but, in and of themselves, do not provide metrics.
D. An increase in the number of violation reports is a possible indication of increased awareness but not as useful as direct testing of awareness levels.

S3-140 Which of the following is the **MOST** important action to take when engaging third-party consultants to conduct an attack and penetration test?

A. Request a list of the software to be used.
B. Provide clear directions to IT staff.
C. Monitor intrusion detection system and firewall logs closely.
D. Establish clear rules of engagement.

D is the correct answer.

Justification:
A. Not as important, but still useful, is to request a list of what software will be used.
B. IT staff should not be alerted in order to maximize effectiveness of the penetration test.
C. Monitoring personnel should not be alerted in order to effectively test their activities.
D. It is critical to establish a clear understanding on what is permissible during the engagement. Otherwise, the tester may inadvertently trigger a system outage or inadvertently corrupt files.

S3-141 Which of the following will **BEST** prevent an employee from using a universal serial bus (USB) drive to copy files from desktop computers?

A. Restrict the available drive allocation on all personal computers.
B. Disable USB ports on all desktop devices.
C. Conduct frequent awareness training with noncompliance penalties.
D. Establish strict access controls to sensitive information.

A is the correct answer.

Justification:
A. Restricting the ability of a personal computer to allocate new drive letters ensures that universal serial bus (USB) drives or even compact disc-writers cannot be attached because they would not be recognized by the operating system.
B. Disabling USB ports on all machines is not practical because mice and other peripherals depend on these connections.
C. Awareness training does not prevent copying of information.
D. Access controls do not prevent copying.

S3-142 What is the **MAIN** drawback of emailing password-protected zip files across the Internet?

A. They all use weak encryption.
B. They are decrypted by the firewall.
C. They may be quarantined by mail filters.
D. They may be corrupted by the receiving mail server.

C is the correct answer.

Justification:
A. Many zip file products are capable of using strong encryption.
B. Firewalls are often unable to decrypt the file to determine if it contains malicious code, thus sending it to quarantine.
C. Often, mail filters will quarantine zip files that are password-protected because the filter (or the firewall) is unable to determine if the file contains malicious code.
D. Such files are not normally corrupted by the mail server.

S3-143 A major trading partner with access to the internal network is unwilling or unable to remediate serious information security exposures within its environment. Which of the following is the **BEST** recommendation?

A. Sign a legal agreement assigning them all liability for any breach.
B. Remove all trading partner access until the situation improves.
C. Set up firewall rules restricting network traffic from that location.
D. Send periodic reminders advising them of their noncompliance.

C is the correct answer.

Justification:
A. Agreements do not protect the integrity of the network.
B. Removing all access will likely result in lost business and be a career-ending solution.
C. It is incumbent on an information security manager to see to the protection of their organization's network but to do so in a manner that does not adversely affect the conduct of business. This can be accomplished by adding specific traffic restrictions for that particular location.
D. Reminders do not protect the integrity of the network.

S3-144 What should documented standards/procedures for the use of cryptography across the enterprise achieve?

A. They should define the circumstances where cryptography should be used.
B. They should define cryptographic algorithms and key lengths.
C. They should describe handling procedures of cryptographic keys.
D. They should establish the use of cryptographic solutions.

A is the correct answer.

Justification:
A. There should be documented standards/procedures for the use of cryptography across the enterprise; they should define the circumstances where cryptography should be used.
B. Procedures should cover the selection of cryptographic algorithms and key lengths but should not define them precisely.
C. Procedures should address the handling of cryptographic keys. However, this is secondary to how and when cryptography should be used.
D. The use of cryptographic solutions should be addressed but, again, this is a secondary consideration.

S3-145 Which of the following is the **MOST** immediate consequence of failing to tune a newly installed intrusion detection system with the threshold set to a low value?

A. The number of false positives increases
B. The number of false negatives increases
C. Active probing is missed
D. Attack profiles are ignored

A is the correct answer.

Justification:
A. Failure to tune an intrusion detection system will result in many false positives, especially when the threshold is set to a low value.
B. An increase in false negatives is less likely given the fact that the threshold for sounding an alarm is set to a low value.
C. Missed active probing is less likely given the fact that the threshold for sounding an alarm is set to a low value.
D. Ignored attack profiles are less likely given the fact that the threshold for sounding an alarm is set to a low value.

S3-146 What is the **MOST** appropriate change management procedure for the handling of emergency program changes?

A. Formal documentation does not need to be completed.
B. Business management approval must be obtained prior to the change.
C. Documentation is completed with approval soon after the change.
D. Emergency changes eliminate certain documentation requirements.

C is the correct answer.

Justification:
A. Formal documentation is still required as soon as possible after the emergency changes have been implemented.
B. Obtaining business approval prior to the change is ideal but not always possible.
C. Even in the case of an emergency change, all change management procedure steps should be completed as in the case of normal changes. The difference lies in the timing of certain events. With an emergency change, it is permissible to obtain certain approvals and other documentation after the emergency has been satisfactorily resolved.
D. Emergency changes require the same process as regular changes, but the process may be delayed until the emergency has been resolved.

S3-147 Who is ultimately responsible for ensuring that information is categorized and that protective measures are taken?

A. Information security officer
B. Security steering committee
C. Data owner
D. Data custodian

B is the correct answer.

Justification:
A. The information security officer supports and implements information security for senior management.
B. Routine administration of all aspects of security is delegated, but senior management must retain overall responsibility.
C. The data owner is responsible for categorizing data security requirements.
D. The data custodian supports and implements information security as directed.

S3-148 Most standard frameworks for information security show the development of an information security program as starting with:

A. policy development and implementation of process.
B. an internal audit and remediation of findings.
C. a risk assessment and control objectives.
D. resource identification and budgetary requirements.

C is the correct answer.

Justification:
A. Policies are written to support objectives, which are determined by business requirements.
B. Audits are conducted to determine compliance with control objectives.
C. An information security program is established to close the gap between the existing state of controls (as identified by a risk assessment) and the state desired on the basis of business requirements, which will be obtained through the meeting of control objectives.
D. A program must have objectives before resources can be allocated in pursuit of those objectives.

X **S3-149** The maturity of an information security program is **PRIMARILY** the result of:

A. a comprehensive risk assessment and analysis.
B. an effective information security strategy.
C. the development of a security architecture.
D. completing a controls statement of applicability.

B is the correct answer.

Justification:

A. Assessing and analyzing risk is required to develop a strategy and will provide some of the information needed to develop it, but will not define the scope and charter of the security program. Also, how the organization chooses to approach identified risk is a business decision that must be made by senior management and identified in a strategy.
B. An effective information security strategy provides clear direction on how the organization will attain security outcomes desired and directed by senior management.
C. A security architecture is ideally a part of implementation after developing the strategy. It is possible to adopt an architecture without a strategy, but its implementation will not necessarily help the organization to attain the security outcomes desired by senior management.
D. The applicability statement is a part of strategy implementation using International Organization for Standardization (ISO) 27001 or 27002 after determining the scope and responsibilities of the program. Like a security architecture, an applicability statement can be adopted without a strategy, but will not necessarily help the organization to attain the security outcomes desired by senior management.

S3-150 A critical device is delivered with a single user ID and password that is required to be shared for multiple users to access the device. An information security manager has been tasked with ensuring all access to the device is authorized. Which of the following would be the **MOST** efficient means to accomplish this?

A. Enable access through a separate device that requires adequate authentication.
B. Implement manual procedures that require a password change after each use.
C. Request the vendor to add multiple user IDs.
D. Analyze the logs to detect unauthorized access.

A is the correct answer.

Justification:

A. Enabling access through a separate device that requires adequate authentication allows authentication tokens to be provisioned and terminated for individuals and also introduces the possibility of logging activity by individual.
B. Implementing manual procedures that require a password change after each use is not effective because users can circumvent the manual procedures.
C. Vendor enhancements may take time and development, and this is a critical device.
D. Analyzing the logs to detect unauthorized access could, in some cases, be an effective complementary control, but because it is detective, it would not be the most effective in this instance.

S3-151 Which of the following documents would be the **BEST** reference to determine whether access control mechanisms are appropriate for a critical application?

A. User security procedures
B. Business process flow
C. IT security standards
D. Regulatory requirements

C is the correct answer.

Justification:
A. Procedures will not indicate the appropriateness of control mechanisms.
B. The business process flow is not relevant to the access control mechanism.
C. IT management should ensure that mechanisms are implemented in line with IT security standards.
D. The organization's own policies, standards and procedures should take into account regulatory requirements.

S3-152 Which of the following is the **MOST** important process that an information security manager needs to negotiate with an outsourced service provider?

A. The right to conduct independent security reviews
B. A legally binding data protection agreement
C. Encryption between the organization and the provider
D. A joint risk assessment of the system

A is the correct answer.

Justification:
A. A key requirement of an outsourced contract involving critical business systems is the establishment of the organization's right to conduct independent security reviews of the provider's security controls.
B. A legally binding data protection agreement is also critical but secondary to conducting independent security reviews, which permits examination of the actual security controls prevailing over the system and, as such, is the more effective risk management tool.
C. Network encryption of the link between the organization and the provider may well be a requirement but by itself will not provide the assurance of independent security reviews.
D. A joint risk assessment of the system in conjunction with the outsourced provider may be a compromise solution should the right to conduct independent security reviews of the controls related to the system prove contractually difficult, but it is not the best option.

S3-153 Which resource is the **MOST** effective in preventing physical access tailgating/piggybacking?

A. Card key door locks
B. Photo identification
C. Awareness training
D. Biometric scanners

C is the correct answer.

Justification:
A. Card key door locks are a physical control that by itself would not be effective against tailgating.
B. Photo identification by itself would not be effective against tailgating.
C. Awareness training would most likely result in any attempted tailgating being challenged by the authorized employee.
D. A biometric scanner is a physical control that by itself would not be effective against tailgating.

X

S3-154 What is the **BEST** approach to implement adequate segregation of duties in business-critical applications, where shared access to elevated privileges by a small group is necessary?

A. Ensure access to individual functions can be granted to individual users only.
B. Implement role-based access control in the application.
C. Enforce manual procedures ensuring separation of conflicting duties.
D. Create service accounts that can only be used by authorized team members.

B is the correct answer.

Justification:

A. Access to individual functions will not ensure appropriate segregation of duties (SoD).
B. Role-based access control is the best way to implement appropriate SoD. Roles will have to be defined once, and then the user could be changed from one role to another without redefining the content of the role each time.
C. Giving a user access to all functions and implementing, in parallel, a manual procedure ensuring SoD is not an effective method, and it would be difficult to enforce and monitor.
D. Creating service accounts that can be used by authorized team members would not provide any help unless their roles are properly segregated.

X

S3-155 Who should approve user access in business-critical applications?

A. The information security manager
B. The data owner
C. The data custodian
D. Business management

B is the correct answer.

Justification:

A. An information security manager will coordinate and execute the implementation of the role-based access control.
B. Data owners are in the best position to validate access rights to users due to their deep understanding of business requirements and of functional implementation within the application. This responsibility should be enforced by the policy.
C. A data custodian will ensure that proper safeguards are in place to protect the data from unauthorized access; it is not the data custodian's responsibility to assign access rights.
D. Business management is not, in all cases, the owner of the data.

S3-156 What is the **MOST** critical success factor of the patch management procedure in an organization where availability is a primary concern?

A. Testing time window prior to deployment
B. Technical skills of the team responsible
C. Certification of validity for deployment
D. Automated deployment to all the servers

A is the correct answer.

Justification:
A. **Having the patch tested prior to implementation on critical systems is an absolute prerequisite where availability is a primary concern because deploying patches that could cause a system to fail could be worse than the vulnerability corrected by the patch.**
B. A high level of technical skills is not required because patches are usually applied via automated tools.
C. Validation of the patch is essential but is unrelated to the testing, which is the primary area of concern.
D. It makes no sense to deploy patches on every system. Vulnerable systems should be the only candidate for patching.

S3-157 Who should be involved in the design of information security procedures to ensure they are functional and accurate?

A. End users
B. Legal counsel
C. Operational units
D. Audit management

C is the correct answer.

Justification:
A. End users are normally not involved in procedure development other than testing.
B. Legal counsel is normally not involved in procedure development.
C. **Procedures at the operational level must be developed by or with the involvement of operational units that will use them. This will ensure that they are functional and accurate.**
D. Audit management generally oversees information security operations but does not get involved at the procedural level.

S3-158 An information security manager reviewed the access control lists and observed that privileged access was granted to an entire department. Which of the following should the information security manager do **FIRST**?

A. Review the procedures for granting access.
B. Establish procedures for granting emergency access.
C. Meet with data owners to understand business needs.
D. Redefine and implement proper access rights.

C is the correct answer.

Justification:
A. Reviewing the procedures for granting access could be correct depending on the priorities set by the business unit, but this would follow understanding the business needs.
B. Procedures for granting emergency access require first understanding business needs.
C. **An information security manager must understand the business needs that motivated the change prior to taking any unilateral action.**
D. Redefining and implementing proper access rights would follow understanding the business needs.

S3-159 Which of the following should be in place before a black box penetration test begins?

A. IT management approval
B. Proper communication and awareness training
C. A clearly stated definition of scope
D. An incident response plan

C is the correct answer.

Justification:
A. IT management approval may not be required based on senior management decisions.
B. Communication, awareness and an incident response plan are not a necessary requirement.
C. Having a clearly stated definition of scope is most important to ensure a proper understanding of risk as well as success criteria.
D. A penetration test could help promote the creation and execution of the incident response plan.

S3-160 Of the following, which is the **MOST** effective way to measure strategic alignment of an information security program?

A. Track audits over time.
B. Evaluate incident losses.
C. Analyze business cases.
D. Interview business owners.

D is the correct answer.

Justification:
A. Audit reports may indicate areas of security activities that do not optimally support the enterprise objectives but will not be as good an indicator as insight from business owners.
B. Losses may or may not be considered acceptable by the enterprise but will not be well correlated with the perception of business support.
C. To the extent that business cases have been developed for particular security activities, they will be a good indication of how well business requirements were considered; however, the perception of business owners will ultimately be the most important factor.
D. It is essential that business owners understand and support the security program and fully understand how its controls impact their activities. This can be most readily accomplished through direct interaction with business leadership.

S3-161 The **MOST** effective way to limit actual and potential impacts of e-discovery in the event of litigation is to:

A. implement strong encryption of all sensitive documentation.
B. ensure segregation of duties and limited access to sensitive data.
C. enforce a policy of not writing or storing potentially sensitive information.
D. develop and enforce comprehensive retention policies.

D is the correct answer.

Justification:
A. Encryption will not prevent the legal requirements to produce documents in the event of legal conflicts.
B. Limiting access to sensitive information based on the need to know may limit which personnel can testify during legal proceedings but will not limit the requirement to produce existing documents.
C. While some organizations have practiced a policy of not committing to writing issues of dubious legality, it is not a sound practice and may violate a variety of laws.
D. Compliance with legally acceptable defined retention policies will limit exposure to the often difficult and costly demands for documentation during legal proceedings such as lawsuits.

S3-162 Which of the following **BEST** protects confidentiality of information?

A. Information classification
B. Segregation of duties
C. Least privilege
D. Systems monitoring

C is the correct answer.

Justification:
A. While classifying information can help focus the assignment of privileges, classification itself does not provide enforcement.
B. Only in very specific situations does segregation of duties safeguard confidentiality of information.
C. Restricting access to information to those who need to have access is the most effective means of protecting confidentiality.
D. Systems monitoring is a detective control rather than a preventive control.

S3-163 Who should determine the appropriate classification of accounting ledger data located on a database server and maintained by a database administrator in the IT department?

A. Database administrator
B. Finance department management
C. Information security manager
D. IT department management

B is the correct answer.

Justification:
A. The database administrator is the custodian of the data who would apply the appropriate security levels for the classification.
B. Data owners are responsible for determining data classification; in this case, management of the finance department would be the owners of accounting ledger data.
C. The security manager would act as an advisor and enforcer.
D. The IT management is the custodian of the data who would apply the appropriate security levels for the classification.

S3-164 A company uses a single employee to update the servers, review the audit logs and maintain access controls. Which of the following choices is the **BEST** compensating control?

A. Verify that only approved changes are made.
B. Perform quarterly penetration tests.
C. Perform monthly vulnerability scans.
D. Implement supervisor review of log files.

A is the correct answer.

Justification:
A. Where segregation of duties is not possible, additional procedures are needed to ensure that the single person with access is not able to abuse that access.
B. Penetration tests do not address insider threat.
C. Vulnerability scans only check hardware and software for changes against set requirements. There is no correlation to unauthorized activities.
D. A sufficiently knowledgeable administrator may be able to manipulate the log files and hide his/her activities from the supervisor.

S3-165 Which of the following would **BEST** assist an information security manager in measuring the existing level of development of security processes against their desired state?

A. Security audit reports
B. Balanced scorecard
C. Capability maturity model
D. Systems and business security architecture

C is the correct answer.

Justification:

A. Security audit reports offer a limited view of the current state of security.
B. Balanced scorecard is a document that enables management to measure the implementation of their strategy and assists in its translation into action.
C. The capability maturity model grades each defined area of security processes on a scale of 0 to 5 based on their maturity and is commonly used by entities to measure their existing state or maturity and then determine the desired one.
D. Systems and business security architecture explain the security architecture of an entity in terms of business strategy, objectives, relationships, risk, constraints and enablers, and provides a business-driven and business-focused view of security architecture, but it is not the best way to determine the existing level of security process development.

S3-166 Who is responsible for raising awareness of the need for adequate funding to support risk mitigation plans?

A. Chief information officer
B. Chief financial officer
C. Information security manager
D. Business unit management

C is the correct answer.

Justification:

A. Even though the chief information officer is involved in the final approval of fund expenditure, it is the information security manager who has the ultimate responsibility for raising awareness.
B. Even though the chief financial officer is involved in the final approval of fund expenditure, it is the information security manager who has the ultimate responsibility for raising awareness.
C. The information security manager is responsible for raising awareness of the need for adequate funding for risk-related mitigation plans.
D. Even though the business unit management is involved in the final approval of fund expenditure, it is the information security manager who has the ultimate responsibility for raising awareness.

S3-167 Which of the following **BEST** describes the key objective of an information security program?

A. Achieve strategic business goals and objectives.
B. Protect information assets using manual and automated controls.
C. Automate information security controls.
D. Eliminate threats to the organization.

A is the correct answer.

Justification:

A. While the activities of the security program are primarily the protection of the organization's assets, the key objective is to support the achievement of the strategic business goals and objectives.
B. An information security program focuses on protecting information assets using manual and automated controls with the objective of supporting the achievement of strategic business goals.
C. Information security is achieved by implementing any type of control; it is achieved not just by using IT controls, but also using manual controls.
D. Threats cannot be eliminated; information security controls help reduce risk to an acceptable level.

S3-168 The effectiveness of segregation of duties may be **MOST** seriously compromised when:

A. user IDs of terminated staff remain active in application systems.
B. access privileges are accumulated based on previous job functions.
C. application role-based access deviates from the organizational hierarchies.
D. role mining tools are used in the access privilege review.

B is the correct answer.

Justification:

A. It is not desirable to leave user IDs of terminated personnel or contractors active in the systems because it increases the potential for unauthorized access. However, the risk related to not effectively managing terminated users is an access management issue, not a segregation of duties issue.
B. When the changing of user roles is not adequately managed, access privileges may cross the boundary of segregation of duties. This often happens when a user's role changes as part of a promotion or transfer and he/she is assigned new system privileges to fulfill the new role, and the privileges of the previous role are not removed.
C. Role-based access is built on the premise that users are granted those privileges that they need to perform their daily job function (role). These may not necessarily be aligned with the organizational hierarchies.
D. Using role mining tools in the access entitlement review may enhance the efficiency and effectiveness of the process, particularly in large and complex environments.

X

S3-169 Which of the following **BEST** mitigates a situation where an application programmer requires access to production data?

A. Create a separate account for the programmer as a power user.
B. Log all of the programmers' activity for review by supervisor.
C. Have the programmer sign a letter accepting full responsibility.
D. Perform regular audits of the application.

B is the correct answer.

Justification:
A. Creating a separate account for the programmer as a power user does not solve the problem.
B. It is not always possible to provide adequate segregation of duties between programming and operations in order to meet certain business requirements. A mitigating control is to record all of the programmers' actions for later review by their supervisor, which would detect any inappropriate action on the part of the programmer.
C. Having the programmer sign a letter accepting full responsibility is not an effective control.
D. Performing regular audits of the application is not relevant to determine if programmer activities are appropriate.

S3-170 What action should be taken in regard to data classification requirements before engaging outsourced providers? Ensure the data classification requirements:

A. are compatible with the provider's own classification.
B. are communicated to the provider.
C. exceed those of the outsourcer.
D. are stated in the contract.

D is the correct answer.

Justification:
A. Ensuring the data classification requirements are compatible with the provider's own classification is an acceptable option but does not provide a requirement for the handling of classified data.
B. Ensuring the data classification requirements are communicated to the provider does not provide a requirement for appropriate handling of classified data.
C. Ensuring the data classification requirements exceed those of the outsourcer is an acceptable option but not as comprehensive or as binding as a legal contract.
D. The most effective mechanism to ensure that the organization's security standards are met by a third party would be a legal agreement stating the handling requirements for classified data and including the right to inspect and audit.

S3-171 What is the **GREATEST** risk when there are an excessive number of firewall rules?

A. One rule may override another rule in the chain and create a loophole.
B. Performance degradation of the whole network may occur.
C. The firewall may not support the increasing number of rules due to limitations.
D. The firewall may show abnormal behavior and may crash or automatically shut down.

A is the correct answer.

Justification:

A. If there are many firewall rules, there is a chance that a particular rule may allow an external connection although other associated rules are overridden. Due to the increasing number of rules, it becomes complex to test them and, over time, a loophole may occur.
B. Excessive firewall rules may impact network performance, but this is a secondary concern.
C. It is unlikely that the capacity to support rules will exceed capacity and is not a significant risk.
D. There is a slight risk that the firewall will behave erratically, but that is not the greatest risk.

S3-172 From an information security perspective, which of the following poses the **MOST** important impact concern in a homogenous network?

A. Increased uncertainty
B. Single points of failure
C. Cascading risk
D. Aggregated risk

D is the correct answer.

Justification:

A. The level of uncertainty is not directly related to the degree of homogeneity. Without proper consideration of a possible collective impact, actual consequences of compromise may not be apparent.
B. Single points of failure are always a consideration but are not related to the degree of homogeneity.
C. Cascading risk is not a function of homogeneity but of how closely systems are coupled.
D. A homogenous network of the same devices is subject to compromise from a common threat vector that, while possibly acceptable in a single device, can create an unacceptable or catastrophic impact in the aggregate (collectively).

S3-173 Which of the following roles is **MOST** responsible for ensuring that information protection policies are consistent with applicable laws and regulations?

A. Executive management
B. The quality manager
C. The board of directors
D. The auditor

C is the correct answer.

Justification:

A. Executive management is responsible for security strategy oversight and alignment and for executing all security program elements.
B. The role of the quality manager is to review security-related documents for accuracy, completeness and comprehension. Ensuring consistency with laws and regulations is not a primary responsibility.
C. The extent of policy compliance with legal and regulatory matters is ultimately a business decision made by the board of directors, who in turn will direct executive management in terms of required policy compliance.
D. The role of the auditor is to review and evaluate policies, procedures, processes, etc., but not to ensure their compliance with laws and regulations.

S3-174 What is the **BEST** way to ensure data protection upon termination of employment?

A. Retrieve identification badge and card keys.
B. Retrieve all personal computer equipment.
C. Erase all of the employee's folders.
D. Ensure all logical access is removed.

D is the correct answer.

Justification:

A. Retrieving identification badge and card keys would only reduce the capability to enter the building.
B. Retrieving the personal computer equipment is a necessary task but does not prevent access to resources.
C. Erasing the employee's folders is not reasonable because the folders may contain information important to the organization.
D. Ensuring that all logical access is removed will guarantee that the former employee will not be able to access company data and that the employee's credentials will not be misused.

S3-175 What is the **MOST** important reason for formally documenting security procedures?

A. Ensure processes are repeatable and sustainable.
B. Ensure alignment with business objectives.
C. Ensure auditability by regulatory agencies.
D. Ensure objective criteria for the application of metrics.

A is the correct answer.

Justification:

A. Without formal documentation, it would be difficult to ensure that security processes are performed correctly and consistently.
B. Alignment with business objectives is not a function of formally documenting security procedures.
C. Processes should not be formally documented merely to satisfy an audit requirement.
D. Although potentially useful in the development of metrics, creating formal documentation to assist in the creation of metrics is a secondary objective.

S3-176 Which of the following is the **BEST** approach for an organization desiring to protect its intellectual property?

A. Conduct awareness sessions on intellectual property policy.
B. Require all employees to sign a nondisclosure agreement.
C. Promptly remove all access when an employee leaves the organization.
D. Restrict access to a need-to-know basis.

D is the correct answer.

Justification:

A. Security awareness regarding intellectual property policy will not prevent violations of this policy.
B. Requiring all employees to sign a nondisclosure agreement is a good control but not as effective as restricting access to a need-to-know basis.
C. Removing all access on termination does not protect intellectual property prior to an employee leaving.
D. Restricting access to a need-to-know basis is the most effective approach to protecting intellectual property.

S3-177 The segregation of duties principle is violated if which of the following individuals has update rights to the database access control list?

A. Data owner
B. Data custodian
C. Systems programmer
D. Security administrator

C is the correct answer.

Justification:

A. The data owner would request and approve updates to the access control list (ACL), but it is not a violation of the segregation of duties principle if the data owner has update rights to the ACL.
B. The data custodian could carry out the updates on the ACL because it is part of the duties as delegated to him/her by the data owner.
C. A systems programmer should not have privileges to modify the ACL because this would give the programmer unlimited control over the system.
D. The security administrator could carry out the updates on the ACL because it is part of their duties as delegated to them by the data owner.

S3-178 An account with full administrative privileges over a production file is found to be accessible by a member of the software development team. This account was set up to allow the developer to download nonsensitive production data for software testing purposes. Assuming all options are possible, which of the following should the information security manager recommend?

A. Restrict account access to read-only.
B. Log all usage of this account.
C. Suspend the account and activate only when needed.
D. Require that a change request be submitted for each download.

A is the correct answer.

Justification:

A. Administrative accounts have permission to change data. This is not required for the developers to perform their tasks. Unauthorized change will damage the integrity of the data. Restricting the account to read-only access will ensure that file integrity can be maintained while permitting access.
B. Logging all usage of the account is a detective control and will not reduce the exposure created by this excessive level of access.
C. Suspending the account and activating only when needed will not reduce the exposure created by this excessive level of access.
D. Requiring that a change request be submitted for each download would be excessively burdensome and will not reduce the exposure created by this excessive level of access.

X **S3-179** Which of the following is the **BEST** indicator that security controls are performing effectively?

A. The monthly service level statistics indicate minimal impact from security issues.
B. The cost of implementing security controls is less than the value of the assets.
C. The percentage of systems that are compliant with security standards is satisfactory.
D. Audit reports do not reflect any significant findings on security.

A is the correct answer.

Justification:
A. The best indicator of effective security control is the evidence of acceptable disruption to business operations.
B. The cost of implementing controls is unrelated to their effectiveness.
C. The percentage of systems that are compliant with security standards is not an indicator of their effectiveness.
D. Audit reports that do not reflect any significant findings on security can support this evidence, but this is generally not sufficiently frequent to be a useful management tool and is only supplemental to monthly service level statistics.

S3-180 An organization's information security manager has been asked to hire a consultant to help assess the maturity level of the organization's information security management. What is the **MOST** important element of the request for proposal?

A. References from other organizations
B. Past experience of the engagement team
C. Sample deliverable provided for review
D. Methodology to be used in the assessment

D is the correct answer.

Justification:
A. References from other organizations are important but not as important as the methodology to be used in the assessment.
B. Past experience of the engagement team is not as important as the methodology to be used.
C. Sample deliverables only tell how the assessment is presented, not the process.
D. Methodology illustrates the process and formulates the basis to align expectations and the execution of the assessment. This also provides a picture of what is required of all parties involved in the assessment.

S3-181 Several business units reported problems with their systems after multiple security patches were deployed. What is the **FIRST** step to handle this problem?

A. Assess the problems and institute rollback procedures, if needed.
B. Disconnect the systems from the network until the problems are corrected.
C. Uninstall the patches from these systems.
D. Contact the vendor regarding the problems that occurred.

A is the correct answer.

Justification:

A. **Assessing the problems and instituting rollback procedures as needed would be the best course of action.**
B. Disconnecting the systems from the network would not identify where the problem was and may make the problem worse.
C. Uninstalling the patches would not identify where the problem was and would recreate the risk the patches were meant to address.
D. Contacting the vendor regarding the problems that occurred is part of the assessment.

S3-182 What is the **BEST** indicator of compliance when defining a service level agreement regarding the level of data confidentiality that is handled by a third-party service provider?

A. Access control matrix
B. Encryption strength
C. Authentication mechanism
D. Data repository

A is the correct answer.

Justification:

A. **The access control matrix is the best indicator of the level of compliance with the service level agreement (SLA) data confidentiality clauses.**
B. Encryption strength might be defined in the SLA but is not a confidentiality compliance indicator.
C. Authentication mechanism might be defined in the SLA but is not a confidentiality compliance indicator.
D. Data repository requirements might be defined in the SLA but is not a confidentiality compliance indicator.

S3-183 In which of the following situations is continuous monitoring the **BEST** option?

A. Where incidents may have a high impact and frequency
B. Where legislation requires strong information security controls
C. Where incidents may have a high impact but low frequency
D. Where e-commerce is a primary business driver

A is the correct answer.

Justification:

A. Continuous monitoring control initiatives are expensive, so they should be used in areas where the risk is at its greatest level. These areas are the ones with high impact and high frequency of occurrence.

B. Regulations and legislations that require tight IT security measures focus on requiring organizations to establish an IT security governance structure that manages IT security with a risk-based approach, so each organization decides which kinds of controls are implemented. Continuous monitoring is not necessarily a requirement.

C. Measures such as contingency planning or insurance are commonly used when incidents rarely happen but have a high impact each time they happen. Continuous monitoring is unlikely to be necessary.

D. Continuous control monitoring initiatives are not needed in all e-commerce environments. There are some e-commerce environments where the impact of incidents is not high enough to support the implementation of this kind of initiative.

S3-184 A third party was engaged to develop a business application. Which of the following is the **BEST** test for the existence of back doors?

A. System monitoring for traffic on network ports
B. Security code reviews for the entire application
C. Reverse engineering the application binaries
D. Running the application from a high-privileged account on a test system

B is the correct answer.

Justification:

A. System monitoring for traffic on network ports would not be able to detect all instances of back doors and is time-consuming and would take much effort.

B. Security code reviews for the entire application is the best measure and will involve reviewing the entire source code to detect all instances of back doors.

C. Reverse engineering the application binaries may not provide any definite clues.

D. Back doors will not surface by running the application on high-privileged accounts because back doors are usually hidden accounts in the applications.

S3-185 What is the **BIGGEST** concern for an information security manager reviewing firewall rules?

A. The firewall allows source routing.
B. The firewall allows broadcast propagation.
C. The firewall allows unregistered ports.
D. The firewall allows nonstandard protocols.

A is the correct answer.

Justification:
A. If the firewall allows source routing, any outsider can carry out spoofing attacks by stealing the internal (private) Internet Protocol addresses of the organization.
B. Broadcast propagation does not create a significant security exposure.
C. Unregistered ports are a poor practice but do not necessarily create a significant security exposure.
D. Nonstandard protocols can be filtered and do not necessarily create a significant security exposure.

S3-186 What is the **MOST** cost-effective means of improving security awareness of staff personnel?

A. Employee monetary incentives
B. User education and training
C. A zero-tolerance security policy
D. Reporting of security infractions

B is the correct answer.

Justification:
A. Incentives perform poorly without user education and training.
B. User education and training is the most cost-effective means of influencing staff to improve security because personnel are the weakest link in security.
C. Unless users are aware of the security requirements, a zero-tolerance security policy would not be as good as education and training.
D. Users would not have the knowledge to accurately interpret and report violations without user education and training.

S3-187 Which of the following is the **MOST** effective at preventing an unauthorized individual from following an authorized person through a secured entrance (tailgating or piggybacking)?

A. Card key door locks
B. Photo identification
C. Biometric scanners
D. Awareness training

D is the correct answer.

Justification:
A. Card key door locks are a physical control, which by itself would not be effective against tailgating.
B. Photo identification is a detective control, which by itself would not prevent tailgating.
C. Biometric scanners would not prevent tailgating.
D. Awareness training is more likely to result in any attempted tailgating being challenged by the authorized employee.

S3-188 How will data owners determine what access and authorizations users will have?

A. Delegating authority to data custodian
B. Cloning existing user accounts
C. Determining hierarchical preferences
D. Mapping to business needs

D is the correct answer.

Justification:
A. Data custodians implement the decisions made by data owners.
B. Access and authorizations are not to be assigned by cloning existing user accounts. By cloning, users may obtain more access rights and privileges than are required to do their job.
C. Access and authorizations should be based on a need-to-know basis. Hierarchical preferences may be based on individual preferences and not on business needs.
D. Access and authorizations should be based on business needs.

S3-189 Which of the following is the **MOST** likely outcome of a well-designed information security awareness course?

A. Increased reporting of security incidents to the incident response function
B. Decreased reporting of security incidents to the incident response function
C. Decrease in the number of password resets
D. Increase in the number of identified system vulnerabilities

A is the correct answer.

Justification:
A. A well-organized information security awareness course informs all employees of existing security policies, the importance of following safe practices for data security and the need to report any possible security incidents to the appropriate individuals in the organization.
B. Decreased reporting of security incidents would not be a likely outcome.
C. A decrease in the number of password resets would not be a likely outcome.
D. An increase in the number of identified system vulnerabilities would not be a likely outcome.

S3-190 Which item would be the **BEST** to include in the information security awareness training program for new general staff employees?

A. Review of various security models
B. Discussion of how to construct strong passwords
C. Review of roles that have privileged access
D. Discussion of vulnerability assessment results

B is the correct answer.

Justification:
A. A review of various security models would not be applicable to general staff employees.
B. All new employees will need to understand techniques for the construction of strong passwords.
C. A review of roles that have privileged access would not be applicable to general staff employees.
D. A discussion of vulnerability assessment results would not be applicable to general staff employees.

S3-191 What is a critical component of a continuous improvement program for information security?

A. Program metrics
B. Developing a service level agreement for security
C. Tying corporate security standards to a recognized international standard
D. Ensuring regulatory compliance

A is the correct answer.

Justification:

A. If an organization is unable to take measurements over time that provide data regarding key aspects of its security program, then continuous improvement is not likely.
B. Although desirable, developing a service level agreement for security is not a critical component for a continuous improvement program.
C. Tying corporate security standards to a recognized international standard is not a critical component for a continuous improvement program.
D. Ensuring regulatory compliance is a separate issue and is not a critical component for a continuous improvement program.

S3-192 An organization has implemented an enterprise resource planning system used by 500 employees from various departments. Which of the following access control approaches is **MOST** appropriate?

A. Rule-based
B. Mandatory
C. Discretionary
D. Role-based

D is the correct answer.

Justification:

A. Rule-based access control needs to define the individual access rules, which is troublesome and error prone in large organizations.
B. In mandatory access control, the individual's access to information resources is based on a clearance level that needs to be defined, which is troublesome in large organizations.
C. In discretionary access control, users have access to resources based on delegation of rights by someone with the proper authority, which requires a significant amount of administration and overhead.
D. Role-based access control is effective and efficient in large user communities because it controls system access by the roles defined for groups of users. Users are assigned to the various roles and the system controls the access based on those roles.

S3-193 An organization plans to contract with an outside service provider to host its corporate web site. The **MOST** important concern for the information security manager is to ensure that:

A. an audit of the service provider uncovers no significant weakness.
B. the contract includes a nondisclosure agreement to protect the organization's intellectual property.
C. the contract should mandate that the service provider will comply with security policies.
D. the third-party service provider conducts regular penetration testing.

C is the correct answer.

Justification:
A. The audit is normally a one-time effort and cannot provide ongoing assurance of the security.
B. A nondisclosure agreement should be part of the contract and would be a part of the policy compliance requirements.
C. It is critical to include the security requirements in the contract based on the company's security policy to ensure that the necessary security controls are implemented by the service provider.
D. Penetration testing alone would not provide total security to the web site; there are many controls that cannot be tested through penetration testing.

X **S3-194** Which of the following is the **MAIN** objective in contracting with an external company to perform penetration testing?

A. To mitigate technical risk
B. To have an independent certification of network security
C. To receive an independent view of security exposures
D. To identify a complete list of vulnerabilities

C is the correct answer.

Justification:
A. Mitigating technical risk is not a direct result of a penetration test.
B. A penetration test would not provide certification of network security.
C. Even though the organization may have the capability to perform penetration testing with internal resources, third-party penetration testing should be performed to gain an independent view of the security exposure.
D. A penetration test would not provide a complete list of vulnerabilities.

S3-195 An organization plans to outsource its customer relationship management to a third-party service provider. Which of the following should the organization do **FIRST**?

A. Request that the third-party provider perform background checks on their employees.
B. Perform an internal risk assessment to determine needed controls.
C. Audit the third-party provider to evaluate their security controls.
D. Perform a security assessment to detect security vulnerabilities.

B is the correct answer.

Justification:
A. A background check should be a standard requirement for the service provider.
B. An internal risk assessment should be performed to identify the risk and determine needed controls.
C. Audit objectives should be determined from the risk assessment results.
D. Security assessment does not cover the operational risk.

S3-196 Which of the following would raise security awareness among an organization's employees?

A. Distributing industry statistics about security incidents
B. Monitoring the magnitude of incidents
C. Encouraging employees to behave in a more conscious manner
D. Continually reinforcing the security policy

D is the correct answer.

Justification:
A. Distributing industry statistics about security incidents would have little bearing on the employee's behavior.
B. Monitoring the magnitude of incidents does not involve the employees.
C. Encouraging employees to behave in a more conscious manner could be an aspect of continual reinforcement of the security policy.
D. Employees must be continually made aware of the policy and expectations of their behavior.

S3-197 Which of the following is the **MOST** appropriate method of ensuring password strength in a large organization?

A. Attempt to reset several passwords to weaker values
B. Install code to capture passwords for periodic audit
C. Sample a subset of users and request their passwords for review
D. Automatic password strength determination on each platform

D is the correct answer.

Justification:
A. Attempting to reset several passwords to weaker values will not ensure adequate password strength.
B. Installing code to capture passwords for periodic audit creates an unnecessary risk.
C. Sampling a subset of users and requesting their passwords for review, would compromise the integrity of the passwords.
D. Automatic testing of password strength and enforcing proper construction is the most effective way of ensuring strong password construction.

S3-198 Which of the following is the **BEST** approach for improving information security management processes? X

A. Conduct periodic security audits.
B. Perform periodic penetration testing.
C. Define and monitor security metrics.
D. Survey business units for feedback.

C is the correct answer.

Justification:
A. Audits will identify deficiencies in established controls; however, they are not effective in evaluating the overall performance for improvement on an ongoing basis.
B. Penetration testing will only uncover technical vulnerabilities and cannot provide a holistic picture of information security management.
C. Defining and monitoring security metrics is a good approach to analyze the performance of the security management process since it determines the baseline and evaluates the performance against the baseline to identify an opportunity for improvement. This is a systematic and structured approach to process improvement.
D. Feedback is subjective and not necessarily reflective of true performance.

S3-199 What should metrics be based on when measuring and monitoring information security programs?

A. Residual risk
B. Levels of security
C. Security objectives
D. Statistics of security incidents

C is the correct answer.

Justification:
A. Metrics are not only used to measure the results of the security controls (residual risk) but also the attributes of the control implementation.
B. Levels of security are only relevant in relation to the security objectives.
C. Metrics should be developed based on security objectives, so they can measure the effectiveness and efficiency of information security controls in relation to the defined objectives.
D. Statistics of security incidents are only a general basis for determining if overall outcomes are meeting expectations, not as a basis for the achievement of individual objectives.

X **S3-200** What is the root cause of a successful cross-site request forgery attack?

A. The application uses multiple redirects for completing a data commit transaction.
B. The application has implemented cookies as the sole authentication mechanism.
C. The application has been installed with a non-legitimate license key.
D. The application is hosted on a server along with other applications.

B is the correct answer.

Justification:
A. Cross-site request forgery (XSRF) is related to an authentication mechanism, not to redirection.
B. XSRF exploits inadequate authentication mechanisms in web applications that rely only on elements such as cookies when performing a transaction. It is a type of web site attack in which unauthorized commands are transmitted from a trusted user.
C. A non-legitimate license key is related to intellectual property rights, not to XSRF vulnerability.
D. Merely hosting multiple applications on the same server is not the root cause of this vulnerability.

S3-201 An organization is entering into an agreement with a new business partner to conduct customer mailings. What is the **MOST** important action that the information security manager needs to perform?

A. A due diligence security review of the business partner's security controls
B. Ensuring that the business partner has an effective business continuity program
C. Ensuring that the third party is contractually obligated to all relevant security requirements
D. Talking to other clients of the business partner to check references for performance

C is the correct answer.

Justification:
A. A due diligence security review is contributory to the contractual agreement but not key.
B. Ensuring that the business partner has an effective business continuity program is contributory to the contractual agreement but not key.
C. The key requirement is that the information security manager ensures that the third party is contractually bound to follow the appropriate security requirements for the process being outsourced. This protects both organizations.
D. Talking to other clients of the business partner is contributory to the contractual agreement but not key.

S3-202 An organization that outsourced its payroll processing needs to perform independent assessments of the security controls of the third party, according to policy requirements. Which of the following is the **MOST** useful requirement to include in the contract?

A. Right to audit
B. Nondisclosure agreement
C. Proper firewall implementation
D. Dedicated security manager for monitoring compliance

A is the correct answer.

Justification:

A. Right to audit would be the most useful requirement because this would provide the company the ability to perform a security audit/assessment whenever there is a business need to examine whether the controls are working effectively at the third party.
B. A nondisclosure agreement is an important requirement and can be examined during the audit.
C. Proper firewall implementation would not be a specific requirement in the contract but part of general control requirements.
D. A dedicated security manager would be a costly solution and not always feasible for most situations.

S3-203 Which of the following is the **MOST** critical activity to ensure the ongoing security of outsourced IT services?

A. Provide security awareness training to the third-party provider's employees.
B. Conduct regular security reviews of the third-party provider.
C. Include security requirements in the service contract.
D. Request that the third-party provider comply with the organization's information security policy.

B is the correct answer.

Justification:

A. Depending on the type of services outsourced, security awareness training may not be relevant or necessary.
B. Regular security audits and reviews of the practices of the provider to prevent potential information security damage will help verify the security of outsourced services.
C. Security requirements should be included in the contract, but what is most important is verifying that the requirements are met by the provider.
D. It is not necessary to require the provider to fully comply with the policy if only some of the policy is related and applicable.

S3-204 A virtual desktop infrastructure enables remote access. The benefit of this approach from a security perspective is to:

A. optimize the IT resource budget by reducing physical maintenance to remote personal computers (PCs).
B. establish segregation of personal and organizational data while using a remote PC.
C. enable the execution of data wipe operations into a remote PC environment.
D. terminate the update of the approved antivirus software list for remote PCs.

B is the correct answer.

Justification:

A. Physical maintenance is reduced in a virtual desktop infrastructure (VDI) environment, but cost reduction is not the benefit of VDI from a security perspective.
B. The major benefit of introducing a VDI is to establish remote desktop hosting while keeping personal areas in a client personal computer (PC) separate. This serves as a control against unauthorized copies of business data on a user PC.
C. Remote data wiping is not possible in a VDI.
D. Termination of antivirus updates may represent a cost savings to the organization, but the presence or absence of antivirus software on a remote PC is irrelevant in a VDI context.

S3-205 Which of the following **BEST** ensures that security risk will be reevaluated when modifications in application developments are made?

A. A problem management process
B. Background screening
C. A change control process
D. Business impact analysis

C is the correct answer.

Justification:

A. Problem management is the general process intended to manage all problems, not those specifically related to security.
B. Background screening is the process to evaluate employee references when they are hired.
C. A change control process is the methodology that ensures that anything that could be impacted by a development change will be reevaluated.
D. Business impact analysis is the methodology used to evaluate impacts and the cost of losing a particular function.

S3-206 In which of the following system development life cycle phases are access control and encryption algorithms chosen?

A. Procedural design
B. Architectural design
C. System design specifications
D. Software development

C is the correct answer.

Justification:

A. The procedural design converts structural components into a procedural description of the software.
B. The architectural design is the phase that identifies the overall system design but not the specifics.
C. The system design specifications phase is when security specifications are identified.
D. Software development is too late a stage because this is the phase when the system is already being coded.

S3-207 Which of the following is generally considered a fundamental component of an information security program?

A. Role-based access control systems
B. Automated access provisioning
C. Security awareness training
D. Intrusion prevention systems

C is the correct answer.

Justification:
A. Role-based access control systems may or may not be necessary but are discretionary.
B. Automated access provisioning may or may not be necessary but are discretionary.
C. Without security awareness training, many components of the security program may not be effectively implemented.
D. Intrusion prevention systems may or may not be necessary but are discretionary.

S3-208 How would an organization know if its new information security program is accomplishing its goals? 

A. Key metrics indicate a reduction in incident impacts.
B. Senior management has approved the program and is supportive of it.
C. Employees are receptive to changes that were implemented.
D. There is an immediate reduction in reported incidents.

A is the correct answer.

Justification:
A. An effective security program will show a trend in impact reduction.
B. Senior management support may result from a performing program but is not as significant as key metrics indicating a reduction in incident impacts.
C. Receptive employees may result from a performing program but are not as significant as key metrics indicating a reduction in incident impacts.
D. An immediate reduction in reported incidents is likely to be from other causes and not a good indicator of the program achieving its goals.

S3-209 A benefit of using a full disclosure (white box) approach as compared to a blind (black box) approach to penetration testing is that:

A. it simulates the real-life situation of an external security attack.
B. human intervention is not required for this type of test.
C. less time is spent on reconnaissance and information gathering.
D. critical infrastructure information is not revealed to the tester.

C is the correct answer.

Justification:
A. Blind (black box) penetration testing is closer to real life than full disclosure (white box) testing.
B. There is no evidence to support that human intervention is not required for this type of test.
C. Data and information required for penetration are shared with the testers, thus eliminating time that would otherwise have been spent on reconnaissance and gathering of information.
D. A full disclosure (white box) methodology requires the knowledge of the subject being tested.

S3-210 An organization is **MOST** likely to include an indemnity clause in a service level agreement because an indemnity clause:

A. reduces the likelihood of an incident.
B. limits impact to the organization.
C. is a regulatory requirement.
D. ensures performance.

B is the correct answer.

Justification:
A. The indemnity clause would not reduce the likelihood of an incident.
B. An indemnity clause is a compensating control that serves to reduce impact if the provider causes financial loss.
C. An indemnity clause is generally not a regulatory requirement.
D. An indemnity clause may provide an incentive to perform but will not ensure it.

S3-211 Which of the following is the **BEST** approach to mitigate online brute force attacks on user accounts?

A. Passwords stored in encrypted form
B. User awareness
C. Strong passwords that are changed periodically
D. Implementation of lockout policies

D is the correct answer.

Justification:
A. Passwords stored in encrypted form will not defeat an online brute force attack if the password itself is easily guessed.
B. User awareness would help to inform users to use strong passwords, but this would not mitigate an online brute force attack.
C. In cases where implementation of account lockout policies is not possible, strong passwords that are changed periodically would be an appropriate choice.
D. Implementation of account lockout policies significantly inhibits brute force attacks.

X **S3-212** Which of the following measures is the **MOST** effective deterrent against disgruntled staff abusing their privileges?

A. Layered defense strategy
B. System audit log monitoring
C. Signed acceptable use policy
D. High-availability systems

C is the correct answer.

Justification:
A. A layered defense strategy would only prevent those activities that are outside of the user's privileges.
B. System audit log monitoring is after the fact and may not be effective.
C. A signed acceptable use policy is often an effective deterrent against malicious activities because of the stated potential for termination of employment and/or legal actions being taken against the individual.
D. High-availability systems do not deter staff abusing privileges.

S3-213 What is an advantage of sending messages using steganographic techniques as opposed to using encryption?

A. The existence of messages is unknown.
B. Required key sizes are smaller.
C. Traffic cannot be sniffed.
D. Reliability of the data is higher in transit.

A is the correct answer.

Justification:

A. The existence of messages is hidden in another file, such as a JPEG image, when using steganography.
B. Some implementations count on security through obscurity and others require keys, which may or may not be smaller in size.
C. Sniffing of steganographic traffic is possible.
D. The reliability of the data is not relevant.

S3-214 There is reason to believe that a recently modified web application has allowed unauthorized access. Which is the **BEST** way to identify an application back door?

A. Black box penetration test
B. Security audit
C. Source code review
D. Vulnerability scan

C is the correct answer.

Justification:

A. Application back doors are almost impossible to identify using a black box penetration test.
B. Security audits will not detect an application back door.
C. Source code review is the typically the only way to find and remove an application back door.
D. A vulnerability scan will only find "known" vulnerability patterns and will, therefore, not find a programmer's application back door.

S3-215 Simple Network Management Protocol v2 (SNMP v2) is used frequently to monitor networks. Which of the following vulnerabilities does it always introduce?

A. Remote buffer overflow
B. Cross-site scripting
C. Cleartext authentication
D. Man-in-the-middle attack

C is the correct answer.

Justification:

A. There have been some isolated cases of remote buffer overflows against Simple Network Management Protocol (SNMP) daemons, but generally that is not a problem.
B. Cross-site scripting is a web application vulnerability that is not related to SNMP.
C. One of the main problems with using SNMP v1 and v2 is the cleartext "community string" that it uses to authenticate. It is easy to sniff and reuse. Most times, the SNMP community string is shared throughout the organization's servers and routers, making this authentication problem a serious threat to security.
D. A man-in-the-middle attack against a User Datagram Protocol makes no sense since there is no active session; every request has the community string and is answered independently.

S3-216 Which of the following is the **FIRST** phase in which security should be addressed in the development cycle of a project?

A. Design
B. Implementation
C. Application security testing
D. Feasibility

D is the correct answer.

Justification:

A. Security requirements must be defined before doing design specification, although changes in design may alter these requirements later on.
B. Security requirements defined during system implementation are typically costly add-ons that are frequently ineffective.
C. Application security testing occurs after security has been implemented.
D. Information security should be considered at the earliest possible stage because it may affect feasibility of the project.

S3-217 A company has installed biometric fingerprint scanners at all entrances in response to a management requirement for better access control. Due to the large number of employees coupled with a slow system response, it takes a substantial amount of time for all workers to gain access to the building and workers are increasingly piggybacking. What is the **BEST** course of action for the information security manager to address this issue?

A. Replace the system for better response time.
B. Escalate the issue to management.
C. Revert to manual entry control procedures.
D. Increase compliance enforcement.

B is the correct answer.

Justification:

A. Upgrading the system is likely to be a costly option and is a management issue.
B. It is a business decision how management wants to deal with the problem, not directly a security issue. Conflicts of this nature are best addressed by management.
C. Given that management has set the requirement, it is unlikely that going back to a manual entry control system will be acceptable.
D. Increasing compliance efforts does not address the underlying issue. Regardless, such a choice should be made by management.

S3-218 Which of the following metrics is the **MOST** useful for the effectiveness of a controls monitoring program?

A. The percentage of key controls being monitored
B. The time between detection and initiating remediation
C. The monitoring cost versus incidents detected
D. The time between an incident and detection

D is the correct answer.

Justification:

A. While the percentage of key controls being monitored is an important metric, it is not an indication of effectiveness.
B. The time between detection and remediation is more of an indication of the effectiveness of the incident response activity.
C. The monitoring cost per incident is an indicator of efficiency rather than effectiveness.
D. The time it takes to detect an incident after it has occurred is a good indication of the effectiveness of the control monitoring effort.

S3-219 At what point should a risk assessment of a new process occur to determine appropriate controls? It should occur:

A. only at the beginning and at the end of the new process.
B. throughout the entire life cycle of the process.
C. at the appropriate point because timing of assessments will differ for processes.
D. depending upon laws and regulations.

B is the correct answer.

Justification:

A. Risk changes at various stages of the life cycle, and if the assessment occurs only at the beginning and end of the process, important issues will be missed.
B. A risk assessment should be conducted throughout the entire life cycle of a new or a changed process. This allows an understanding of how implementation of an early control will affect control needs later on in a process.
C. The timing of assessments should occur at each stage of the life cycle regardless of the process.
D. Laws and regulations are not relevant to when risk should be assessed.

S3-220 What would be the **MOST** significant security risk when using wireless local area network technology?

A. Man-in-the-middle attack
B. Spoofing of data packets
C. Rogue access point
D. Session hijacking

C is the correct answer.

Justification:

A. Man-in-the-middle attacks are can occur in any media and are not dependent on the use of a wireless local area network (WLAN) technology.
B. Spoofing of data packets is not dependent on the use of a WLAN technology.
C. A rogue access point masquerades as a legitimate access point. The risk is that legitimate users may connect through this access point and have their traffic monitored.
D. Session hijacking is not dependent on the use of a WLAN technology.

S3-221 Which of the following will **BEST** prevent external security attacks?

A. Static Internet Protocol addressing
B. Network address translation
C. Background checks for temporary employees
D. Securing and analyzing system access logs

B is the correct answer.

Justification:

A. Static Internet Protocol addressing is helpful to an attacker.
B. Network address translation is helpful by having internal addresses that are nonroutable.
C. Background checks of temporary employees are more likely to prevent an attack launched from within the enterprise.
D. Writing all computer logs to removable media does not prevent an attack.

S3-222 How should an information security manager determine the selection of controls required to meet business objectives?

A. Prioritize the use of role-based access controls.
B. Focus on key controls.
C. Restrict controls to only critical applications.
D. Focus on automated controls.

B is the correct answer.

Justification:

A. Prioritizing the use of role-based access controls could be an example of possible key controls but is only one of the typical key controls.
B. Key controls are the essential controls to reduce risk and are most effective for the protection of information assets.
C. Controls cannot be restricted to just critical applications because, in many cases, noncritical applications can provide access to critical ones.
D. Focusing on automated controls would eliminate many essential nonautomated key controls such as policies, standards, procedures and necessary physical controls.

S3-223 What is the purpose of a corrective control?

A. To reduce adverse events
B. To identify a compromise
C. To mitigate impact
D. To ensure compliance

C is the correct answer.

Justification:

A. Preventive controls, such as firewalls, reduce the occurrence of adverse events.
B. Compromise can be detected by detective controls, such as intrusion detection systems.
C. Corrective controls serve to reduce or mitigate impacts, such as providing recovery capabilities.
D. Compliance could be ensured by preventive controls, such as access controls.

S3-224 What does the following statement reflect: "All desktops are required to use Windows 7 Service Pack 1, and all servers are required to use Windows Server 2008 R2 Service Pack 1."

A. The statement is a policy.
B. The statement is a guideline.
C. The statement is a standard.
D. The statement is a procedure.

C is the correct answer.

Justification:
A. A policy would not be so detailed.
B. A guideline is not mandatory and is more like a recommendation.
C. A standard can include required hardware and software mechanisms. A standard sets the allowable boundaries for software or hardware as well as people, processes and technologies.
D. Procedures are usually detailed, step-by-step required actions.

S3-225 What is the **BEST** method for detecting and monitoring a hacker's activities without exposing information assets to unnecessary risk?

A. Firewalls
B. Bastion hosts
C. Decoy files
D. Screened subnets

C is the correct answer.

Justification:
A. Firewalls attempt to keep the hacker out.
B. Bastion hosts attempt to keep the hacker out.
C. Decoy files, often referred to as honeypots, are the best choice for diverting a hacker away from critical files and alerting security of the hacker's presence.
D. Screened subnets or demilitarized zones provide a middle ground between the trusted internal network and the external untrusted Internet but does not help detect hacker activities.

S3-226 Which of the following would be the **BEST** way to improve employee attitude toward, and commitment to, information security?

A. Implement restrictive controls.
B. Customize methods training to the audience.
C. Apply administrative penalties.
D. Initiate stronger supervision.

B is the correct answer.

Justification:

A. Implementing restrictive controls may improve compliance but is not likely to improve attitude.
B. Cultural differences will dictate the best behavior modification techniques to customize training. For example, some cultures value relationships over monetary rewards.
C. Applying administrative penalties may also increase compliance but is likely to have a negative effect on attitudes.
D. Initiating stronger supervision may improve attitudes in certain circumstances, enterprises and geographic locations, but not in others.

S3-227 When considering outsourcing services, at what point should information security become involved in the vendor management process?

A. During contract negotiation
B. Upon request for assistance from the business unit
C. When requirements are being established
D. When a security incident occurs

C is the correct answer.

Justification:

A. Waiting until later in the process can lead to vendors having to re-bid and can disrupt negotiations.
B. There may be situations where information security involvement is not required, but those situations would be established by conducting an initial risk assessment.
C. Information security should be involved in the vendor or third-party management process from the beginning of the selection process, when the business is defining what it needs. This will ensure that all bids for the service take into consideration, and reflect in bid prices, the security requirements.
D. Waiting until after the contract is signed when an incident occurs can expose the enterprise to significant security risk, with little recourse to correct, because the contract has already been executed.

S3-228 Which of the following should be performed **EXCLUSIVELY** by the information security department?

A. Monitoring unauthorized access to operating systems
B. Configuring user access to operating systems
C. Approving operating system access standards
D. Configuring the firewall to protect operating systems

C is the correct answer.

Justification:

A. Monitoring unauthorized access may or may not be performed by the information security department.
B. Configuring user access is normally performed by the IT department.
C. The approval of standards to meet the requirements of policies should be performed by the information security department. Segregation of duties will be required to ensure that operational constraints do not result in standards not being met. The implementation of the standards may be performed in conjunction with the IT department. Approving security standards is performed exclusively by the information security department.
D. Configuring firewalls is normally a function of the IT department in accordance with the standards set by the information security department.

S3-229 Which of the following is the **MOST** critical success factor of an information security program?

A. Developing information security policies and procedures
B. Senior management commitment
C. Conducting security training and awareness for all users
D. Establishing an information security management system

B is the correct answer.

Justification:

A. Developing policies and procedures is important, but without senior management commitment, implementation will be difficult.
B. Without senior management commitment, it would be difficult to implement a successful information security program.
C. Conducting training and awareness exercises is not the most critical success factor.
D. Establishing an information security management system is essential, but without management support and commitment, it is unlikely to be successful.

S3-230 How does the development of an information security program begin?

A. Risk is assessed and analyzed.
B. The security architecture is developed.
C. The controls statement of applicability is completed.
D. Required outcomes are defined.

D is the correct answer.

Justification:

A. Assessing and analyzing risk is required to develop a strategy and will provide some of the information needed to develop the strategy that will achieve the desired outcomes, but it will not define the scope and charter of the security program.
B. A security architecture is a part of implementation subsequent to developing the strategy.
C. The applicability statement is a part of strategy implementation using International Organization for Standardization (ISO) 27001 or 27002 subsequent to determining the scope and responsibilities of the program.
D. After management has determined the desired outcomes of the information security program, development of a strategy can begin as well as initiating the process of developing information security governance structures, achieving organizational adoption and developing an implementation strategy, which will define the scope and responsibilities of the security program.

S3-231 Which of the following constitutes the **MAIN** project activities undertaken in developing an information security program?

A. Controls design and deployment
B. Security organization development
C. Logical and conceptual architecture design
D. Development of risk management objectives

A is the correct answer.

Justification:

A. The majority of program development activities will involve designing, testing and deploying controls that achieve the risk management objectives.
B. The security organization should be fairly well developed prior to attempting to implement a security program.
C. Conceptual and logical architecture designs should have been completed as a part of strategy and road map development.
D. Risk management objectives are part of strategy development.

S3-232 A financial institution plans to allocate information security resources to each of its business divisions. What areas should security activities focus on?

A. Areas where strict regulatory requirements apply
B. Areas that require the shortest recovery time objective
C. Areas that can maximize return on security investment
D. Areas where threat likelihood and impact are greatest

D is the correct answer.

Justification:

A. While regulatory requirements may be a major consideration, there may be other areas of greater threat and impact to the enterprise.
B. Watching the recovery time objective (RTO) requirement is very important from a business continuity perspective, but this only illustrates a part of the information security framework. Regulatory compliance may also touch upon RTO initiatives.
C. It is difficult to set up a single formula so that the most profitable business line always has the most critical information security initiatives in the enterprise.
D. Security activities should focus on the areas where threat, likelihood and impact are the greatest.

S3-233 Which of the following should be responsible for final approval of security patch implementation?

A. The application development manager
B. The business asset owner
C. The information security officer
D. The business continuity coordinator

B is the correct answer.

Justification:

A. When business logic has been modified, the application development team may be involved in testing; however, the team's involvement will be less necessary when a security patch is being released.
B. In order to ensure that no serious business interruption takes place due to any unexpected problems, it is important to bring business asset owners into the final sign-off loop when a security patch is being released.
C. The information security officer is informed of security patches currently being released; however, the information security officer's approval may not always be required for patch release.
D. A business continuity coordinator would not be involved in approving security patches in normal day-to-day operations.

X S3-234 Which is the **FIRST** thing that should be determined by the information security manager when developing an information security program?

A. The control objectives
B. The strategic aims
C. The desired outcomes
D. The logical architecture

C is the correct answer.

Justification:

A. Control objectives cannot be determined until desired outcomes have been determined and subsequent specific objectives defined.
B. Without determining the desired outcomes of the security program, the strategic aims that would lead to the desired outcomes cannot be determined.
C. Without determining the desired outcomes of the security program, it will be difficult or impossible to determine a viable strategy, control objectives and logical architecture.
D. Architecture is the physical manifestation of policy which is developed subsequent to and in support of strategy.

X S3-235 Which of the following is the **BEST** way to mitigate the risk of the database administrator reading sensitive data from the database?

A. Log all access to sensitive data.
B. Employ application-level encryption.
C. Install a database monitoring solution.
D. Develop a data security policy.

B is the correct answer.

Justification:

A. Access logging can be easily turned off by the database administrator.
B. Data encrypted at the application level that is stored in a database cannot be viewed in cleartext by the database administrator.
C. A database monitoring solution can be bypassed by the database administrator.
D. A security policy will only be effective if the database administrator chooses to adhere to the policy.

S3-236 In a large enterprise, what makes an information security awareness program **MOST** effective?

A. The program is developed by a professional training company.
B. The program is embedded into the orientation process.
C. The program is customized to the audience using the appropriate delivery channel.
D. The program is required by the information security policy.

C is the correct answer.

Justification:

A. It does not have to be developed by a professional training company to make it effective.
B. The awareness program should be embedded into the orientation process for new employees, but that does not necessarily indicate efficacy.
C. An awareness program should be customized for different types of audiences (e.g., for new employees, system administration, sales and delivery channels such as posters or e-learning).
D. Being required by policy does not make the program more effective.

S3-237 What is the **PRIMARY** basis for the prioritization of security spending and budgeting?

A. The identified levels of risk
B. Industry trends
C. An increased cost of services
D. The allocated revenue of the enterprise

A is the correct answer.

Justification:

A. The first required action is to conduct a risk assessment of the enterprise's key processes to identify control gaps and determine where investments should be made to mitigate risk and to determine order of prioritization. This must be conducted with consideration of enterprise goals and strategy.

B. Prioritization should not be based on the trends at other organizations because each organization has unique requirements and business objectives.

C. Prioritization by cost alone is not aligned with a risk-based approach.

D. Although the revenue may increase, it is not wise to link the IT budget to a fixed percentage of revenue because this could lead to spending more or less than is necessary to effectively address risk.

S3-238 Which of the following will the data backup policy contain?

A. Criteria for data backup
B. Personnel responsible for backup
C. A data backup schedule
D. A list of systems to be backed up

A is the correct answer.

Justification:

A. A policy is a high-level statement of management intent and will essentially contain the criteria to be followed for backing up any data such as critical data, confidential data and project data, and the frequency of backup.

B. Personnel responsible for backup are a procedural detail and will not be included in the data backup policy.

C. A data backup schedule is a procedural detail and will not be included in the data backup policy.

D. A list of systems to be backed up is a procedural detail and will not be included in the data backup policy.

S3-239 Which of the following roles is **MOST** appropriately responsible for ensuring that security awareness and training material is effectively deployed to reach the intended audience?

A. The human resources department
B. The business manager
C. The subject matter experts
D. The information security department

D is the correct answer.

Justification:

A. The human resources department may assist in disseminating security awareness material but the primary responsibility rests with the information security department.

B. The business manager may also assist in information dissemination but is not primarily responsible.

C. Subject matter experts are not normally involved with security awareness activities.

D. The information security department oversees the information security program. This includes ensuring that training reaches the intended audience.

X **S3-240** Which of the following should be done **FIRST** when making a decision to allow access to the information processing facility of an enterprise to a new external party?

A. A contract language review
B. A risk assessment
C. The exposure factor
D. Vendor due diligence

B is the correct answer.

Justification:

A. A contract language review is part of the risk assessment.
B. A risk assessment identifies the risk involved in allowing access to an external party and the required controls.
C. The exposure factor is part of the risk assessment.
D. Vendor due diligence is part of the risk assessment.

S3-241 An enterprise has a network of suppliers that it allows to remotely access an important database that contains critical supply chain data. What is the **BEST** control to ensure that the individual supplier representatives who have access to the system do not improperly access or modify information within this system?

A. User access rights
B. Biometric access controls
C. Password authentication
D. Two-factor authentication

A is the correct answer.

Justification:

A. User access rights limit the access and rights that users have to a network, file system or database once they have been authenticated.
B. Biometric access controls is a method of user access control that manages user access to an overall system, not generally to a specific set of files or records.
C. Password authentication controls access but not rights once the system is accessed.
D. Two-factor authentication controls access but not rights once the system is accessed.

S3-242 Which of the following aspects is **MOST** important to include in the service level agreement to promote resolution of operational issues with a cloud computing vendor?

A. The court of jurisdiction
B. A process description
C. Audit requirements
D. Defined responsibilities

D is the correct answer.

Justification:

A. The court of jurisdiction may be defined in the agreement, and in fact may be a benefit or a detriment to a satisfactory solution of operational issues, but seeking court remedies is generally costly and time-consuming and is not the best way to resolve operational issues with a vendor.
B. A process description has a minimal impact on issue resolution.
C. Audits may help identify and determine the nature of issues but by themselves will not help resolve the issues.
D. When issues arise with cloud vendors, it is most important to identify responsibility ownership. This will promptly determine the next action to be taken for follow-up.

S3-243 When securing wireless access points, which of the following controls would **BEST** assure confidentiality?

A. Implementing wireless intrusion prevention systems
B. Not broadcasting the service set identifier
C. Implementing wired equivalent privacy authentication
D. Enforcing a virtual private network over wireless

D is the correct answer.

Justification:
A. A wireless intrusion prevention system is a detective system and would not prevent wireless sniffing.
B. Not broadcasting the service set identifier does not reduce the risk of wireless packets being captured.
C. Wired equivalent privacy authentication is known to be weak and does not protect individual confidentiality.
D. Enforcing a virtual private network over wireless is the best option to enforce strong authentication and encryption of the sessions.

S3-244 Which of the following terms and conditions represent a significant deficiency if included in a commercial hot site contract?

A. The facility will be shared in multiple disaster declarations.
B. All equipment is provided "at time of disaster, not on floor."
C. The facility is subject to a "first-come, first-served" policy.
D. Equipment may be substituted with equivalent model.

B is the correct answer.

Justification:
A. Many commercial providers require sharing facilities in cases where there are multiple simultaneous declarations.
B. Equipment provided "at time of disaster, not on floor" means that the equipment is not available but will be acquired by the commercial hot site provider on a best effort basis. This does not meet the requirements of a hot site.
C. Many commercial providers require sharing facilities in cases where there are multiple simultaneous declarations, and that priority may be established on a first-come, first-served basis.
D. It is common for the provider to substitute equivalent or better equipment, as they are frequently upgrading and changing equipment.

S3-245 Which of the following would **PRIMARILY** provide the potential for users to bypass a form-based authentication mechanism in an application with a back-end database?

A. A weak password of six characters
B. A structured query language (SQL) injection
C. A session time-out of long duration
D. Lack of an account lockout after multiple wrong attempts

B is the correct answer.

Justification:
A. Weak passwords can make it easy to access the application, but there is no bypass of authentication.
B. Although structured query language injection is well understood and preventable, it still is a significant security risk for many enterprises writing code. Using SQL injection, one can pass SQL statements in a manner that bypasses the logon page and allows access to the application.
C. Long time-out duration is not relevant to the authentication mechanism.
D. Because the authentication mechanism is bypassed, account lockout is not initiated.

S3-246 From a security perspective, which of the following is the **MOST** important step when an employee is transferred to a different function?

A. Reviewing and modifying access rights
B. Assigning new security responsibilities
C. Conducting specific training for the new role
D. Knowledge of security weaknesses in the previous department

A is the correct answer.

Justification:
A. When an employee is transferred from one function to another, it is very important to review and update the logical access rights to ensure that any access no longer needed is removed and appropriate access for the new position is granted.
B. Assigning new security responsibilities may or may not be required.
C. Training for a new role has no direct security implications.
D. Having knowledge of security weaknesses in the previous department is not relevant.

S3-247 Which of the following is the **BEST** way to erase confidential information stored on magnetic tapes?

A. Performing a low-level format
B. Rewriting with zeros
C. Burning them
D. Degaussing them

D is the correct answer.

Justification:
A. Performing a low-level format may be adequate but is a slow process, and with the right tools, data can still be recovered.
B. Rewriting with zeroes will not overwrite information located in the disk slack space.
C. Burning destroys the tapes and does not allow their reuse.
D. Degaussing the magnetic tapes would quickly dispose of all information because the magnetic domains are thoroughly scrambled and would not allow reuse.

S3-248 For which of the following purposes would ethical hacking **MOST** likely be used? As a:

A. process resiliency test at an alternate site.
B. substitute for substantive testing.
C. control assessment of legacy applications.
D. final check in a cyberattack recovery process.

C is the correct answer.

Justification:
A. It is not common to conduct ethical hacking as part of disaster recovery testing at an alternate site.
B. Substantive testing involves obtaining audit evidence on the completeness, accuracy or existence of activities or transactions during the audit period. Ethical hacking would not be used as a substitute for substantive testing.
C. The problem with legacy applications is that there is typically not enough documentation to study their functionalities, including security controls. To assess control effectiveness, ethical hacking could be an efficient way to find out weaknesses rather than reviewing program code.
D. It is not necessarily a recommended practice to engage in ethical hacking in the last phase of a system recovery process after a cyberattack.

S3-249 A contract has just been signed with a new vendor to manage IT support services. Which of the following tasks should the information security manager ensure is performed **NEXT**?

A. Establish vendor monitoring.
B. Define reporting relationships.
C. Create a service level agreement.
D. Have the vendor sign a nondisclosure agreement.

A is the correct answer.

Justification:

A. **Once the contract is signed, the security manager should ensure that continuous vendor monitoring is established and operational. This control will help identify and provide alerts on security events and minimize potential losses.**
B. The reporting relationships will have been defined prior to the contract being signed.
C. The service level agreement will be part of the contract.
D. Nondisclosure agreements will have been signed prior to entering in contract discussions.

S3-250 Which of the following will be **MOST** important in calculating accurate return on investment in information security?

A. Excluding qualitative risk for accuracy in calculated figures
B. Establishing processes to ensure cost reductions
C. Measuring monetary values in a consistent manner
D. Treating security investment as a profit center

C is the correct answer.

Justification:

A. If something is an important risk factor, an attempt should be made to quantify it even though it may not be highly accurate.
B. Establishing processes to ensure cost reductions is not relevant to calculating return on investment (ROI).
C. **There must be consistency in metrics in order to have reasonably accurate and consistent results. In assessing security risk, it is not a good idea to simply exclude qualitative risk because of the difficulties in measurement.**
D. Whether or not security investment is treated as a profit center does not affect ROI calculations.

S3-251 Which of the following roles performs the day-to-day duties required to ensure the protection and integrity of data?

A. Data owners
B. Data users
C. Steering committees
D. Data custodians

D is the correct answer.

Justification:

A. Data owners decide the level of classification based upon business needs for the protection of data and periodically review the classification assignments and make changes as necessary. Information owners may be executives or managers responsible for the protection of data and may be held liable for negligence if there is a failure to protect the data.
B. Data users follow procedures set out in the enterprise's security policy and are expected to adhere to privacy and security regulations that are often specific to sensitive application fields (e.g., health care, finance, legal).
C. Steering committees serve as an effective communication channel for management's aims and directions and provide an ongoing basis for ensuring alignment of the security program with enterprise objectives. Steering committees are also instrumental in achieving behavior change toward a culture that promotes good security practices and policy compliance.
D. Data custodians oversee the day-to-day duties required to ensure the protection and integrity of data. A custodian, such as IT systems personnel, may be responsible for performing regular backups, testing the validity of backups and maintaining records in accordance with classification policies. In addition, a custodian may be the administrator for the enterprise's information classification scheme.

S3-252 When outsourcing to an offshore provider, the **MOST** difficult element to determine during a security review will be:

A. technical competency.
B. incompatible culture.
C. defense in depth.
D. adequate policies.

B is the correct answer.

Justification:

A. Technical competency is a usual area for review to ensure that the offshore provider meets acceptable standards.
B. Individuals in different cultures often have different perspectives on what information is considered sensitive or confidential and how the information should be handled. Those perspectives may not be consistent with the enterprise's requirements. Cultural norms are not usually an area of consideration in a security review or during an onsite inspection.
C. Defense in depth is a usual area for review to ensure that the offshore provider meets acceptable standards.
D. Policies are a usual area for review to ensure that the offshore provider meets acceptable standards.

S3-253 Addressing production risk is **PRIMARILY** a function of:

A. release management.
B. incident management.
C. change management.
D. configuration management.

C is the correct answer.

Justification:
A. Release management is the specific process to manage risk of production system deployment.
B. Incident management is not directly relevant to life cycle stages.
C. Change management is the overall process to assess and control risk introduced by changes.
D. Configuration management is the specific process to manage risk associated with system configuration.

S3-254 Which of the following is the **BEST** approach to dealing with inadequate funding of the security program?

A. Eliminate low-priority security services.
B. Require management to accept the increased risk.
C. Prioritize risk mitigation and educate management.
D. Reduce monitoring and compliance enforcement activities.

C is the correct answer.

Justification:
A. Prioritizing security activities is always useful, but eliminating even low-priority security services should be a last resort.
B. If budgets are seriously constrained, management is already addressing increases in other risk and is likely to be aware of the issue; a proactive approach to doing more with less will be well received.
C. Allocating resources to the areas of highest risk and benefit and educating management on the potential consequences of underfunding is the best approach.
D. Reducing monitoring activities may unnecessarily increase risk when lower-cost options to perform those functions may be available.

S3-255 During an audit, an information security manager discovered that sales representatives are sending sensitive customer information through email messages. Which of the following is the **BEST** course of action to address the issue?

A. Review the finding with the sales manager to evaluate the risk and impact.
B. Report the issue to senior management immediately.
C. Request that the sales representatives stop emailing sensitive information.
D. Provide security awareness training to the sales representatives.

A is the correct answer.

Justification:
A. It is always good practice to engage the management of the business unit when addressing security threats and risk. The input from business unit management is critical in formulating the next step.
B. The issue should not be escalated until gaining an understanding of the risk and business issues from the business unit manager.
C. Requesting the representatives stop sending sensitive information can be a temporary remediation but does not solve the underlying problem.
D. Awareness training may help but does not resolve the problem.

S3-256 Which of the following is **BEST** used to define minimum requirements for database security settings?

A. Procedures
B. Guidelines
C. Baselines
D. Policies

C is the correct answer.

Justification:
A. Procedures determine the steps, not the configuration requirements.
B. Guidelines are not enforceable.
C. Baselines set the minimum security controls required for safeguarding an IT system based on its identified needs for confidentiality, integrity and/or availability protection.
D. Policies determine direction, but not detailed configurations.

X **S3-257** Who would be the **PRIMARY** user of metrics regarding the number of email messages quarantined due to virus infection versus the number of infected email messages that were not caught?

A. The security steering committee
B. The board of directors
C. IT managers
D. The information security manager

D is the correct answer.

Justification:
A. Metrics support decisions. Knowing the number of email messages blocked due to viruses would not on its own be an actionable piece of information for the steering committee.
B. The board of directors would have no use for the information.
C. IT managers would be interested, but it would not be in their purview to address the issue.
D. Information regarding the effectiveness of the current email antivirus control is most useful to the information security manager and staff because they can use the information to initiate an investigation to determine why the control is not performing as expected and to determine whether there are other factors contributing to the failure of the control. When these determinations are made, the information security manager can use these metrics, along with data collected during the investigation, to support decisions to alter processes or add to (or change) the controls in place.

S3-258 To ensure that all employees follow procedures regarding the integrity and confidentiality of personal identifiable information, a hospital required that policies and procedures be put in place for data access and that all data stored should be encrypted. This is an example of what type of controls?

A. Administrative and technical controls
B. Administrative and deterrent controls
C. Technical and physical controls
D. Administrative and corrective controls

A is the correct answer.

Justification:
A. Administrative and technical controls are the only correct answers.
B. Encryption is not a deterrent control.
C. Encryption is not a physical control.
D. Encryption is not a corrective control.

S3-259 From an information security manager's perspective, which of the following is the **MOST** important element of a third-party contract to outsource a sensitive business process?

A. Security service level agreements
B. Background checks for key personnel
C. Specific system requirements
D. A right to audit

D is the correct answer.

Justification:

A. While service level agreements are an important consideration, without the ability to audit the provider, it is very difficult to validate compliance with the contract.
B. Background checks would not be the security manager's job.
C. Specific system requirements are more of an operational issue.
D. A right to audit is essential to ensure contract compliance.

S3-260 An information security manager has implemented procedures for monitoring specific activities on the network. The system administrator has been trained to analyze the network events, take appropriate action and provide reports to the information security manager. What additional monitoring should be implemented to give a more accurate, risk-based view of network activity?

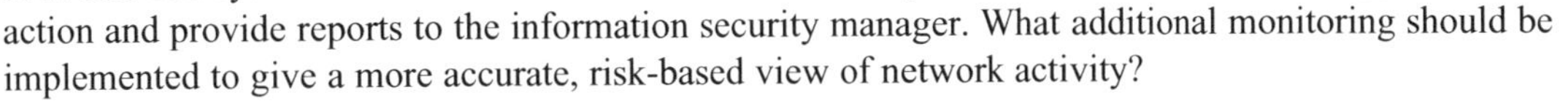

A. The system administrator should be monitored by a separate reviewer.
B. All activity on the network should be monitored.
C. No additional monitoring is needed in this situation.
D. Monitoring should be done only by the information security manager.

A is the correct answer.

Justification:

A. The system administrator needs to be monitored to ensure that the administrator is in compliance with the information security program. Normally, an administrator will have more rights on the network than an end user, and while an administrator can monitor others, administrators must be monitored as well. The primary objective is to ensure that risk is managed appropriately, balancing operational efficiency against adequate safety.
B. To simply monitor all network activity would be excessive and is not a risk-based approach to protecting the enterprise.
C. Additional monitoring is needed in this situation. The system administrator needs to be monitored for the specific activities.
D. The information security manager needs to use the resources available within the enterprise to assist in monitoring compliance. Using expertise for monitoring is an efficient method and should be used when possible.

S3-261 Which of the following defines requirements for securing the technical infrastructure?

A. Information security strategy
B. Information security guidelines
C. Information security model
D. Information security architecture

D is the correct answer.

Justification:
A. A strategy is a broad, high-level document and not a standard.
B. A guideline is advisory in nature.
C. A security model shows the relationships between components.
D. Minimum standards for securing the technical infrastructure should be defined in a security architecture document. This document defines how components are secured and the security services that should be in place.

S3-262 It is **MOST** important that information security architecture be based on which of the following?

A. Industry good practices
B. Information technology plans
C. Information security good practices
D. Organizational policies and standards

D is the correct answer.

Justification:
A. Good practices suppose that what is best for one is best for all. For some organizations it will be overkill and in others, insufficient.
B. Information technology plans should be based on the architecture.
C. Information security good practices suppose that what is best for one is best for all. For some organizations it will be overkill and in others, insufficient.
D. Information security architecture is a manifestation of policy and must implement the technical standards.

S3-263 Which of the following **BEST** defines the relationships among security technologies?

A. Security metrics
B. Network topology
C. Security architecture
D. Process improvement models

C is the correct answer.

Justification:
A. Security metrics measure functions within the security practice but do not explain the use and relationships of security technologies.
B. Network topology diagrams do not describe the use and relationships of these technologies.
C. Security architecture explains the use and relationships of security mechanisms.
D. Process improvement models deal with processes, not the relationship between technologies.

S3-264 Which of the following are likely to be updated **MOST** frequently?

A. Procedures for hardening database servers
B. Standards for password length and complexity
C. Policies addressing information security governance
D. Standards for document retention and destruction

A is the correct answer.

Justification:
A. **Procedures, especially with regard to the hardening of operating systems, will be subject to constant change; as operating systems change and evolve, the procedures for hardening will have to keep pace.**
B. Standards should generally be more static and less subject to frequent change.
C. Well-conceived, mature policies will rarely require change.
D. Standards regarding document retention and destruction will rarely need to be changed.

S3-265 Which of the following is the **MOST** important information to include in an information security standard?

A. Creation date
B. Author name
C. Initial draft approval date
D. Last review date

D is the correct answer.

Justification:
A. The creation date is not that important.
B. The name of the author is not that important.
C. The initial draft date is not that important.
D. **The last review date confirms the currency of the standard, affirming that management has reviewed the standard to assure that nothing in the environment has changed that would necessitate an update to the standard.**

S3-266 What is the **BEST** approach to manage regulatory and legal compliance in a global organization operating in multiple governmental jurisdictions with differing requirements?

A. Bring all locations into conformity with the aggregate requirements of all governmental jurisdictions.
B. Establish baseline standards for all locations and add supplemental standards as required.
C. Bring all locations into conformity with a generally accepted set of industry good practices.
D. Establish a baseline standard incorporating those requirements that all jurisdictions have in common.

B is the correct answer.

Justification:
A. Forcing all locations to be in compliance with all the regulations places an undue burden on those locations and may result in contradictory requirements.
B. **It is more efficient to establish a baseline standard and then develop additional standards for locations that must meet specific requirements.**
C. Using industry good practices may cause certain locations to fail regulatory compliance.
D. Seeking a lowest common denominator may cause certain locations to fail regulatory compliance.

S3-267 Which of the following should be reviewed to ensure that security controls are effective?

A. Risk assessment policies
B. Return on security investment
C. Security metrics
D. User access rights

C is the correct answer.

Justification:

A. Reviewing risk assessment policies would not ensure that the controls are actually working.
B. Reviewing return on security investment provides business justifications in implementing controls but does not measure effectiveness of the control itself.
C. Reviewing security metrics provides senior management a snapshot view and trends of an organization's security posture.
D. Reviewing user access rights is a joint responsibility of the data custodian and the data owner and does not measure control effectiveness.

S3-268 Which of the following should be included in a good privacy statement?

A. A notification of liability on accuracy of information
B. A notification that information will be encrypted
C. A statement of what the company will do with information it collects
D. A description of the information classification process

C is the correct answer.

Justification:

A. A notification of liability on accuracy of information should be located in the web site's disclaimer.
B. Although encryption may be applied, this is not generally disclosed.
C. Most privacy laws and regulations require disclosure on how information will be used.
D. Information classification is unrelated to privacy statements and would be contained in a separate policy.

S3-269 Which of the following would be **MOST** effective in successfully implementing restrictive password policies?

A. Regular password audits
B. Single sign-on system
C. Security awareness program
D. Penalties for noncompliance

C is the correct answer.

Justification:

A. Regular password audits would not be as effective on their own; people would go around them unless forced by the system.
B. Single sign-on is a technology solution that would enforce password complexity but would not help the implementation of the policy.
C. To be successful in implementing restrictive password policies, it is necessary to obtain the buy-in of the end users by promoting the need for the change. The best way to accomplish this is through a security awareness program.
D. Penalties for noncompliance are likely to cause resentment and will not be helpful in a successful implementation.

S3-270 When an organization is setting up a relationship with a third-party IT service provider, which of the following is one of the **MOST** important topics to include in the contract from a security standpoint?

A. Compliance with international security standards
B. Use of a two-factor authentication system
C. Existence of an alternate hot site in case of business disruption
D. Compliance with the organization's information security requirements

D is the correct answer.

Justification:
A. Requiring compliance only with this security standard does not guarantee that a service provider complies with the organization's security requirements.
B. The requirement to use a specific kind of control methodology is not usually stated in the contract with third-party service providers.
C. The requirement for a hot site is not usually stated in the contract with third-party service providers.
D. From a security standpoint, compliance with the organization's information security requirements is one of the most important topics that should be included in the contract with third-party service provider.

S3-271 When developing an information security program, what is the **MOST** useful source of information for determining available human resources?

A. Proficiency test
B. Job descriptions
C. Organization chart
D. Skills inventory

D is the correct answer.

Justification:
A. Proficiency testing is useful but only with regard to specific technical skills.
B. Job descriptions would not be as useful because they may be out of date or not sufficiently detailed.
C. An organization chart would not provide the details necessary to determine the resources required for this activity.
D. A skills inventory would help identify the available human resources, any gaps and the training requirements for developing resources.

S3-272 Which of the following is an advantage of a centralized information security organizational structure?

A. It is easier to promote security awareness.
B. It is easier to manage and control.
C. It is more responsive to business unit needs.
D. It provides a faster turnaround for security requests.

B is the correct answer.

Justification:

A. Decentralization allows the of use field security personnel as security missionaries or ambassadors to spread the security awareness message.
B. It is easier to manage and control a centralized structure. Promoting security awareness is an advantage of decentralization.
C. Decentralized operations allow security administrators to be more responsive.
D. Being close to the business allows decentralized security administrators to achieve a faster turnaround than that achieved in a centralized operation.

S3-273 The organization has decided to outsource the majority of the IT department with a vendor that is hosting servers in a foreign country. Of the following, which is the **MOST** critical security consideration?

A. Laws and regulations of the country of origin may not be enforceable in the foreign country.
B. A security breach notification might get delayed due to the time difference.
C. Additional network intrusion detection sensors should be installed, resulting in an additional cost.
D. The company could lose physical control over the server and be unable to monitor the physical security posture of the servers.

A is the correct answer.

Justification:

A. A company is held to the local laws and regulations of the country in which the company resides, even if the company decides to place servers with a vendor that hosts the servers in a foreign country. A potential violation of local laws applicable to the company might not be recognized or rectified (i.e., prosecuted) due to the lack of knowledge of the local laws that are applicable and the inability to enforce the laws.
B. Time difference does not play a role in a 24/7 environment. Pagers, cellular phones, telephones, etc., are usually available to communicate notifications.
C. Installation of additional network intrusion detection sensors is a manageable problem that requires additional funding, but it can be addressed.
D. Most hosting providers have standardized the level of physical security that is in place. Regular physical audits can address such concerns.

S3-274 What is the **PRIMARY** purpose of installing an intrusion detection system?

A. To identify weaknesses in network security
B. To identify patterns of suspicious access
C. To identify how an attack was launched on the network
D. To identify potential attacks on the internal network

D is the correct answer.

Justification:
A. An intrusion detection system is not designed to identify weaknesses in network security.
B. An intrusion detection system is not designed to identify patterns of suspicious logon attempts.
C. Identifying how an attack was launched is secondary.
D. The most important function of an intrusion detection system is to identify potential attacks on the network.

S3-275 What is the **GREATEST** benefit of decentralized security management?

A. Reduction of the total cost of ownership
B. Improved compliance with organizational policies and standards
C. Better alignment of security to business needs
D. Easier administration

C is the correct answer.

Justification:
A. Reduction of the total cost of ownership is a benefit of centralized security management.
B. Improved compliance is a benefit of centralized security management.
C. Better alignment of security to business needs is the only answer that fits because the other choices are benefits of centralized security management.
D. Easier administration is a benefit of centralized security management.

S3-276 Which of the following is one of the **BEST** metrics an information security manager can employ to effectively evaluate the results of a security program?

A. Number of controls implemented
B. Percent of control objectives accomplished
C. Percent of compliance with the security policy
D. Reduction in the number of reported security incidents

B is the correct answer.

Justification:
A. Number of controls implemented does not have a direct relationship with the results of a security program.
B. Control objectives are directly related to business objectives; therefore, they would be the best metrics.
C. Percent in compliance with the security policy is a useful metric but says nothing about achieving control objectives.
D. A reduction in the number of security incidents has no direct bearing on whether control objectives are being achieved.

S3-277 After obtaining commitment from senior management, which of the following should be completed **NEXT** when establishing an information security program?

A. Define security metrics.
B. Conduct a risk assessment.
C. Perform a gap analysis.
D. Procure security tools.

B is the correct answer.

Justification:

A. Defining security metrics is a subsequent consideration after control objectives are determined and a strategy is developed.
B. When establishing an information security program, conducting a risk assessment is key to identifying the needs of the organization and developing a security strategy.
C. A gap analysis would be used after the desired state of security and the current state are determined to assess what needs to happen to fill the gap.
D. Procuring security tools is a subsequent consideration.

S3-278 Obtaining another party's public key is required to initiate which of the following activities?

A. Authorization
B. Digital signing
C. Authentication
D. Nonrepudiation

C is the correct answer.

Justification:

A. Authorization is not a public key infrastructure function.
B. A private key is used for signing.
C. The counterparty's public key is used for authentication.
D. The private key is used for nonrepudiation.

S3-279 Which one of the following combinations offers the **STRONGEST** encryption and authentication method for 802.11 wireless networks?

A. Wired equivalent privacy with 128-bit preshared key authentication
B. Temporal Key Integrity Protocol-Message Integrity Check with the RC4 cipher
C. Wi-Fi Protected Access 2 (WPA2) and preshared key authentication
D. WPA2 and 802.1x authentication

D is the correct answer.

Justification:

A. Wired Equivalent Privacy (WEP) with 128-bit preshared key authentication can be easily cracked with open source tools. WEP is easily compromised and is no longer recommended for secure wireless networks.
B. Temporal Key Integrity Protocol-Message Integrity Check (TKIP-MIC) with the RC4 cipher is not as strong as WPA2 with 802.1x authentication.
C. Wi-Fi Protected Access 2 (WPA2) with preshared keys uses the strongest level of encryption, but the authentication is more easily compromised.
D. WPA2 and 802.1x authentication is the strongest form of wireless authentication currently available. WPA2 combined with 802.1x forces the user to authenticate using strong Advanced Encryption Standard encryption.

S3-280 Which one of the following types of detection is **NECESSARY** to mitigate a denial or distributed denial-of-service attack?

A. Signature-based detection
B. Deep packet inspection
C. Virus detection
D. Anomaly-based detection

D is the correct answer.

Justification:

A. Signature-based detection cannot react to a distributed denial-of-service (DDoS) attack because it does not have any insight into increases in traffic levels.
B. Deep packet inspection allows a protocol to be inspected and is not related to denial-of-service (DoS) attacks.
C. Virus detection would have no effect on DDoS detection or mitigation.
D. Anomaly-based detection establishes normal traffic patterns and then detects any deviation from that baseline. Traffic baselines are greatly exceeded when under a DDoS attack and are quickly identified by anomaly-based detection.

S3-281 The **MOST** effective technical approach to mitigate the risk of confidential information being disclosed in outgoing email attachments is to implement:

A. content filtering.
B. data classification.
C. information security awareness.
D. encryption for all attachments.

A is the correct answer.

Justification:

A. Content filtering provides the ability to examine the content of attachments and prevent information containing certain words or phrases, or of certain identifiable classifications, from being sent out of the enterprise.
B. Data classification helps identify the material that should not be transmitted via email attachments but by itself will not prevent it.
C. Information security awareness training also helps limit confidential material from being disclosed via email as long as personnel are aware of what information should not be exposed and willingly comply with the requirements, but it is not as effective as outgoing content filtering.
D. Encrypting all attachments is not effective because it does not limit the content and may actually obscure confidential information contained in the email.

S3-282 An organization is planning to deliver subscription-based educational services to customers online that will require customers to log in with their user IDs and passwords. Which of the following is the **BEST** method to validate passwords entered by a customer before access to educational resources is granted?

A. Encryption
B. Content filtering
C. Database hardening
D. Hashing

D is the correct answer.

Justification:

A. Encryption is the application of an algorithm that converts the plaintext password to the encrypted form, but using encrypted passwords requires that they be decrypted for authentication—this would expose the actual password. Also, the authentication mechanism would need to have access to the encryption key in order to decrypt the password for authentication. This would allow anyone with the appropriate access to the server to decrypt user passwords, which is not typically acceptable and is not a secure practice.
B. Content filtering is not a component of password validation.
C. Database hardening helps in enhancing the security of a database but does not assist with password validation.
D. Hashing refers to a one-way algorithm that always creates the same output if applied to the same input. When hashing passwords, only the password's hash value (output) is stored, not the actual password (input). When a user logs in and enters the password, the hash is applied to the password by the authentication mechanism and compared to the stored hash. If the hash matches, then access is granted. The actual password cannot be derived from the hash (because it is a one-way algorithm), so there is no chance of the password being compromised from the hash values stored on the server.

S3-283 Integrating a number of different activities in the development of an information security infrastructure is **BEST** achieved by developing:

A. a business plan.
B. an architecture.
C. requirements.
D. specifications.

B is the correct answer.

Justification:

A. A business plan may address some issues of integrating activities, but that is not its main purpose.
B. An architecture allows different activities to be integrated under one design authority.
C. Requirements do not generally address integration.
D. Specifications do not address integration.

X

S3-284 After deciding to acquire a security information and event management system, it is **MOST** important for the information security manager to:

A. perform a comparative analysis of available systems.
B. develop a comprehensive business case for the system.
C. utilize the organization's existing acquisition process.
D. ensure that there is adequate network capacity for the system.

C is the correct answer.

Justification:
A. A comparative analysis should have been accomplished prior to the decision to purchase.
B. Development of a business case should have been accomplished prior to the decision to purchase.
C. The information security manager should always use existing organization practices and processes whenever possible to minimize potential issues with other departments.
D. Ensuring adequate capacity should have been accomplished prior to the decision to purchase.

X

S3-285 Application level controls are **MOST** likely to be employed when:

A. general controls are not sufficient.
B. detective controls are required.
C. preventive controls are required.
D. corrective controls are the only option.

A is the correct answer.

Justification:
A. Application controls are employed when general system controls do not provide an adequate level of security.
B. Detective controls exist at both general and application levels.
C. Preventive controls exist at both general and application levels.
D. Corrective controls exist at both general and application levels.

S3-286 An organization has commissioned an information security expert to perform network penetration testing and has provided the expert with information about the infrastructure to be tested. The benefit of this approach is:

A. more time is devoted to exploitation than to fingerprinting and discovery.
B. this accurately simulates an external hacking attempt.
C. the ability to exploit Transmission Control Protocol/Internet Protocol vulnerabilities.
D. the elimination of the need for penetration testing tools.

A is the correct answer.

Justification:
A. When information is provided to the penetration tester (white box testing), less time is spent on discovering and understanding the target to be penetrated.
B. A black box approach, where no information is provided, better simulates an actual hacking attempt.
C. Both white box and black box approaches could exploit Transmission Control Protocol/Internet Protocol vulnerabilities.
D. Both white box and black box approaches would require use of penetration testing tools.

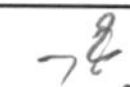

S3-287 Which of the following is the **BEST** way to verify that all critical production servers are utilizing up-to-date virus signature files?

A. Verify the date that signature files were last pushed out.
B. Use a recently identified benign virus to test if it is quarantined.
C. Research the most recent signature file and compare to the console.
D. Check a sample of servers that the signature files are current.

D is the correct answer.

Justification:

A. The fact that an update was pushed out to a server does not guarantee that it was properly loaded onto that server.
B. Personnel should never release a virus, no matter how benign.
C. Checking the vendor information to the management console would still not be indicative as to whether the file was properly loaded on the server.
D. The only accurate way to check the signature files is to look at a sample of servers.

S3-288 Which one of the following measures will **BEST** indicate the effectiveness of an incident response process?

A. Number of open incidents
B. Reduction of the number of security incidents
C. Reduction of the average response time to an incident
D. Number of incidents handled per month

C is the correct answer.

Justification:

A. The total number of open incidents is not an indicator of incident response effectiveness because the team does not have direct control over the number of incidents it must handle at any given time.
B. Reduction of the number of security incidents generally cannot be attributed to the effectiveness of the response team but rather to improved controls.
C. Reduction of response time helps minimize the impact of the incident and is the best indicator of the effectiveness of the incident response process.
D. The number of incidents handled per month would not be a direct indicator of team effectiveness.

S3-289 A security baseline can **BEST** be used for:

A. securing unstable environments.
B. establishing uniform system hardening.
C. prioritizing security objectives.
D. establishing a corporate security policy.

B is the correct answer.

Justification:

A. The stability of an environment is not necessarily related to baselines; the application of a security baseline can sometimes even destabilize an environment by conflicting with existing software.
B. A security baseline establishes a uniform security standard to be applied across similar systems.
C. A baseline does not prioritize security objectives.
D. Baselines are established as the result of a policy; they are not part of the policy development.

S3-290 The newly appointed chief information security officer (CISO) of a pharmaceutical company is given the task of creating information security procedures for all departments in the company. Which one of the following groups should the CISO initially approach to write the procedures?

A. Legal department
B. End users
C. Senior management
D. Operations department

D is the correct answer.

Justification:
A. The legal department is not typically involved in writing procedures, except for its own procedures.
B. End users are not typically involved in writing procedures.
C. Senior management would not be directly involved in the writing of security procedures.
D. The operations group has firsthand knowledge of organizational processes and responsibilities and should ensure that all procedures that are written are functionally sound.

S3-291 Which of the following is the **MOST** appropriate control to address compliance with specific regulatory requirements?

A. Policies
B. Standards
C. Procedures
D. Guidelines

B is the correct answer.

Justification:
A. Policies are a statement of management intent, expectations and direction and should not address the specifics of regulatory compliance.
B. Standards set the allowable boundaries for technologies, procedures and practices and thus are the appropriate documentation to define compliance requirements.
C. Procedures are developed in order to provide instruction for meeting standards, but cannot be developed without established standards.
D. Guidelines are not mandatory and will not normally address issues of regulatory compliance.

S3-292 Which of the following is the **MOST** effective method to enhance information security awareness?

A. Timely emails that address actual security threats
B. Security training from specialized external experts for key IT personnel
C. Role-specific awareness training
D. General online security awareness training for all staff

C is the correct answer.

Justification:
A. Email is not a strong communication medium to enhance information security awareness.
B. Training for IT personnel is important, but information security awareness training needs to be provided to all employees.
C. Role-based training that includes simulation of actual information security incidents is the most effective method to teach employees how their specific function can impact information security.
D. Well-developed general awareness training can be an acceptable method to enhance information security awareness if resources are not available for role-specific training, but it is not typically as effective.

X **S3-293** Which of the following is the **MOST** effective way to ensure that noncompliance to information security standards is resolved?

A. Periodic audits of noncompliant areas
B. An ongoing vulnerability scanning program
C. Annual security awareness training
D. Regular reports to executive management

D is the correct answer.

Justification:

A. Periodic audits can be effective but only when combined with reporting.
B. Vulnerability scanning has little to do with noncompliance with standards.
C. Training can increase management's awareness regarding information security, but awareness training is generally not as compelling to management as having their names highlighted on a compliance report.
D. The concern of having their area of responsibility reported as noncompliant to their peers and executives is generally the most effective motivation for management to take action.

S3-294 Which one of the following phases of the application development life cycle for in-house development represents the **BEST** opportunity for an information security manager to influence the outcome of the development effort?

A. System design for a new application
B. User acceptance testing and sign-off
C. Requirements gathering and analysis
D. Implementation

C is the correct answer.

Justification:

A. The design phase helps determine how the requirements will be implemented; however, if an information security manager first becomes involved in the design phase, the manager will likely find that influencing the outcome of the development effort will be more difficult.
B. The user acceptance testing and sign-off phase is too late in the life cycle to effectively influence the outcome.
C. An information security manager should be involved in the earliest phase of the application development life cycle to effectively influence the outcome of the development effort. Of the choices listed, the requirements gathering and analysis phase represents the earliest opportunity for an information security manager to have such influence. During this phase, both functional and nonfunctional requirements, including security, should be considered.
D. The implementation phase is too late in the life cycle to effectively influence the outcome.

S3-295 An information security manager has instructed a system database administrator to implement native database auditing in order to meet regulatory requirements for privileged user monitoring. Which of the following is the **PRIMARY** reason that the database administrator would be concerned? Native database auditing:

A. interferes with policy-driven event logging.
B. affects production database performance.
C. requires development of supplementary tools.
D. impairs flexibility in configuration management.

B is the correct answer.

Justification:
A. Interference with policy-driven event logging is a potential concern but secondary to performance impact.
B. Many database products come with a native audit log function. Although it can be easily activated, there is a risk that it may negatively impact the performance of the database.
C. The need to develop supplementary tools is a potential concern but secondary to performance impact.
D. Impaired flexibility in configuration management is not an issue.

S3-296 Controls that fail closed (secure) will present a risk to:

A. confidentiality.
B. integrity.
C. authenticity.
D. availability.

D is the correct answer.

Justification:
A. The blocked access will not generally impact confidentiality.
B. The blocked access will not generally impact integrity.
C. The blocked access will not generally impact authenticity.
D. A control (such as a firewall) that fails in a closed condition will typically prevent access to resources behind it, thus impacting availability.

S3-297 Which of the following is the **MOST** important aspect that needs to be considered from a security perspective when payroll processes are outsourced to an external service provider?

A. A cost-benefit analysis has been completed.
B. Privacy requirements are met.
C. The service provider ensures a secure data transfer.
D. No significant security incident occurred at the service provider.

B is the correct answer.

Justification:
A. A cost-benefit analysis should be undertaken from a business perspective but not from a security perspective.
B. Applicable privacy requirements may be a matter of law or policy and will require consideration when outsourcing processes that involve personal information.
C. When data are transferred, it may be necessary to ensure data security, but there are many other privacy and security issues to consider.
D. Past incidents may not reflect the current security posture of the service provider nor do they reflect applicable security requirements.

S3-298 When initially establishing an information security program, it is **MOST** important that managers:

A. examine and understand the culture within the organization.
B. analyze and understand the control system of the organization.
C. identify and evaluate the overall risk exposure of the organization.
D. examine and assess the security resources of the organization.

C is the correct answer.

Justification:

A. Examining and understanding the culture within the organization is an important step in the overall evaluation process.
B. Analyzing and understanding the control system is an essential step to determine what risk is addressed and what control objectives are currently in place.
C. Identifying and evaluating the overall risk is most important, because it includes the other three elements, in addition to others.
D. Examining and assessing security resources is important information in determining and evaluating overall risk and exposure of an organization.

S3-299 The requirement for due diligence is **MOST** closely associated with which of the following?

A. The right to audit
B. Service level agreements
C. Appropriate standard of care
D. Periodic security reviews

C is the correct answer.

Justification:

A. The right to audit is an important consideration when evaluating an enterprise but is not as closely related to the concept of due diligence.
B. Service level agreements are an important consideration when evaluating an enterprise but are not as closely related to the concept of due diligence.
C. The standard of care is most closely related to due diligence. It is based on the legal notion of the steps that would be taken by a person of similar competency in similar circumstances.
D. Periodic security reviews is not as closely related to due diligence.

S3-300 Which one of the following factors affects the extent to which controls should be layered?

A. Impact on productivity
B. Common failure modes
C. Maintenance cost of controls
D. Controls that fail in a closed condition

B is the correct answer.

Justification:

A. Negative impacts on productivity could indicate that controls may be too restrictive, but it is not a consideration for layering.
B. Common failure modes in existing controls must be addressed by adding or modifying controls so that they fail under different conditions. This is done to manage the aggregate risk of total control failure.
C. Excessive maintenance costs will probably increase and not be addressed by layering additional controls.
D. Controls that fail closed pose a risk to availability, but layering would not always address this risk.

S3-301 When recommending a control to protect organization applications against structured query language injection, the information security manager is **MOST** likely to suggest:

A. hardening of web servers.
B. consolidating multiple sites into a single portal.
C. coding standards and reviewing code.
D. using Hypertext Transfer Protocol Secure (HTTPS) in place of HTTP.

C is the correct answer.

Justification:
A. Hardening of web servers does not reduce this type of vulnerability.
B. Consolidating multiple sites into a single portal does not reduce this type of vulnerability.
C. Implementing secure coding standards and peer review as part of the enterprise's system development life cycle (SDLC) are controls that address structured query language injection.
D. Using Hypertext Transfer Protocol Secure (HTTPS) instead of HTTP does not reduce this type of vulnerability.

S3-302 What is the most significant attribute of a good information security metric?

A. It is meaningful to the recipient.
B. It is reliable and accurate.
C. It impacts productivity.
D. It is scalable and cost-effective.

A is the correct answer.

Justification:
A. Information provided by metrics that are not meaningful to the recipient is of little value.
B. While reliability and accuracy are important criteria for selecting information security metrics, it must first be determined that the information provided helps recipients accomplish their tasks.
C. The impact on productivity must be balanced against the usefulness of the metric; however, it is a valid consideration.
D. Cost-effectiveness must be balanced against the usefulness of the metric; however, it is a valid consideration. Scalability of metrics—in most situations—is more of a nice-to-have criterion than a selection criterion.

S3-303 Why is a certificate authority needed in a public key infrastructure?

A. It provides a proof of the integrity of data.
B. It prevents the denial of specific transactions.
C. It attests to the validity of a user's public key.
D. It stores a user's private key.

C is the correct answer.

Justification:
A. The certificate authority (CA) does not provide proof of message integrity.
B. Nonrepudiation prevents a party from denying that the party originated a specific transaction and is provided by a user's private key signing communication.
C. The CA is a trusted third party that attests to the authenticity of a user's public key by digitally signing it with the CA's private key.
D. A conventional CA does not store a user's private key.

S3-304 In what circumstances should mandatory access controls be used?

A. When the organization has a high risk tolerance
B. When delegation of rights is contrary to policy
C. When the control policy specifies continuous oversight
D. When access is permitted, unless explicitly denied

B is the correct answer.

Justification:
A. Mandatory access controls (MACs) are a restrictive control employed in situations of low risk tolerance.
B. With MAC, the security policy is centrally controlled by a security policy administrator, and users do not have the ability to delegate rights.
C. A requirement for continuous oversight is not related to MACs.
D. MACs do not allow access as a default condition.

S3-305 Which of the following are the **MOST** important criteria when selecting virus protection software?

A. Product market share and annualized cost
B. Ability to interface with intrusion detection system software and firewalls
C. Alert notifications and impact assessments for new viruses
D. Ease of maintenance and frequency of updates

D is the correct answer.

Justification:
A. Market share and annualized cost is secondary in nature.
B. Ability to interface with the intrusion detection system is secondary in nature.
C. Automatic notifications are very useful but not the most important criteria.
D. For the software to be effective, it must be easy to maintain and keep current.

S3-306 What is the **BEST** evidence of a mature information security program?

A. A comprehensive risk assessment and analysis exists.
B. A development of a physical security architecture exists.
C. A controls statement of applicability exists.
D. An effective information security strategy exists.

D is the correct answer.

Justification:
A. Assessing and analyzing risk is required to develop a strategy and will provide some of the information needed to develop it but will not define the scope and charter of the security program.
B. A physical security architecture is a part of an implementation.
C. The applicability statement is a part of the strategy implementation using International Organization for Standardization/International Electrotechnical Commission (ISO/IEC) 27001 or 27002 subsequent to determining the scope and responsibilities of the program.
D. The process of developing information security governance structures, achieving organizational adoption and developing a strategy to implement will define the scope and responsibilities of the security program.

S3-307 The IT department has been tasked with developing a new transaction processing system for online account management. At which stage should the information security department become involved?

A. Feasibility
B. Requirements
C. Design
D. User acceptance testing

A is the correct answer.

Justification:

A. Involve the security department as early as possible. Security considerations will affect feasibility. Security that is added later in the process often is not nearly as effective as security that is considered from end to end.
B. The requirements stage is too late in the process, and the introduction of security requirements will potentially cause delays and/or incur other costs that are neither budgeted nor anticipated by stakeholders.
C. The design stage is too late in the process, and the introduction of security requirements will potentially cause delays and/or incur other costs that are neither budgeted nor anticipated by stakeholders.
D. The user acceptance testing stage is too late in the process, and the introduction of security requirements will potentially cause delays and/or incur other costs that are neither budgeted nor anticipated by stakeholders.

S3-308 Which of the following is an example of a corrective control?

A. Diverting incoming traffic as a response to a denial-of-service attack
B. Filtering network traffic
C. Examining inbound network traffic for viruses
D. Logging inbound network traffic

A is the correct answer.

Justification:

A. Diverting incoming traffic helps correct the situation and, therefore, is a corrective control.
B. Filtering network traffic is a preventive control.
C. Examining inbound network traffic for viruses is a detective control.
D. Logging inbound network traffic is a detective control.

X **S3-309** What is the **MAIN** objective of integrating the information security process into the system development life cycle?

A. It ensures audit compliance.
B. It ensures that appropriate controls are implemented.
C. It delineates roles and responsibilities.
D. It establishes the foundation for development or acquisition.

B is the correct answer.

Justification:

A. Simply integrating information security processes into the system development life cycle (SDLC) will not ensure audit success; it is merely a piece of the compliance puzzle that must be reviewed by the auditor.
B. Establishing information security processes at the front end of any development project and using the process at each stage of the SDLC ensures that the appropriate security controls are implemented based on the review and assessment completed by security staff.
C. The purpose of integrating the information security process at the front end of any SDLC project is to reduce the risk of delays or rework rather than to identify roles and responsibilities for information security in the project.
D. The information security process should be performed at each phase of the SDLC to ensure that appropriate controls are in place. However, integration of information security does not establish the foundation for the make versus buy decision.

S3-310 Which of the following is the **MAIN** reason for implementing a corporate information security education and awareness program?

A. To achieve commitment from the board and senior management
B. To assign roles and responsibility for information security
C. To establish a culture that is conducive to effective security
D. To meet information security policy and regulatory requirements

C is the correct answer.

Justification:

A. Implementing such a program is an ongoing process that supports senior management's commitment toward information security.
B. Assigning roles and responsibilities for information security is achieved largely by implementing information security policies. However, to be effective, it is important that the policy be supported by a corporate information security education and awareness program.
C. Education, training and awareness help in the dissemination of information on the necessity of information security and in building a conducive environment for secure and reliable business operations.
D. The information security policy and regulatory requirements contribute content to an education and awareness program.

S3-311 What is the **PRIMARY** reason an information security manager should have a sound understanding of information technology?

A. To prevent IT personnel from misleading the information security manager
B. To implement supplemental information security technologies
C. To understand requirements of a conceptual information security architecture
D. To understand the IT issues related to achieving adequate information security

D is the correct answer.

Justification:

A. This is not the main reason for the information security manager to have technical knowledge, but it is helpful.
B. The information security manager is not generally responsible for implementing IT.
C. Technical knowledge is not required for developing a conceptual information security architecture.
D. The information security manager must understand information technology in enough depth to make informed decisions about technologies and the risk that must be addressed.

S3-312 Which of the following activities is **MOST** effective for developing a data classification schema?

A. Classifying critical data based on protection levels
B. Classifying data based on the possibility of leakage
C. Aligning the schema with data leak prevention tools
D. Building awareness of the benefit of data classification

D is the correct answer.

Justification:

A. Data protection levels are decided based on classification or business value.
B. Data are classified on business value and not on the possibility of leakage. Protection of the data may well be based on the possibility of leakage.
C. Aligning the schema with data leak prevention (DLP) tools may help while automating protection, but the data classification schema already has to exist for it to align with DLP.
D. While developing a data classification schema, it is most important that all users are made aware of the need for accurate data classification to reduce the cost of overprotection and the risk of underprotection of information assets.

X **S3-313** Which of the following choices is a **MAJOR** concern with using the database snapshot of the audit log function?

A. Degradation of performance
B. Loss of data integrity
C. Difficulty maintaining consistency
D. Inflexible configuration change

A is the correct answer.

Justification:

A. **Evidential capability increases if data are taken from a location that is close to the origination point. For database auditing, activation of a built-in log may be ideal. However, there is a trade-off. The more elaborate logging becomes, the slower the performance. It is important to strike a balance.**
B. If database recovery log is impaired, there is a chance that data integrity may be lost. However, it is unlikely that audit logging will impair the integrity of the database.
C. Database replication functionality will control the consistency between database instances.
D. It is difficult to judge whether configuration change will become complex as the result of audit log activation. It depends on many different factors. Therefore, this is not the best option.

X **S3-314** What is the **PRIMARY** goal of developing an information security program?

A. To implement the strategy
B. To optimize resources
C. To deliver on metrics
D. To achieve assurance

A is the correct answer.

Justification:

A. **The development of an information security program is usually seen as a manifestation of the information security strategy. Thus, the goal of developing the information security program is to implement the strategy.**
B. Optimizing resources can be achieved in an information security program once the program has been aligned to the strategy.
C. Delivery of the metrics is a subset of strategic alignment with the information security program in an organization.
D. Assurance of information security occurs upon the strategic alignment of the information security program.

S3-315 Which of the following is the **GREATEST** success factor for effectively managing information security?

A. An adequate budget
B. Senior level authority
C. A robust technology
D. Effective business relationships

D is the correct answer.

Justification:

A. An adequate budget is important, but without cooperation and support from senior managers, it is unlikely that the security program will be effective.
B. Senior level authority can be helpful in communicating at the right organizational levels, but effective security requires persuasion, cooperation and operating in a collaborative manner.
C. Good technology and a robust network will certainly help security be effective, but they are only one part of what is required.
D. Support for information security from senior managers is essential for an effective security program. This requires developing good relationships throughout the organization and particularly with influential managers.

X

S3-316 When should a request for proposal be issued?

A. At the project feasibility stage
B. Upon management project approval
C. Prior to developing a project budget
D. When developing the business case

C is the correct answer.

Justification:

A. Assessing project feasibility involves a variety of factors that must be determined prior to issuing a request for proposal (RFP).
B. An RFP is a document distributed to software vendors requesting them to submit a proposal to develop or provide a software solution. Final management approval is likely to occur subsequent to receiving responses to an RFP.
C. Development of a project budget depends on the responses to an RFP.
D. The business case will be developed as a part of determining feasibility, which occurs prior to issuing an RFP.

X **S3-317** Which of the following is the **MOST** effective method for ensuring that outsourced operations comply with the company's information security posture?

A. The vendor is provided with audit documentation.
B. A comprehensive contract is written with service level metrics and penalties.
C. Periodic onsite visits are made to the vendor's site.
D. An onsite audit and compliance review is performed.

D is the correct answer.

Justification
A. Audit documentation may not show whether the vendor meets the company's needs; the company needs to know the testing procedures.
B. While comprehensive contracts set minimum service levels, contracts do not ensure that vendors will perform without confirming oversight.
C. Onsite visits to the vendor's site are not sufficient by themselves; they should be coupled with an audit approach to gauge information security compliance.
D. Audits and compliance reviews are the most effect way to ensure compliance.

S3-318 Why is public key infrastructure the preferred model when providing encryption keys to a large number of individuals?

A. It is computationally more efficient.
B. It is more scalable than a symmetric key.
C. It is less costly to maintain than a symmetric key approach.
D. It provides greater encryption strength than a secret key model.

B is the correct answer.

Justification:
A. Public key cryptography is computationally intensive due to the long key lengths required.
B. Symmetric or secret key encryption requires a separate key for each pair of individuals who wish to have confidential communication, resulting in an exponential increase in the number of keys as the number of users increase, creating an intractable distribution and storage problems. Public key infrastructure keys increase arithmetically, making it more practical from a scalability point of view.
C. Public key cryptography typically requires more maintenance and is more costly than a symmetric key approach in small scale implementations.
D. Secret key encryption requires shorter key lengths to achieve equivalent strength.

S3-319 The implementation of an effective change management process is an example of a:

A. corrective control.
B. deterrent control.
C. preventative control.
D. compensating control.

C is the correct answer.

Justification:

A. A corrective control is designed to correct errors, omissions and unauthorized uses and intrusions, once they are detected.
B. Deterrent controls are intended to discourage individuals from intentionally violating information security policy or procedures. Change management is intended to reduce the introduction of vulnerability by unauthorized changes.
C. An effective change management process can prevent (and detect) unauthorized changes. It requires formal approval, documentation and testing of all changes by a supervisory process.
D. Compensating controls are meant to mitigate impact when existing controls fail. Change management is the primary control for preventing or detecting unauthorized changes. It is not compensating for another control that has that function.

S3-320 Serious security incidents typically lead to renewed focus by management on information security that then usually fades over time. What opportunity should the information security manager seize to **BEST** use this renewed focus?

A. To improve the integration of business and information security processes
B. To increase information security budgets and staffing levels
C. To develop tighter controls and stronger compliance efforts
D. To acquire better supplemental technical security controls

A is the correct answer.

Justification:

A. Close integration of information security governance with overall enterprise governance is likely to provide better long-term information security by institutionalizing activities and increasing visibility in all organizational activities.
B. Increased budgets and staff may improve information security but will not have the same beneficial impact as incorporating security into the strategic levels of the organization's operations.
C. Control strength and compliance efforts must be balanced against business requirements, culture and other organizational factors and are best accomplished at the governance level.
D. While technical security controls may improve some aspects of security, they will not address management issues nor provide the enduring organizational changes needed for improved maturity levels.

S3-321 An information security manager has received complaints from senior management about the level of security delivered by a third-party service provider. The service provider is a long-standing vendor providing services based on a service agreement that has been renewed regularly without much change over the last four years. Which of the following actions is the **FIRST** one the information security manager should take in this situation?

A. Ensure that security requirements in the service agreement meet current business requirements.
B. Review security metrics to determine whether the vendor is meeting the terms of the service agreement.
C. Conduct a formal assessment of the vendor's capability to deliver security services.
D. Automate the incident reporting process to ensure timely reporting and monitoring.

A is the correct answer.

Justification:

A. Because the service agreement has not been significantly revised in four years, it is entirely likely that the vendor is delivering exactly what was purchased and that the disappointment shown by senior management is the result of the agreement not reflecting current business requirements.
B. Knowing whether the vendor is meeting the terms of the agreement is actionable only after the information security manager is certain that the terms of the agreement align with the business requirements of the company.
C. If the vendor has committed to a level of security services that metrics indicate are consistently not being met, it may be worthwhile to conduct a formal assessment of the vendor's capabilities to determine whether a new vendor is needed. However, knowing how what was contracted aligns with business requirements needs to be the first step.
D. Automation of the incident reporting process to ensure timely reporting and monitoring is only a reporting mechanism and does not resolve the issues faced.

S3-322 How should an information security manager proceed when selecting a public cloud vendor to provide outsourced infrastructure and software?

A. Insist on strict service level agreements to guarantee application availability.
B. Verify that the vendor's security posture meets the organization's requirements.
C. Update the organization's security policies to reflect the vendor agreement.
D. Consult a third party to provide an audit report to assess the vendor's security program.

B is the correct answer.

Justification:

A. Agreements that address availability do not address other aspects of the organization's security policy.
B. When considering a cloud implementation, an information security manager must verify that a chosen vendor will meet the organization's security requirements.
C. An organization defines its security policies with its business risk in mind. Changing internal policy requirements to reflect what a vendor can deliver may not be sufficient and could raise risk to the organization making it an inappropriate approach.
D. Third-party audit reports are snapshots that tell what was true at a particular time and address only those items that were within the audit scope. Each organization has its own security policy considerations, and verification with the vendor should be accomplished with the organization's specific considerations and requirements in mind.

S3-323 Which of the following approaches is the **BEST** for designing role-based access controls?

A. Create a matrix of work functions.
B. Apply persistent data labels.
C. Enable multifactor authentication.
D. Use individual logon scripts.

A is the correct answer.

Justification:

A. A matrix that documents the functions associated with particular kinds of work, typically referred to as a segregation of duties matrix, shows which roles are required or need various permissions.

B. Persistent data labels apply to mandatory access control environments where permissions are brokered by the classification levels of objects themselves. They do not factor into role-based access controls.

C. Multifactor authentication deals with how users authenticate their identities, which helps to ensure that people are who they claim to be. It does not determine the permissions that they are assigned, particularly in a role-based access control model, where permissions are assigned to roles rather than individual users.

D. Using automated logon scripts is practical in some environments, but assigning permissions to individual accounts is contrary to the intent of role-based access controls.

S3-324 Which of the following change management process steps can be bypassed to implement an emergency change?

A. Documentation
B. Authorization
C. Scheduling
D. Testing

C is the correct answer.

Justification:

A. Emergency changes require documentation, although it may occur after implementation.

B. Emergency changes require formal authorization, although it may occur after implementation.

C. When a change is being made on an emergency basis, it generally is implemented outside of the normal schedule. However, it should not bypass other aspects of the change management process.

D. Emergency changes require testing.

S3-325 Which of the following criteria is the **MOST** essential for operational metrics?

A. Timeliness of the reporting
B. Relevance to the recipient
C. Accuracy of the measurement
D. The cost of obtaining the metrics

B is the correct answer.

Justification:
A. Timeliness of reporting is important, but secondary to relevance.
B. Unless the metric is relevant to the recipient and the recipient understands what the metric means and what action to take, if any, all other criteria are of little importance.
C. A high degree of accuracy is not essential as long as the metric is reliable and indications are within an acceptable range.
D. Cost is always a consideration, but secondary to the others.

S3-326 The information security manager can **BEST** ensure that critical information resources are adequately protected by aligning the information security program with:

A. external audit.
B. vulnerability scan results.
C. formal reporting.
D. change management.

D is the correct answer.

Justification:
A. External audit is a backward-looking function. It can help to highlight deficiencies, but it cannot ensure protection of information assets.
B. Vulnerability scanning is an identification process that occurs within the information security program. The protection of critical information resources involves more than identification of vulnerabilities.
C. Reporting is a backward-looking function. It can help to highlight deficiencies, but it cannot ensure protection of information assets.
D. Aligning the information security program with change management ensures that the information security manager is immediately aware of changes as they are considered, so that gap analysis between current controls and the target state can be considered and any identified gaps addressed before changes are made, including gaps that may be created or expanded by the proposed changes.

S3-327 What is the **MOST** important reason that an information security manager must have an understanding of information technology?

A. To ensure the proper configuration of the devices that store and process information
B. To understand the risk of technology and its contribution to security objectives
C. To assist and advise on the acquisition and deployment of information technology
D. To improve communication between information security and business functions

B is the correct answer.

Justification:
A. The configuration of the devices is not the primary responsibility of the information security manager. The security manager will work through technical staff to ensure that configurations are appropriate.
B. Knowledge of information technology helps the information security manager understand how changes in the technical environment affect the security posture and its contribution to control objectives.
C. Advising on acquisition and deployment in regard to security issues is a secondary function of the information security manager.
D. Information security decisions can be made most effectively when they are understood by people in business functions, but this is secondary to understanding the relationship between technology and information security.

S3-328 Which of the following conditions is **MOST** likely to require that a corporate standard be modified?

A. The standard does not conform to procedures.
B. IT staff does not understand the standard.
C. The standard is inconsistent with guidelines.
D. Control objectives are not being met.

D is the correct answer.

Justification:
A. If a procedure does not meet the standard, the procedure must be changed, not the standard.
B. IT staff not understanding the standard may require clarification and/or training.
C. Inconsistencies with the guidelines require that the guidelines be changed to conform to the standard.
D. If conformance with the standard does not achieve control objectives, the standard requires modification.

S3-329 The protection of sensitive data stored at a third-party location requires:

A. assurances that the third party will comply with the requirements of the contract.
B. commitments to completion of periodic independent security audits.
C. security awareness training and background checks of all third-party employees.
D. periodic review of third-party contracts and policies to ensure compliance.

A is the correct answer.

Justification:

A. When storing data with a third party, the ownership and responsibility for the adequate protection of the data remains with the outsourcing organization. The outsourcing organization should have measures in place to provide assurance of compliance with the terms of the contract, which should be written on the basis of the organizational risk appetite.

B. Independent security audits are one assurance mechanism that an organization may use to verify compliance with contractual requirements, but whether these are appropriate is situational and based on the organizational risk appetite.

C. Awareness training and background checks are assurance mechanisms, but may or may not be appropriate or important in all cases.

D. Review of contracts and policies is important, but it does not assure compliance.

S3-330 A new business application requires deviation from the standard configuration of the operating system (OS). Which of the following steps should the security manager take **FIRST**?

A. Contact the vendor to modify the application.
B. Assess risk and identify compensating controls.
C. Approve an exception to the policy to meet business needs.
D. Review and update the OS baseline configuration.

B is the correct answer.

Justification:

A. The security manager would contact the vendor to modify the application only after assessing the risk and identifying compensating controls.

B. Before approving any exception, the security manager should first check for compensating controls and assess the possible risk due to deviation.

C. The security manager may make a case for deviation from the policy, but this would be based on a risk assessment and compensating controls. The deviation itself would be approved in accordance with a defined process.

D. Updating the baseline configuration is not associated with requests for deviations.

S3-331 Which of the following training mechanisms is the **MOST** effective means of promoting an organizational security culture?

A. Choose a subset of influential people to promote the benefits of the security program.
B. Hold structured training in small groups on an annual basis.
C. Require each employee to complete a self-paced training module once per year.
D. Deliver training to all employees across the organization via streaming video.

A is the correct answer.

Justification:

A. Certain people are either individually inclined or required by their positions to have greater interest in promoting security than others. By selecting these people and offering them broad, diverse opportunities for security education, they are able to act as ambassadors to their respective teams and departments, imparting a gradual and significant change in an organizational culture toward better security.
B. Structured training rarely aligns with the interests of individual employees when chosen at random to fill a small-group setting.
C. Computer-based training is a common approach to annual information awareness, but there is limited evidence that employees retain the information or adopt it into their regular activities.
D. Streaming-video "webinars" are among the least effective means of presenting information, requiring very little interaction from end users.

S3-332 Where should resource requirements for information security initially be identified? X

A. In policies
B. In the architecture
C. In the strategy
D. In procedures

C is the correct answer.

Justification:

A. Policies are developed to implement the strategy and may specify some requirements, but they are subsequent to, and a part of, implementing the strategy.
B. The architecture must implement the policies and standards.
C. The strategy must define the requirements for the resources necessary to implement the program. This is different from the tactical detail level necessary to identify specific resources.
D. Procedures will define resource acquisition processes but will not specify requirements.

X **S3-333** In a financial institution, under which of the following circumstances will policies **MOST** likely need modification?

A. Current access controls have been insufficient to prevent a series of serious network breaches.
B. The information security manager has determined that compliance with configuration standards is inadequate.
C. The results of an audit have identified a going concern issue with the organization.
D. Management has mandated compliance with a newly enacted set of information security requirements.

D is the correct answer.

Justification:
A. Necessary modifications to access controls are most likely going to be reflected in standards, not policy.
B. Compliance with existing standards is not likely to require a policy change but rather better enforcement.
C. If the viability of the organization is in doubt (going concern), it is not likely that a change in policy will solve the problem.
D. A new set of regulations requiring significant changes to the information security program is most likely going to be reflected in modifications of policy.

X **S3-334** Which of the following tools should a newly hired information security manager review to gain an understanding of how effectively the current set of information security projects are managed?

A. A project database
B. A project portfolio database
C. Policy documents
D. A program management office

B is the correct answer.

Justification:
A. A project database may contain information for one specific project and updates to various parameters pertaining to the current status of that single project.
B. A project portfolio database is the basis for project portfolio management. It includes project data such as owner, schedules, objectives, project type, status and cost. Project portfolio management requires specific project portfolio reports.
C. Policy documents on project management set direction for the design, development, implementation and monitoring of the project.
D. A program management office is the team that oversees the delivery of the project portfolio. Review of the office may provide meaningful insights into the skill set and organizational structure but not on how effectively the current set of information security projects is managed.

S3-335 What is the **MAIN** objective for developing an information security program?

A. To create the information security policy
B. To maximize system uptime
C. To develop strong controls
D. To implement the strategy

D is the correct answer.

Justification:

A. The policy should not be written for its own sake. To be effective, the policy must address the threat and risk landscape that is usually the basis for strategy development.
B. The degree of uptime required will be defined as a part of strategy development balanced against costs.
C. Not all controls need to be strong, and the degree of control must be determined by cost-effectiveness, impact on productivity and other factors.
D. The information security strategy provides a development road map to which the program is built.

S3-336 An information security manager has implemented an automated process to compare physical access using swipe cards operated by the physical security department with logical access in the single sign-on (SSO) system. What is the **MOST** likely use for this information?

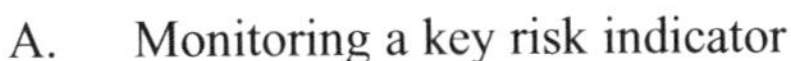

A. Monitoring a key risk indicator
B. Determining whether staff is piggybacking
C. Overseeing the physical security department
D. Evaluating the SSO process

A is the correct answer.

Justification:

A. Discrepancies between physical and logical access can occur for a variety of reasons, but all are indications that something is wrong and risk is elevated. Discrepancies could indicate piggybacking, shared passwords or attempts at unauthorized access, and therefore, this monitoring can serve as a key risk indicator (KRI).
B. Potential piggybacking can be flagged if more individuals log in from within the network than physically enter the facility; however, this is just one KRI.
C. Although this information could indicate that the physical access control is not functioning properly, the responsibility for oversight of the physical security department is not usually a function of the information security manager.
D. Comparing physical access and logical access is not an effective way to monitor the single sign-on (SSO) system, and there are other methods more specific and useful for this purpose.

S3-337 What activity should the information security manager perform **FIRST** after finding that compliance with a set of standards is weak?

A. Initiate the exception process.
B. Modify policy to address the risk.
C. Increase compliance enforcement.
D. Perform a risk assessment.

D is the correct answer.

Justification:

A. The exception process can be used after assessing the noncompliance risk and determining whether compensating controls are required.
B. Modifying policy is not necessary, unless there is no applicable standard and policy.
C. It is not appropriate to increase compliance enforcement until the information security manager has determined the extent of the risk posed by weak compliance.
D. The first action after finding noncompliance with particular standards should be to determine the risk to the organization and the potential impact.

S3-338 When is the access control process **MOST** effective?

A. When it integrates access controls for all IT assets
B. When it ensures that all user activities are uniquely identifiable
C. When it validates user identities via authentication mechanisms
D. When it is based on a role-based access control model

B is the correct answer.

Justification:

A. Integrating access controls for all IT assets improves efficiency and operational ease, but does not impact effectiveness. The same results can be accomplished without integration at the expense of additional time and effort.
B. The primary objective of the access control process is to prevent unauthorized access, which requires user activities to be uniquely identifiable for accountability purposes.
C. Validation of identity through authentication supports the access control process, but knowing that users are who they claim to be does not translate into control of the actions of those users unless individual user activities can be traced back to their sources.
D. The role-based access control model increases efficiency, but does not increase effectiveness.

S3-339 The relationship between policies and corporate standards can **BEST** be described by which of the following associations?

A. Standards and policies have only an indirect relationship.
B. Standards provide a detailed description of the meaning of a policy.
C. Standards provide direction on achieving compliance with policy intent.
D. Standards can exist without a relationship to any particular policy.

C is the correct answer.

Justification:
A. In most cases, there is a direct relationship between policy and corporate standards.
B. Corporate standards generally do not provide details on the meaning of policy, rather on the acceptable limits needed to comply with policy intent.
C. Corporate standards set the allowable limits and boundaries for people, processes and technology as an expression of policy intent and, therefore, provide direction on policy compliance.
D. It would be a poor practice to have corporate standards not directly expressing the intent of a particular policy. To the extent that they exist, they should rely on an implicit policy.

S3-340 A person working at a bank receives a call on a voice-over Internet protocol line from a person claiming to be an employee of the bank at another branch office. He is requesting customer information. The **FIRST** action to take when receiving this type of call is to:

A. obtain the email address of the caller and have the recipient transmit the information using message encryption.
B. advise the employee who received the call to hang up and then return the call to the other branch using the number in the office phone directory.
C. pose business-related questions to the caller, and if a proper reply is received, the recipient may forward the information to the caller.
D. ask the person to call back later and notify regulatory officers of a possible fraud attempt.

B is the correct answer.

Justification:
A. Sensitive information should not be sent to an third-party who has not been validated even if encryption is used, because the organization cannot guarantee that the recipient will be unable to decipher information in a time period during which the information can still be used.
B. If the call recipient suspects any chance of social engineering over the phone, the callback option is quite effective. The best approach to identifying the caller is who they say they are is to call them back using the legitimate phone number listed in the office phone directory. The recipient should not use a phone number or email address provided by the caller. Once the legitimacy of the call has been reasonably verified, the information may be transmitted using message encryption. Even after voice verification, it is essential that encryption be used because voice verification might be subject to additional attacks (e.g., man-in-the-middle).
C. Even when there is a strong suspicion of fraud, the recipient should not indicate this to the caller over the phone. Instead, the recipient should hang up and call back using the phone number from the office phone directory.
D. The person attempting social engineering will attempt to pass business (or non-business) related questions to the caller. If proper answer is obtained, the recipient will continue the conversation. If the caller feels uneasy, he/she will have the control to end the conversation. Because this technique puts the control of the conversation on the attacker, it is not the best answer.

S3-341 What must change management achieve from a risk management perspective?

A. It must be operated by information security to ensure that security is maintained.
B. It must be overseen by the steering committee because of its importance.
C. It must be secondary to release and configuration management.
D. It must include mandatory notification of the information security department.

D is the correct answer.

Justification:

A. It is not important who oversees the change management process provided notification occurs and a consistent process is in place.
B. Change management oversight may or may not be the responsibility of the steering committee.
C. Change management is just as essential as release and configuration management to properly manage risk. Release and configuration management may be included as part of the change management process.
D. In some organizations, information security is represented on the change control board. At a minimum, the change management function must have a process that ensures notification to information security of proposed changes in order to manage the risk that the change may affect.

S3-342 Which of the following reasons is **MOST** likely why an organization has decided to outsource intrusion detection services?

A. As a response to audit recommendations
B. Due to the complexity of interpreting attacks
C. As a result of a cost-benefit analysis
D. Due to lack of competent staff

C is the correct answer.

Justification:

A. Audit recommendations may lead to a cost-benefit analysis, but generally do not direct a particular approach to solving an identified problem.
B. Technology for intrusion detection that reduces complexity to manage the levels is available, but may not be cost-effective.
C. A cost-benefit analysis addresses the trade-offs between in-house and outsourced services. If outsourcing is chosen, it is generally chosen on the basis of cost-effectiveness.
D. Hiring staff with the proper skill set for intrusion detection is generally possible but may not be cost-effective.

S3-343 Which of the following would be an appropriate control for ensuring the authenticity of orders received in an electronic data interchange system application?

A. Acknowledge receipt of electronic orders with a confirmation message
B. Perform reasonableness checks on quantities ordered before filling orders
C. Encrypt electronic orders
D. Verify the identity of senders and determine whether orders correspond to contract terms

D is the correct answer.

Justification:

A. Acknowledging the receipt of electronic orders with a confirmation message is good practice but will not authenticate orders from customers.
B. Performing reasonableness checks on quantities ordered before placing orders is a control for ensuring the correctness of the company's orders, not the authenticity of its customers' orders.
C. Encrypting sensitive messages is an appropriate step but does not apply to messages received.
D. An electronic data interchange system is subject not only to the usual risk exposures of computer systems but also to those arising from the potential ineffectiveness of controls on the part of the trading partner and the third-party service provider, making authentication of users and messages a major security concern.

S3-344 What is the initial step that an information security manager would take during the requirements gathering phase of an IT project to avoid project failure?

A. Develop a comprehensive methodology that defines and documents project needs.
B. Build security requirements into the design of the system with consideration of enterprise security needs.
C. Ensure that the business problem is clearly understood before working on the solution.
D. Create a project plan based on the principles of agile development methodology.

C is the correct answer.

Justification:

A. Developing a methodology is a step separate from defining requirements.
B. The question relates to requirements gathering phase of the project, not the design phase. Therefore, it would be too early to start building the requirement into the design.
C. The key to successful requirements gathering is to focus initially on the business problem before trying to develop a solution. Otherwise, the solution may address the wrong problem.
D. An agile development methodology first requires the determination of business requirements.

S3-345 What is the **PRIMARY** benefit of a security awareness training program?

A. To reduce the likelihood of an information security event
B. To encourage compliance with information security policy
C. To comply with the local and industry-specific regulation and legislation
D. To provide employees with expectations for information security

A is the correct answer.

Justification:

A. **Each person/employee should know how information security is related to his/her job role and why work tasks should be performed in an appropriate way that protects the organization and its assets.**
B. Although compliance with the information security policy is important, security awareness training goes beyond to include cultural and behavioral elements of information security.
C. Industry-specific regulation and legislation are not the primary drivers of security awareness training programs.
D. Employee expectations do not necessarily ensure understanding of information security or influence cultural or behavioral attitudes directly.

S3-346 Who has the inherent authority to grant an exception to information security policy?

A. The business process owner
B. The departmental manager
C. The policy approver
D. The information security manager

C is the correct answer.

Justification:

A. The business process owner is typically required to enforce the policy and would not normally have the authority to grant an exception.
B. The departmental manager cannot approve an exception to policy because he/she is not responsible for the policy delivering its promised results.
C. **The person or body empowered to approve a policy is empowered to grant exceptions to it because in approving it, he/she assumed responsibility for the results that it promises to deliver.**
D. The information security manager cannot approve an exception to policy because he/she is not responsible for the policy delivering its promised results.

S3-347 What human resources (HR) activity is **MOST** crucial in managing mobile devices supplied by the organization?

A. HR provides termination notices.
B. HR provides background checks.
C. HR provides reporting structures.
D. HR provides awareness support.

A is the correct answer.

Justification:

A. When the human resources (HR) department provides staff termination notices, security management can perform de-provisioning of mobile devices.
B. Background checks generally do not help the management of mobile devices.
C. Reporting structures generally do not affect the management of mobile devices.
D. HR could support information security awareness programs. However, from the management perspective, device de-provisioning upon staff termination will be more important.

S3-348 A newly appointed information security manager has been asked to redefine information security requirements because senior management is unhappy with the current state of information security. Which of the following choices would the information security manager consider **MOST** critical?

A. An industry framework
B. The business strategy
C. The technology infrastructure
D. User competencies

B is the correct answer.

Justification:

A. Industry frameworks are useful in improving security implementation to the extent that they align with and support business objectives.
B. The most critical factor to be considered in defining information security requirements is the business strategy because everything that the business does—including information security—is only done for the sake of pursuing the business strategy.
C. Security requirements are driven by the information security policy, procedures and practices. The technology infrastructure needs to be considered while implementing security, but if the current infrastructure cannot support information security requirements that are aligned to the business strategy, then the infrastructure will also need to be reevaluated.
D. User competencies reflect a current state and may be useful in mapping a path forward for the lowest cost, but competencies can be enhanced to those required by providing training to bring users to the required level. The business strategy is the driver of information security requirements (and all other activities).

X

S3-349 The extent to which senior management supports the implementation of the strategy and risk management activities of an information security program will **FIRST** determine:

A. the charter.
B. the budget.
C. policy.
D. the reporting structure.

A is the correct answer.

Justification:

A. Without management support, the program will never be able to establish a charter that will allow it to function within the environment. All of the other choices follow the charter.
B. Without a charter for the program, there will be no budget because the program will not exist.
C. A charter is needed to establish the program before policy can be developed.
D. The reporting structure will not be established until the program is chartered.

S3-350 When implementing a cloud computing solution that will provide Software as a Service (SaaS) to the organization, what is the **GREATEST** concern for the information security manager?

A. The lack of clear regulations regarding the storage of data with a third party
B. The training of the users to access the new technology properly
C. The risk of network failure and the resulting loss of application availability
D. The possibility of disclosure of sensitive data in transit or storage

D is the correct answer.

Justification:

A. Many jurisdictions have regulations regarding data privacy. The concern of the information security manager is compliance with those regulations, not the lack of regulations.
B. The training of how to use Software as a Service (SaaS) is no different than the need for training required for more traditional solutions. In most cases, the use of SaaS is fairly simple and requires minimal technology, but is not within the scope of the information security manager's responsibility in any case.
C. Loss of application availability as a result of network failure is an inherent risk associated with SaaS and must be taken into account by the organization as part of the decision to move to cloud computing, but this is a business decision rather than a principle concern of the information security manager.
D. Disclosure of sensitive data is a primary concern of the information security manager.

S3-351 An advantage of using a cloud computing solution over a locally hosted solution is:

A. the ability to obtain storage and bandwidth on demand.
B. reduced requirements for training of users and managers.
C. increased security as a result of encrypting data in transit.
D. the opportunity to control changes to applications and data.

A is the correct answer.

Justification:

A. One key advantage of cloud computing is the ability to rapidly adjust storage and network bandwidth needs as required. This is generally not possible in locally hosted environments.
B. The amount of training required for users and managers is not substantially different between a cloud and a local solution.
C. Sensitive data can be encrypted in transit regardless of whether it is locally hosted or hosted on a cloud provider.
D. Access controls may be established in both local and cloud solutions.

S3-352 To establish the contractual relationship between entities using public key infrastructure, the certificate authority must provide which of the following? X

A. A registration authority
B. A digital certificate
C. A nonrepudiation capability
D. A certification practice statement

D is the correct answer.

Justification:

A. The registration authority is responsible for authentication of users prior to the issuance of a certificate.
B. A digital certificate is the electronic credentials of individual entities but does not provide the contractual relationship of users and the certificate authority.
C. Nonrepudiation is an inherent capability of a public key infrastructure by the virtue of the signing capability.
D. The certification practice statement provides the contractual requirements between the relying parties and the certificate authority.

S3-353 Which of the following control practices represents the **FIRST** layer of the defense-in-depth strategy? X

A. Data privacy
B. Authentication
C. Incident response
D. Backup

B is the correct answer.

Justification:

A. Data privacy is part of the second layer, which is containment.
B. Authentication is part of prevention, which is the first layer of defense in depth.
C. Incident response is part of the fourth layer of defense, which is reaction.
D. A backup policy is part of the last layer of defense, which is recovery/restoration.

S3-354 Which of the following is **MOST** likely to initiate a review of an information security standard? Changes in the:

A. effectiveness of security controls.
B. responsibilities of department heads.
C. information security procedures.
D. results of periodic risk assessments.

D is the correct answer.

Justification:

A. Changes in the effectiveness of security controls will require a review of the controls, not necessarily the standards.
B. Changes in the roles and responsibilities of department heads will not require a change to security standards, which will be captured during risk review.
C. Standards set the requirements for procedures, so a change in procedures is not likely to affect the standard.
D. Security policies need to be reviewed regularly in order to ensure that they appropriately address the organization's security objectives. A review of a security standard is prompted by changes in external and internal risk factors that are captured during risk assessment.

S3-355 The **MOST** likely reason to segment a network by trust domains is to:

A. limit consequences of a compromise.
B. reduce vulnerability to a breach.
C. facilitate automated network scanning.
D. implement a data classification scheme.

A is the correct answer.

Justification:

A. Segmentation by trust domain limits the potential consequences of a successful compromise by constraining the scope of impact.
B. Segmentation by trust domain does not substantially change vulnerability.
C. Automated network scanning can treat a network as logically segmented without reliance on trust domains.
D. Segmentation is not implemented primarily to facilitate data classification.

S3-356 Effective strategic alignment of the information security program requires:

A. active participation by a steering committee.
B. creation of a strategic planning business unit.
C. regular interaction with business owners.
D. acceptance of cultural and technical limitations.

C is the correct answer.

Justification:
A. Active participation by a steering committee made up of business owners or their delegates is one way to accomplish strategic alignment, but a steering committee is not the only way to achieve this goal.
B. If an organization has a strategic planning business unit, active participation in its activities may provide insight into future business directions and ensure that security considerations are included in the planning progress, but strategic alignment of the information security program does not require creation of such a unit.
C. Alignment of the information security program requires an understanding of business plans and objectives as determined by business owners. Although the method of achieving regular interaction with business owners can vary based on the size and structure of an organization, the interaction itself is a requirement.
D. Alignment of an information security program must take into account culture and existing technology, but information security supports the business objectives of the organization, which may include changes to the culture and technology currently in place. These aspects of the organization should not be accepted as foundations for program alignment when they are misaligned with business objectives.

S3-357 Business management is finalizing the contents of a segregation of duties matrix to be loaded in a purchase order system. Which of the following should the information security manager recommend in order to **BEST** improve the effectiveness of the matrix?

A. Ensure approvers are aligned with the organizational chart
B. Trace approvers' paths to eliminate routing deadlocks
C. Set triggers to go off in the event of exceptions
D. Identify conflicts in the approvers' authority limits

D is the correct answer.

Justification:
A. The approver's structure in a purchase order system may not necessarily be in sync with the organizational structure. Depending on business requirements, modified hierarchy is acceptable purely in terms of approving certain transactions.
B. It is rare that the structure of an approver's routing path will end up with deadlocks. If a highly complicated approval structure is developed, something similar to deadlock may occur (e.g., it takes very long time until request is approved). Even so, it is unlikely that routing effectiveness becomes a primary driver for quality improvement.
C. Setting triggers to go off in the event of exceptions is a technical feature to be implemented inside the database. It is not relevant advice to be given to business management.
D. In order to make the segregation of duties matrix complete, it is best to ensure that no conflicts exist in approvers' authorities. If there are any, it will introduce a flaw in the control, resulting the successful execution of unauthorized transactions.

X

S3-358 Active information security awareness programs **PRIMARILY** influence:

A. acceptable risk.
B. residual risk.
C. control objectives.
D. business objectives.

B is the correct answer.

Justification:

A. The level of risk that an organization deems acceptable is a business decision. Controls, including active security awareness programs, are implemented to reduce risk to acceptable levels and do not influence what level of risk is acceptable.

B. An information security awareness program is an administrative control that reduces vulnerability, thereby yielding lower residual risk.

C. Security awareness may be a control objective, depending on the information security strategy of the organization, but such a program does not primarily influence the objectives of other controls.

D. Security awareness does not primarily influence business objectives.

X

S3-359 With which of the following business functions is integration of information security **MOST** likely to result in risk being addressed as a standard part of production processing?

A. Quality assurance
B. Procurement
C. Compliance
D. Project management

A is the correct answer.

Justification:

A. Quality assurance uses metrics as indicators to identify systemic problems in processes that may result in unacceptable levels of output quality. Because this monitoring is intended to be effectively continuous as a matter of statistical sampling, integrating information security with quality assurance helps to ensure that risk is addressed as a standard part of production processing.

B. Procurement approves initial acquisitions, but it has no involvement in implementation or production monitoring.

C. Compliance focuses on legal and regulatory requirements, which represent a subset of overall risk.

D. The involvement of the project management office is typically limited to planning and implementation.

S3-360 When considering outsourcing technical or business processes, one of the **MAIN** concerns of the information security manager is whether the third-party service provider will:

A. deliver a level of quality acceptable to the organization's established customer base.
B. agree to service level agreements with penalties sufficient to offset potential losses.
C. provide technical services at a lower cost than would be possible on an in-house basis.
D. meet the organization's security requirements on an ongoing and verifiable basis.

D is the correct answer.

Justification:

A. Quality assurance is an area of concern when dealing with third-party service providers, but it is not a primary focus of the information security manager.
B. Penalties written into service level agreements are a form of risk transfer (sharing) that may be appropriate for an organization's business objectives, but the sufficiency of such arrangements are not a primary focus of the information security manager.
C. Reducing or controlling cost is typically one of the main reasons that organizations choose to enter into third-party service agreements, but whether or not such agreements deliver their expected cost savings is not a primary focus of the information security manager.
D. When an organization enters into an outsourcing agreement with a third-party service provider, the information security manager becomes responsible for ensuring that the provider adheres to the same security requirements as apply to the organization itself and that any variances are documented and presented to senior management for an appropriate risk response. The challenge of being able to assess a provider's security behaviors on an ongoing and verifiable basis is one of the main concerns of the information security manager in any outsourcing arrangement.

S3-361 IT management has standardized the Internet browser used within the organization. This practice is **MOST** effective in meeting which of the following objectives? ✗

A. Prevent attacks designed to exploit known vulnerabilities.
B. Ensure the subscription count is aligned with contract.
C. Invalidate illegal browser script program development.
D. Guarantee compatibility with internal web-based applications.

D is the correct answer.

Justification:

A. Standardizations of browsers within organizations typically results in delays to upgrades and patching, making it more likely that the standardized browser will be susceptible to exploitation of known vulnerabilities.
B. Controlling the version of the Internet browser used may not support the reconciliation between the subscription count and contract license number.
C. Browser script development is generally not constrained by browser standardization. Information security managers seeking to prevent script execution would typically do so by standardizing configurations rather than versions.
D. Internal web applications typically depend on particular versions of a web browser. Many organizations choose to retain versions of their browsers beyond periods of support in order to maintain compatibility with their deployed applications.

S3-362 It is **MOST** important that the information security program be integrated with:

A. internal audit.
B. quality assurance.
C. information technology.
D. business unit managers.

C is the correct answer.

Justification:

A. A good relationship with internal audit can provide considerable support for achieving information security objectives, but an information security program can be effective with or without this relationship.
B. Understanding the quality assurance (QA) process offers an opportunity for information security because security-related controls may be integrated into QA for greater efficiency, but this integration is less important than the link with IT.
C. While the information security program will achieve greater success by integrating assurance activities with all of the listed departments, effective integration with IT is essential, because IT is responsible for the hands-on implementation of and operation of information processing systems.
D. Business unit managers are responsible for front-line business operations and help to identify and escalate security incidents and other risks, so integration with them is valuable. However, business requirements implemented in technology must necessarily be shared with IT, so integration with IT is the most important link for an information security program.

S3-363 Which of the following would be the **BEST** metric for an information security manager to use to support a request to fund new controls?

A. Adverse yearly incident trends
B. Audit findings of poor compliance
C. Results of a vulnerability scan
D. Increased external port scans

A is the correct answer.

Justification:

A. Security incidents occur because either a control failed or there was no control in place. Trends are a metric providing their own points of reference.
B. Failures of compliance with existing controls are not likely to be solved by additional controls. Also, an audit finding absent any prior findings of compliance or other reference point is a measure, not a metric.
C. Without knowing exposure, threat and potential impact, risk cannot be determined and will be poor support for new controls. Also, results of a vulnerability scan constitute a measure, not a metric.
D. Port scans are common and generally will not support funding of new controls.

DOMAIN 4—INFORMATION SECURITY INCIDENT MANAGEMENT (19%)

S4-1 Which of the following should be determined **FIRST** when establishing a business continuity program?

A. Cost to rebuild information processing facilities
B. Incremental daily cost of the unavailability of systems
C. Location and cost of offsite recovery facilities
D. Composition and mission of individual recovery teams

B is the correct answer.

Justification:
A. The cost to rebuild information processing facilities would not be the first thing to determine.
B. Prior to creating a detailed business continuity plan, it is important to determine the incremental daily cost of losing different systems. This will allow recovery time objectives to be determined.
C. Location and cost of a recovery facility cannot be addressed until the potential losses are calculated, which will determine the type of recovery site that is needed—and this will affect cost.
D. Individual recovery team requirements will occur after the requirements for business continuity are determined.

S4-2 A company has a network of branch offices with local file/print and mail servers; each branch individually contracts a hot site. Which of the following would be the **GREATEST** weakness in recovery capability?

A. Exclusive use of the hot site is limited to six weeks.
B. The hot site may have to be shared with other customers.
C. The time of declaration determines site access priority.
D. The provider services all major companies in the area.

D is the correct answer.

Justification:
A. Access to a hot site is not indefinite; the recovery plan should address a long-term outage.
B. Sharing a hot site facility is common practice and sometimes necessary in the case of a major disaster and not a significant weakness.
C. First come, first served is a standard practice in hosted facilities and does not constitute a major weakness.
D. In case of a disaster affecting a localized geographical area, the vendor's facility and capabilities could be insufficient for all of its clients, which will all be competing for the same resource. Preference will likely be given to the larger corporations, possibly delaying the recovery of a branch that will likely be smaller than other clients based locally.

S4-3 Which of the following actions should be taken when an online trading company discovers a network attack in progress?

A. Shut off all network access points
B. Dump all event logs to removable media
C. Isolate the affected network segment
D. Enable trace logging on all events

C is the correct answer.

Justification:

A. Shutting off all network access points would create a denial of service that could result in loss of revenue.
B. Dumping event logs, while useful, would not mitigate the immediate threat posed by the network attack.
C. Isolating the affected network segment will mitigate the immediate threat while allowing unaffected portions of the business to continue processing.
D. Enabling trace logging, while useful, would not mitigate the immediate threat posed by the network attack.

S4-4 Which of the following choices should be assessed after the likelihood of a loss event has been determined?

A. The magnitude of impact
B. Risk tolerance
C. The replacement cost of assets
D. The book value of assets

A is the correct answer.

Justification:

A. Disaster recovery is driven by risk, which is a combination of likelihood and consequences. Once likelihood has been determined, the next step is to determine the magnitude of impact.
B. Risk tolerance is the acceptable deviation from acceptable risk. This is taken into account once risk has been quantified, which is dependent on determining the magnitude of impact.
C. Replacement cost is needed only when replacement is required.
D. Book value does not represent actual asset value and cannot be used to measure magnitude of impact.

S4-5 What is the **FIRST** priority when responding to a major security incident?

A. Documentation
B. Monitoring
C. Restoration
D. Containment

D is the correct answer.

Justification:

A. Documentation is important, but it should follow containment.
B. Monitoring is important and should be ongoing but does not limit the impact of the incident.
C. Restoration follows containment.
D. The first priority in responding to a security incident is to contain it to limit the impact.

S4-6 Although control effectiveness has recently been tested, a serious compromise occurred. What is the **FIRST** action that the information security manager should take?

A. Evaluate control objectives.
B. Develop more stringent controls.
C. Perform a root cause analysis.
D. Repeat the control test.

C is the correct answer.

Justification:

A. Control objectives cannot be evaluated until the exact nature of the compromise is understood, and therefore, it is not clear how to best provide a solution.
B. Increasing the restrictiveness of controls should only take place if it is determined by root cause analysis to be necessary to solve the problem.
C. Assessing the root cause is the first step in understanding whether control objectives and controls are inadequate or there was some other cause that must be addressed.
D. Repeating the control test does not provide a root cause of the compromise that occurred.

S4-7 Which of the following is the **MOST** important element to ensure the success of a disaster recovery test at a vendor-provided hot site?

A. Tests are scheduled on weekends.
B. Network Internet Protocol addresses are predefined.
C. Equipment at the hot site is identical.
D. Business management actively participates.

D is the correct answer.

Justification:

A. Testing on weekends can be advantageous, but this is not the most important choice.
B. Because vendor-provided hot sites are in a state of constant change, it is not always possible to have network addresses defined in advance.
C. Although it would be ideal to provide for identical equipment at the hot site, this is not always practical because multiple customers must be served and equipment specifications will vary.
D. Disaster recovery testing requires the allocation of sufficient resources to be successful. Without the support of management, these resources will not be available, and testing will suffer as a result.

S4-8 At the conclusion of a disaster recovery test, which of the following should **ALWAYS** be performed prior to leaving the vendor's hot site facility?

A. Erase data and software from devices.
B. Conduct a meeting to evaluate the test.
C. Complete an assessment of the hot site provider.
D. Evaluate the results from all test scripts.

A is the correct answer.

Justification:

A. For security and privacy reasons, all organizational data and software should be erased prior to departure.
B. Evaluations can occur back at the office after everyone is rested.
C. An assessment of the hot site provider should be included in the post-mortem.
D. Results of the test are a part of the post-mortem.

S4-9 Which of the following is a key component of an incident response policy?

A. Updated call trees
B. Escalation criteria
C. Press release templates
D. Critical backup files inventory

B is the correct answer.

Justification:

A. Call trees are too detailed, change too frequently and are not a part of policy.
B. Escalation criteria, indicating the circumstances under which specific actions are to be undertaken, should be contained within an incident response policy.
C. Press release templates are too detailed to be included in a policy document.
D. Lists of critical backup files are too detailed to be included in a policy document.

S4-10 How does a security information and event management solution **MOST** likely detect the existence of an advanced persistent threat in its infrastructure?

A. Through analysis of the network traffic history
B. Through stateful inspection of firewall packets
C. Through identification of zero-day attacks
D. Through vulnerability assessments

A is the correct answer.

Justification:

A. Advanced persistent threat (APT) refers to stealthy attacks not easily discovered without detailed analysis of behavior and traffic flows. Security information and event management (SIEM) solutions analyze network traffic over long periods of time to identify variances in behavior that may reveal APTs.
B. Stateful inspection is a function of some firewalls, but is not part of a SIEM solution. A stateful inspection firewall keeps track of the destination Internet Protocol address of each packet that leaves the organization's internal network. Whenever the response to a packet is received, its record is referenced to ascertain and ensure that the incoming message is in response to the request that went out from the organization.
C. Zero-day attacks are not APTs because they are unknown until they manifest for the first time and cannot be proactively detected by SIEM solutions.
D. A vulnerability assessment identifies areas that may potentially be exploited, but does not detect attempts at exploitation, so it is not related to APT.

S4-11 Why should an incident management team conduct a postincident review?

A. To identify relevant electronic evidence
B. To identify lessons learned
C. To identify the hacker
D. To identify affected areas

B is the correct answer.

Justification:

A. Evaluating the relevance of evidence is not the primary purpose for a postincident review because this should have been already established during the response to the incident.
B. Postincident reviews are beneficial in determining ways to improve the response process through lessons learned from the attack.
C. Identifying who launched the attack is not the primary purpose for a postincident review because this should have been already established during the response to the incident.
D. Identifying what areas were affected is not the primary purpose for a postincident review because this should have been already established during the response to the incident.

S4-12 What is the **MOST** important concern when an organization with multiple data centers designates one of its own facilities as the recovery site?

A. Communication line capacity between data centers
B. Current processing capacity loads at data centers
C. Differences in logical security at each center
D. Synchronization of system software release versions

B is the correct answer.

Justification:

A. Although line capacity is important from a mirroring perspective, this is secondary to having the necessary capacity to restore critical systems.
B. If data centers are operating at or near capacity, it may prove difficult to recover critical operations at an alternate data center.
C. Differences in logical security is a much easier issue to overcome and is, therefore, of less concern.
D. Synchronization of system software releases is a much easier issue to overcome and is, therefore, of less concern.

S4-13 Which of the following is **MOST** important in determining whether a disaster recovery test is successful?

A. Only business data files from offsite storage are used.
B. IT staff fully recovers the processing infrastructure.
C. Critical business processes are duplicated.
D. All systems are restored within recovery time objectives.

C is the correct answer.

Justification:

A. Although ensuring that only materials taken from offsite storage are used in the test is important, this is not as critical in determining a test's success.
B. While full recovery of the processing infrastructure is a key recovery milestone, it does not ensure the success of a test.
C. To ensure that a disaster recovery test is successful, it is most important to determine whether all critical business functions were successfully recovered and duplicated.
D. Achieving the recovery time objectives is an important milestone, but it does not necessarily prove that the critical business functions can be conducted, due to interdependencies with other applications and key elements such as data, staff, manual processes, materials and accessories, etc.

X **S4-14** Which of the following is **MOST** important when deciding whether to build an alternate facility or subscribe to a third-party hot site?

A. Cost to build a redundant processing facility and location
B. Daily cost of losing critical systems and recovery time objectives
C. Infrastructure complexity and system sensitivity
D. Criticality results from the business impact analysis

A is the correct answer.

Justification:

A. Location is critical since the recover site must not be subject to the same disaster as the primary site and then cost is the second main consideration.
B. The cost of losing critical systems is not affected by a buy or build choice.
C. Infrastructure complexity and system sensitivity is the same whether in a third-party facility or not.
D. Criticality is the same regardless of the alternate site choice.

X **S4-15** A new email virus that uses an attachment disguised as a picture file is spreading rapidly over the Internet. Which of the following should be performed **FIRST** in response to this threat?

A. Quarantine all picture files stored on file servers.
B. Block all emails containing picture file attachments.
C. Quarantine all mail servers connected to the Internet.
D. Block incoming Internet mail but permit outgoing mail.

B is the correct answer.

Justification:

A. There is no indication of infection and quarantining all picture files is unnecessary.
B. Until signature files can be updated, incoming email containing picture file attachments should be blocked.
C. Quarantine of all mail servers is unnecessary because only those emails containing attached picture files are in question.
D. Blocking all incoming mail is unnecessary as long as picture files are blocked.

S4-16 When a large organization discovers that it is the subject of a network probe, which of the following actions should be taken?

A. Reboot the router connecting the demilitarized zone (DMZ) to the firewall.
B. Power down all servers located on the DMZ segment.
C. Monitor the probe and isolate the affected segment.
D. Enable server trace logging on the affected segment.

C is the correct answer.

Justification:
A. Rebooting the router is not warranted.
B. Powering down the demilitarized zone servers is not warranted.
C. In the case of a probe, the situation should be monitored and the affected network segment isolated.
D. Enabling server trace routing is not warranted.

S4-17 An organization determined that if its email system failed for three days, the cost to the organization would be eight times greater than if it could be recovered in one day. This determination **MOST** likely was the result of:

A. disaster recovery planning.
B. business impact analysis.
C. site proximity analysis.
D. full interruption testing.

B is the correct answer.

Justification:
A. A disaster recovery plan does not include impact of system loss. A business impact analysis must be completed prior to disaster recovery planning.
B. A business impact analysis is used to establish the escalation of loss over time in addition to other elements.
C. Site proximity is a consideration during disaster recovery planning for locating your recovery site. Where the site is located does not indicate the business impact.
D. Full interruption testing is used to validate disaster recovery plans. A business impact analysis must be completed prior to disaster recovery planning.

X **S4-18** Which of the following should be performed **FIRST** in the aftermath of a denial-of-service (DoS) attack?

A. Restore servers from backup media stored offsite.
B. Conduct an assessment to determine system status.
C. Perform an impact analysis of the outage.
D. Isolate the screened subnet.

B is the correct answer.

Justification:
A. Servers may not have been affected, so it is not necessary at this point to rebuild any servers.
B. An assessment should be conducted to determine the overall system status and whether any permanent damage occurred.
C. An impact analysis of the outage will not provide any immediate benefit.
D. Isolating the screened subnet is after the fact and will not provide any benefit.

S4-19 Which of the following is the **MOST** important element to ensure the successful recovery of a business during a disaster?

A. Detailed technical recovery plans are maintained offsite.
B. Network redundancy is maintained through separate providers.
C. Hot site equipment needs are recertified on a regular basis.
D. Appropriate declaration criteria have been established.

A is the correct answer.

Justification:
A. In a major disaster, staff can be injured or can be prevented from traveling to the hot site, so technical skills and business knowledge can be lost. It is, therefore, critical to maintain an updated copy of the detailed recovery plan at an offsite location. In a disaster situation, without the detailed technical plan, business recovery will be seriously impaired.
B. Continuity of the business requires adequate network redundancy. Ideally, the business continuity program addresses this satisfactorily.
C. Continuity of the business requires hot site infrastructure that is certified as compatible and clear criteria. Ideally, the business continuity program addresses this satisfactorily.
D. Continuity of the business requires clear criteria for declaring a disaster. Ideally, the business continuity program addresses this satisfactorily.

X **S4-20** Addressing the root cause of an incident is one aspect of which of the following incident management processes?

A. Eradication
B. Recovery
C. Lessons learned
D. Containment

A is the correct answer.

Justification:
A. Determining the root cause of an incident and eliminating it are key activities that occur as part of the eradication process.
B. Recovery focuses on restoring systems or services to conditions specified in service delivery objectives (SDOs) or business continuity plans (BCPs).
C. Lessons learned are documented at the end of the incident response process, after the root cause has been identified and remediated.
D. Containment focuses on preventing the spread of damage associated with an incident, typically while the root cause either is still unknown or is known but cannot yet be remediated.

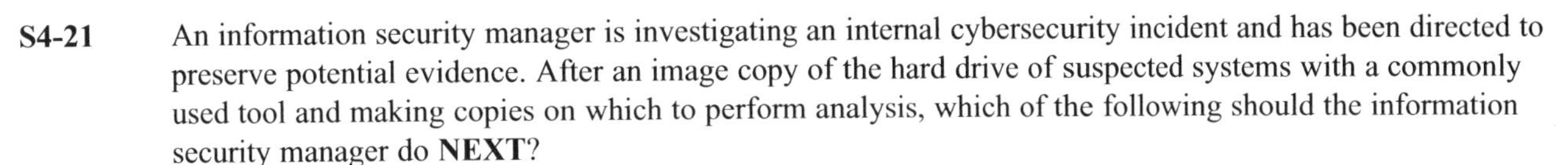

S4-21 An information security manager is investigating an internal cybersecurity incident and has been directed to preserve potential evidence. After an image copy of the hard drive of suspected systems with a commonly used tool and making copies on which to perform analysis, which of the following should the information security manager do **NEXT**?

A. Encrypt the primary and backup hard drive images.
B. Use an alternate tool to make an image copy of the hard drive.
C. Generate hashes for the primary and backup hard drive images.
D. Document the process used to make an image copy of the hard drive.

C is the correct answer.

Justification:

A. Ensuring the confidentiality of the memory dumps is not a primary concern during forensic analysis. Encrypted memory dumps cannot be analyzed.
B. If a memory dump made with an alternate tool is desired, it should be made only after the existing primary and backup dumps have been hashed, so that their authenticity can be established if necessary.
C. Generating hashes for the primary and backup memory dumps provides a means of demonstrating that the dump used for analysis is identical to the one stored for reference. It is essential that this step be performed before anything might happen to corrupt the original memory source, so it should be done as soon as possible.
D. Documentation of the process should exist as part of the incident response procedures, but if it does not, the middle of an incident is not the best time to create it.

S4-22 When an organization is using an automated tool to manage and house its business continuity plans, which of the following is the **PRIMARY** concern?

A. Ensuring accessibility should a disaster occur
B. Versioning control as plans are modified
C. Broken hyperlinks to resources stored elsewhere
D. Tracking changes in personnel and plan assets

A is the correct answer.

Justification:

A. If all of the plans exist only in electronic form, this presents a serious weakness if the electronic version is dependent on restoration of the intranet or other systems that are no longer available.
B. Versioning control is actually easier with an automated system.
C. Broken hyperlinks are a concern, but less serious than plan accessibility.
D. Tracking changes in personnel and plan assets is actually easier with an automated system.

S4-23 The information security manager identifies a vulnerability in a publicly exposed business application during risk assessment activities. The **NEXT** step he/she should take is:

A. containment.
B. eradication.
C. analysis.
D. recovery.

C is the correct answer.

Justification

A. Containment is necessary when an incident is found to have occurred. Prior to analysis, the information security manager has no way of knowing whether an incident may have occurred in the past or might even still be underway, so analysis should precede containment.
B. Eradication is undertaken once an incident has been contained, which requires that it first be analyzed to determine its scope.
C. Identification of a vulnerability does not necessarily mean that an incident has occurred, but reliance on automated detection mechanisms when a vulnerability has been identified may allow any compromises that have already occurred to continue unimpeded. Analysis is appropriate to determine whether a threat actor may have already exploited the vulnerability and, if so, to determine the scope of the compromise.
D. Recovery is the last step taken before concluding an incident. At the time that a vulnerability is detected, there is no apparent impact, so recovery is not yet needed. Eradication and recovery will take place if an incident has occurred. However, it is important to first determine if an incident has taken place.

S4-24 Which of the following actions should be taken when an information security manager discovers that a hacker is footprinting the network perimeter?

A. Reboot the border router connected to the firewall.
B. Check intrusion detection system logs and monitor for any active attacks.
C. Update IDS software to the latest available version.
D. Enable server trace routing on the demilitarized zone segment.

B is the correct answer.

Justification:

A. Rebooting the router would not be relevant.
B. Information security should check the intrusion detection system (IDS) logs and continue to monitor the situation. It would be inappropriate to take any action beyond that.
C. Updating the IDS could create a temporary exposure until the new version can be properly tuned.
D. Enabling server trace routing is of no use.

S4-25 The triage phase of the incident response plan provides:

A. a snapshot of the current status of all incident activity reported.
B. a global, high-level view of the open incidents.
C. a tactical review of incident's progression and resolution.
D. a comprehensive basis for changes to the enterprise architecture.

A is the correct answer.

Justification:

A. Triage gives a snapshot based on both strategic and tactical reviews for the purposes of assigning limited resources to where they can be most effective.
B. Triage addresses the tactical level of the incident to be able to determine the best path to resolution and does not focus exclusively on the high-level view.
C. Triage provides a view on both the tactical and strategic levels and occurs prior to resolution.
D. Triage occurs before root-cause analysis, so it does not provide a comprehensive basis for changes to the enterprise architecture.

S4-26 Which of the following is the **MOST** serious exposure of automatically updating virus signature files on every desktop each Friday at 11:00 p.m. (2300 hours)?

A. Most new viruses' signatures are identified over weekends.
B. Technical personnel are not available to support the operation.
C. Systems are vulnerable to new viruses during the intervening week.
D. The update's success or failure is not known until Monday.

C is the correct answer.

Justification:

A. The fact that most new viruses' signatures are identified over weekends is secondary to leaving systems vulnerable during the intervening week.
B. The fact that technical personnel are not available is secondary to leaving systems vulnerable during the intervening week.
C. Updating virus signature files on a weekly basis carries the risk that the systems will be vulnerable to viruses released during the week; far more frequent updating is essential.
D. The fact that success or failure is not known until Monday is secondary to leaving systems vulnerable during the intervening week.

S4-27 When performing a business impact analysis, which of the following should calculate the recovery time and cost estimates?

A. Business continuity coordinator
B. Information security manager
C. Business process owners
D. IT management

C is the correct answer.

Justification:

A. The business continuity coordinator will not be able to provide the correct level of detailed knowledge.
B. The information security manager will not have the level of detailed knowledge needed.
C. Business process owners are in the best position to understand the true impact on the business that a system outage would create.
D. IT management would not be able to provide the required level of detailed knowledge.

S4-28 Which of the following is **MOST** closely associated with a business continuity program?

A. Confirming that detailed technical recovery plans exist
B. Periodically testing network redundancy
C. Updating the hot site equipment configuration every quarter
D. Developing recovery time objectives for critical functions

D is the correct answer.

Justification:
A. Technical recovery plans are associated with infrastructure disaster recovery.
B. Network redundancy is associated with infrastructure disaster recovery.
C. Equipment needs is associated with infrastructure disaster recovery.
D. Of the choices, only recovery time objectives directly relate to business continuity.

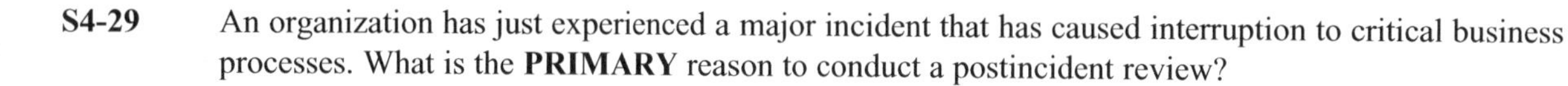

S4-29 An organization has just experienced a major incident that has caused interruption to critical business processes. What is the **PRIMARY** reason to conduct a postincident review?

A. To identify responsible parties and apply disciplinary action or reward as appropriate
B. To document the event and ensure that all steps taken and issues encountered are listed
C. To ensure that all damage that resulted from the crisis is being fixed and that systems are stable
D. To determine the root cause of the crisis and take steps to prevent re-occurrence

D is the correct answer.

Justification:
A. The postincident review should address any training needs and personnel that played a role in the event, but this is not the primary objective.
B. Documenting the event should have been conducted during the incident the postincident review will rely on the documentation to learn what happened and the chronology of events.
C. It is important to ensure that all damage has been fixed, but this should have been done before the postincident review.
D. During the incident, the efforts may have been focused on getting the business back up and running; the postincident review should ensure that the root cause of the incident is identified and addressed.

S4-30 A computer incident response team manual should **PRIMARILY** contain which of the following documents?

A. Risk assessment results
B. Severity criteria
C. Emergency call tree directory
D. Table of critical backup files

B is the correct answer.

Justification:
A. Risk assessment results are a document that would not likely be included in a computer incident response team (CIRT) manual.
B. Quickly ranking the severity criteria of an incident is a key element of incident response.
C. The emergency call tree directory is a document that would not likely be included in a CIRT manual.
D. A table of critical backup files is a document that would not likely be included in a CIRT manual.

S4-31 Which of the following would represent a violation of the chain of custody when a backup tape has been identified as evidence in a fraud investigation? The tape was:

A. removed into the custody of law enforcement investigators.
B. kept in the tape library pending further analysis.
C. sealed in a signed envelope and locked in a safe under dual control.
D. handed over to authorized independent investigators.

B is the correct answer.

Justification:

A. Removing the tape into the custody of law enforcement provides clear indication of who was in custody of the tape at all times.
B. Because a number of individuals would have access to the tape library and could have accessed and tampered with the tape, the chain of custody could not be verified.
C. Sealing the tape and locking it in a safe provides clear indication of who was in custody of the tape at all times.
D. Handing the tape over to authorized independent investigators provides clear indication of who was in custody of the tape at all times.

S4-32 When properly tested, which of the following would **MOST** effectively support an information security manager in handling a security breach?

A. Business continuity plan
B. Disaster recovery plan
C. Incident response plan
D. Vulnerability management plan

C is the correct answer.

Justification:

A. A business continuity plan would be triggered during the execution of the incident response plan in case it developed into a disaster causing serious business interruption.
B. A disaster recovery plan would be triggered during the execution of the incident response plan if it developed into a disaster.
C. An incident response plan documents the step-by-step process to follow, as well as the related roles and responsibilities pertaining to all parties involved in responding to an information security breach.
D. A vulnerability management plan is a procedure to address technical vulnerabilities and mitigate the risk through configuration changes (patch management).

S4-33 The **PRIMARY** way in which incident management adds value to an organization is by:

A. reducing the overall threat level.
B. optimizing risk management efforts.
C. eliminating redundant recovery plans.
D. streamlining the reporting structure.

B is the correct answer.

Justification:

A. Incident management focuses on prevention, containment and restoration activities and does not reduce the threat level.
B. Incident management is a component of risk management that can provide an optimal balance between prevention, containment and restoration.
C. Recovery plans are created by business and process owners. Incident management should ideally be integrated with continuity and recovery plans, but an organization does not seek to evaluate these plans for redundancy.
D. Reporting structures are typically created for business reasons. Incident management may play a role in clarifying or modifying the structures used for reporting incidents in particular, but streamlining the reporting structure is not the primary way in which incident management adds value to an organization.

X S4-34 Why is "slack space" of value to an information security manager as part of an incident investigation?

A. Hidden data may be stored there.
B. The slack space contains login information.
C. Slack space is encrypted.
D. It provides flexible space for the investigation.

A is the correct answer.

Justification:

A. "Slack space" is the unused space between where the file data end and the end of the cluster the data occupy.
B. Login information is not typically stored in the slack space.
C. Encryption for the slack space is no different from the rest of the file system.
D. Slack space is not a viable means of storage during an investigation.

S4-35 What is the **PRIMARY** objective of a postincident review in incident response?

A. To adjust budget provisioning
B. To preserve forensic data
C. To improve the response process
D. To ensure the incident is fully documented

C is the correct answer.

Justification:

A. Adjusting budget provisioning is secondary.
B. Forensic data should already be preserved and is not part of postincident review.
C. The primary objective is to find any weakness in the current process and improve it.
D. The incident should have been fully documented prior to conducting the postincident review and ensuring its completeness is secondary.

S4-36 What is the **PRIMARY** basis for a detailed business continuity plan?

A. Consideration of different alternatives
B. The solution that is least expensive
C. Strategies that cover all applications
D. Strategies validated by senior management

D is the correct answer.

Justification:
A. Senior management should select the most appropriate strategy from the alternatives provided.
B. All recovery strategies have associated costs, including costs of preparing for disruptions and putting them to use in the event of a disruption. The latter can be insured against, but not the former. The best recovery option need not be the least expensive.
C. The selection of strategy depends on criticality of the business process and applications supporting the processes. It need not necessarily cover all applications.
D. A recovery strategy identifies the best way to recover a system in case of disaster and provides guidance based on detailed recovery procedures that can be developed. Different strategies should be developed and all alternatives presented to senior management. Senior management should select the most appropriate strategy from the alternatives provided. The selected strategy should be used for further development of the detailed business continuity plan.

S4-37 A web server in a financial institution that has been compromised using a super-user account has been isolated, and proper forensic processes have been followed. What is the most appropriate next step?

A. Rebuild the server from the last verified backup.
B. Place the web server in quarantine.
C. Shut down the server in an organized manner.
D. Rebuild the server with original media and relevant patches.

D is the correct answer.

Justification:
A. Rebuilding from the last known verified backup poses the risk that the verified backup may have been compromised by the super-user at a different time.
B. Placing the web server in quarantine should have already occurred in the forensic process.
C. The step of shutting down in an organized manner is out of sequence and no longer a problem. The forensic process is already finished and evidence has already been acquired.
D. The original media should be used because one could never find and eliminate all the changes a super-user may have made or the time lines in which these changes were made.

S4-38 Evidence from a compromised server must be acquired for a forensic investigation. What would be the **BEST** source?

A. A bit-level copy of the hard drive
B. The last verified backup stored offsite
C. Data from volatile memory
D. Backup servers

A is the correct answer.

Justification:

A. The bit-level copy image file ensures forensic quality evidence that is admissible in a court of law.
B. The last verified backup will not copy everything and will not provide a forensic quality image for investigative work.
C. Dumping memory runs the risk that swap files or other disk activities will alter disk-based evidence. Standard advice from law enforcement is to pull the power plug on the compromised server to maximize preservation of evidence.
D. Backup servers may not have been compromised.

S4-39 Which of the following choices is **MOST** useful to an incident response team determining the severity level of reported security incidents?

A. Reviewing past incidents to determine impact
B. Integrating incident management with business continuity
C. Maintaining an inventory of assets and resources
D. Involving managers from affected operational areas

D is the correct answer.

Justification:

A. Past incidents can be a useful guide to the types and severity of incidents but will not necessarily provide any information on a current incident.
B. Integrating incident management with business continuity facilitates response to high-severity incidents, but severity level must be determined prior to invoking the business continuity plan.
C. Maintaining an inventory of assets and resources may be helpful when determining the severity of incidents but is not a requirement.
D. The incident response team is likely not as well informed regarding each operational area impacted by a security incident as the managers from those areas, so it makes sense to consult with the managers to get their estimates.

S4-40 What is the **FIRST** action an information security manager should take when a company laptop is reported stolen?

A. Evaluate the impact of the information loss.
B. Update the corporate laptop inventory.
C. Initiate appropriate incident response procedures.
D. Disable the user account immediately.

C is the correct answer.

Justification:

A. Evaluating the impact of the information loss would be a part of incident response procedures.
B. Updating inventory is of minor significance and can be done anytime.
C. The first step is to initiate incident response procedures.
D. Disabling the user account would be addressed as a part of incident response.

S4-41 Which of the following actions should take place immediately after a security breach is reported to an information security manager?

A. Confirm the incident.
B. Determine impact.
C. Notify affected stakeholders.
D. Isolate the incident.

A is the correct answer.

Justification:

A. **Before performing analysis of impact, notification or isolation of an incident, it must be validated as a real security incident.**
B. Before performing analysis of the impact of an incident, it must be validated as a real security incident.
C. Before notification of stakeholders, it must be validated as a real security incident.
D. Before isolation of an incident, it must be validated as a real security incident.

S4-42 What is the **PRIMARY** factor that should be taken into consideration when designing the technical solution for a disaster recovery site?

A. Service delivery objective
B. Recovery time objective
C. Allowable interruption window
D. Maximum tolerable outage

C is the correct answer.

Justification:

A. The service delivery objective is the required level of functionality that must be supported during the alternate process mode until the normal situation is restored, which is directly related to business needs.
B. The recovery time objective (RTO) is commonly agreed to be the time frame between a disaster and the return to normal or acceptable operations defined by the service level objective. The RTO must be shorter than the allowable interruption window (AIW).
C. **The length of the AIW is defined by business management and determines the acceptable time frame between a disaster and the restoration of critical services/applications. AIW is generally based on the downtime before the organization suffers major financial damage. The technical implementation of the disaster recovery site will be based on this constraint, especially the choice between a mirrored, hot, warm or cold site.**
D. Maximum tolerable outage is the amount of time the organization can operate in alternate mode based on various factors such as accessibility and performance levels.

X

S4-43 What is the **PRIMARY** factor to be taken into account when designing a backup strategy that will be consistent with a disaster recovery strategy?

A. Volume of sensitive data
B. Recovery point objective
C. Recovery time objective
D. Interruption window

B is the correct answer.

Justification:

A. The volume of data will be used to determine the capacity of the backup solution.
B. The recovery point objective defines the maximum loss of data acceptable by the business (i.e., age of data to be restored). It will directly determine the basic elements of the backup strategy—frequency of the backups and what kind of backup is the most appropriate (disk-to-disk, on tape, mirroring).
C. The recovery time objective—the time between disaster and return to normal operation—will not have any impact on the backup strategy.
D. The availability to restore backups in a time frame consistent with the interruption window will have to be checked and will influence the strategy (e.g., full backup versus incremental), but this will not be the primary factor.

S4-44 Which of the following actions is **MOST** important when a server is infected with a virus?

A. Isolate the infected server(s) from the network.
B. Identify all potential damage caused by the infection.
C. Ensure that the virus database files are current.
D. Establish security weaknesses in the firewall.

A is the correct answer.

Justification:

A. The priority in this event is to minimize the effect of the virus infection and to prevent it from spreading by removing the infected server(s) from the network.
B. After the network is secured from further infection, the damage assessment can be performed.
C. The virus signature files should be updated on a regular basis regardless of when a server was infected.
D. An undetected virus infection is a function of the antivirus software and generally unrelated to weakness in the firewall.

S4-45 Which of the following provides the **BEST** confirmation that the business continuity plan/disaster recovery plan (BCP/DRP) objectives have been achieved?

A. The recovery time objective was not exceeded during testing.
B. Objective testing of the BCP/DRP has been carried out consistently.
C. The recovery point objective was proved inadequate by DRP testing.
D. Information assets have been valued and assigned to owners according to the BCP/DRP.

A is the correct answer.

Justification:

A. Consistent achievement of recovery time objectives during testing provides the most objective evidence that business continuity plan/disaster recovery plan (BCP/DRP) objectives have been achieved.
B. Objective testing of the BCP/DRP will not serve as a basis for evaluating the alignment of the risk management process in business continuity/disaster recovery planning.
C. If the recovery point objective is inadequate, the objectives of BCPs have not been achieved.
D. Mere valuation and assignment of information assets to owners (according to the BCP/DRP) will not serve as a basis for evaluating the alignment of the risk management process in business continuity/disaster recovery planning.

S4-46 Which of the following situations would be of the **MOST** concern to a security manager?

A. Audit logs are not enabled on a production server.
B. The logon ID for a terminated systems analyst still exists on the system.
C. The help desk has received numerous results of users receiving phishing emails.
D. A Trojan was found to be installed on a system administrator's laptop.

D is the correct answer.

Justification:

A. Audit logs not enabled on a production server, although important, do not pose as immediate or as critical a threat as a Trojan installed on a system administrator's laptop.
B. The logon ID for a terminated employee existing on the system poses a risk, but unless it is a disgruntled or malicious employee, it is not likely to be a critical threat.
C. Numerous reports of phishing emails are a risk. But in this situation, employees recognize the threat and are responding appropriately, so it is not a critical threat.
D. The discovery of a Trojan installed on a system's administrator's laptop is a highly significant threat from an attacker and may mean that privileged user accounts and passwords have been compromised.

S4-47 A customer credit card database has been reported as being breached by hackers. What is the **FIRST** step in dealing with this attack?

A. Confirm the incident.
B. Notify senior management.
C. Start containment.
D. Notify law enforcement.

A is the correct answer.

Justification:

A. Validating that the condition is a true security incident is the necessary first step in determining the correct response.
B. Notifying senior management could be part of the incident response process that takes place after confirming an incident.
C. The containment stage would follow confirming the incident.
D. Notifying law enforcement by the appropriate party could be part of the incident response process that takes place after confirming an incident.

S4-48 A root kit was used to capture detailed accounts receivable information. What is the next step to ensure admissibility of evidence from a legal standpoint, once the incident has been identified and the server isolated?

A. Document how the attack occurred.
B. Notify law enforcement.
C. Take an image copy of the media.
D. Close the accounts receivable system.

C is the correct answer.

Justification:

A. Documentation is subsequent to taking an image copy and may be supplementary.
B. Notifying law enforcement is subsequent to taking an image copy, preserving evidence and maintaining the chain of custody.
C. Taking an image copy of the media along with preserving any other evidence and maintaining the chain of custody is a recommended practice to ensure legal admissibility.
D. Closing the accounts receivable system is not a practical solution.

S4-49 Which of the following is **MOST** important when collecting evidence for forensic analysis?

A. Ensure the assignment of qualified personnel.
B. Request the IT department do an image copy.
C. Disconnect from the network and isolate the affected devices.
D. Ensure law enforcement personnel are present before the forensic analysis commences.

A is the correct answer.

Justification:

A. Without the initial assignment of forensic expertise, the required levels of evidence may not be preserved properly.
B. The IT department is unlikely to have that level of expertise and should, therefore, be prevented from taking action.
C. Disconnecting from the network may be a prudent step prior to collecting evidence but does not eliminate the requirement for properly qualified forensic personnel.
D. Notifying law enforcement will likely occur after the forensic analysis has been completed.

S4-50 What is the **BEST** method for mitigating against network denial-of-service (DoS) attacks?

A. Ensure all servers are up to date on OS patches.
B. Employ packet filtering to drop suspect packets.
C. Implement network address translation to make internal addresses nonroutable.
D. Implement load balancing for Internet-facing devices.

B is the correct answer.

Justification:
A. In general, patching servers will not affect network traffic.
B. Packet filtering techniques are the only ones which reduce network congestion caused by a network denial-of-service (DoS) attack.
C. Implementing network address translation would not be effective in mitigating most network DoS attacks.
D. Load balancing would not be as effective in mitigating most network DoS attacks.

S4-51 To justify the establishment of an incident management team, an information security manager would find which of the following to be the **MOST** effective?

A. Assessment of business impact of past incidents
B. Need for an independent review of incident causes
C. Need for constant improvement on the security level
D. Possible business benefits from incident impact reduction

D is the correct answer.

Justification:
A. The assessment of business impact of past incidents would need to be completed to articulate the benefits.
B. Having an independent review benefits the incident management process.
C. The need for constant improvement on the security level is a benefit to the organization.
D. Business benefits from incident impact reduction would be the most important goal for establishing an incident management team.

S4-52 A database was compromised by guessing the password for a shared administrative account and confidential customer information was stolen. The information security manager was able to detect this breach by analyzing which of the following?

A. Invalid logon attempts
B. Write access violations
C. Concurrent logons
D. Firewall logs

A is the correct answer.

Justification:
A. Because the password for the shared administrative account was obtained through guessing, it is probable that there were multiple unsuccessful logon attempts before the correct password was deduced. Searching the logs for invalid logon attempts could, therefore, lead to the discovery of this unauthorized activity.
B. Write access violations would not necessarily be observed because the information was merely copied and not altered.
C. Because the account is shared, reviewing the logs for concurrent logons would not reveal unauthorized activity because concurrent usage is common in this situation.
D. Firewall logs would not necessarily contain information regarding logon attempts.

S4-53 The postincident review of a security incident revealed that there was a process that was not monitored. As a result monitoring functionality has been implemented. Which of the following may **BEST** be expected from this remediation?

A. Reduction in total incident duration
B. Increase in risk tolerance
C. Improvement in identification
D. Facilitation of escalation

C is the correct answer.

Justification:

A. Monitoring may cause incident durations to become longer as each event is investigated and possibly escalated for further remediation.
B. Risk tolerance is a determination made by senior management based on the results of a risk analysis and the amount of risk senior management believes the organization can manage effectively. Risk tolerance will not change from implementation of a monitoring process.
C. When a key process is not monitored, that lack of monitoring may lead to a security vulnerability or threat going undiscovered resulting in a security incident. Once consistent monitoring is implemented, identification of vulnerabilities and threats will improve.
D. Monitoring itself is simply an identification and reporting tool; it has little bearing on how information is escalated to other staff members for investigation and resolution.

X **S4-54** To determine how a security breach occurred on the corporate network, a security manager looks at the logs of various devices. Which of the following **BEST** facilitates the correlation and review of these logs?

A. Database server
B. Domain name server
C. Time server
D. Proxy server

C is the correct answer.

Justification:

A. The database server would not assist in the correlation and review of the logs.
B. The domain name server would not assist in the correlation and review of the logs.
C. To accurately reconstruct the course of events, a time reference is needed, and that is provided by the time server.
D. The proxy server would not assist in the correlation and review of the logs.

S4-55 An organization has been experiencing a number of network-based security attacks that all appear to originate internally. What is the **BEST** course of action?

A. Require the use of strong passwords.
B. Assign static Internet Protocol addresses.
C. Implement centralized logging software.
D. Install an intrusion detection system.

D is the correct answer.

Justification:

A. Requiring the use of strong passwords will not be sufficiently effective against an internal network-based attack.
B. Assigning Internet Protocol (IP) addresses would not be effective since these can be spoofed.
C. Implementing centralized logging software will not necessarily provide information on the source of the attack.
D. Installing an intrusion detection system (IDS) will allow the information security manager to better pinpoint the source of the attack so that countermeasures may then be taken. An IDS is not limited to detection of attacks originating externally. Proper placement of agents on the internal network can be effectively used to detect an internally based attack.

S4-56 A serious vulnerability is reported in the firewall software used by an organization. Which of the following should be the immediate action of the information security manager?

A. Ensure that all operating system patches are up to date.
B. Block inbound traffic until a suitable solution is found.
C. Obtain guidance from the firewall manufacturer.
D. Commission a penetration test.

C is the correct answer.

Justification:

A. Ensuring that all operating system patches are up to date is a good practice, in general, but it will not necessarily address the reported vulnerability.
B. Blocking inbound traffic may not be practical or effective from a business perspective.
C. The best source of information is the firewall manufacturer because the manufacturer may have a patch to fix the vulnerability or a work-around solution.
D. Commissioning a penetration test will take too much time and will not necessarily provide a solution for corrective actions.

S4-57 A virus incident has been reported and eradicated. The information security manager is **MOST** interested in knowing the:

A. intrusion detection system configuration.
B. type and payload of the virus.
C. virus entry path.
D. origin of the virus.

C is the correct answer.

Justification:

A. Because the virus was reported and eradicated, there is no reason to suspect that the intrusion detection system is misconfigured. The first step is to determine the entry path so that the investigation can identify what controls failed.
B. Information on type and payload of the virus is a secondary consideration because eradication has been concluded.
C. To prevent the recurrence, the security manager must find out how the virus entered the system and implement required controls.
D. The origin of the virus is not immediately actionable information and is not necessarily relevant.

S4-58 Which of the following processes is critical for deciding prioritization of actions in a business continuity plan?

A. Business impact analysis
B. Risk assessment
C. Vulnerability assessment
D. Business process mapping

A is the correct answer.

Justification:

A. A business impact analysis (BIA) provides results, such as impact from a security incident and required response times. The BIA is the most critical process for deciding which part of the information system/business process should be given prioritization in case of a security incident.
B. Risk assessment is a very important process for the creation of a business continuity plan. Risk assessment provides information on the likelihood of occurrence of security incidents and assists in the selection of countermeasures but not in the prioritization.
C. A vulnerability assessment provides information regarding the security weaknesses of the system, supporting the risk analysis process.
D. Business process mapping facilitates the creation of the plan by providing mapping guidance on actions after the decision on critical business processes has been made—translating business prioritization to IT prioritization. Business process mapping does not help in making a decision but in implementing a decision.

S4-59 In addition to backup data, which of the following is the **MOST** important to store offsite in the event of a disaster?

A. Copies of critical contracts and service level agreements
B. Copies of the business continuity plan
C. Key software escrow agreements for the purchased systems
D. List of emergency numbers of service providers

B is the correct answer.

Justification:

A. Copies of contracts and service level agreements would not be as immediately critical as the business continuity plan (BCP) itself.
B. Without a copy of the BCP, recovery efforts would be severely hampered or may not be effective. The BCP would contain a list of the emergency numbers of service providers.
C. Key software escrow agreements would not be as immediately critical as the BCP itself.
D. A list of emergency numbers would be a part of the BCP.

S4-60 Which of the following is the **MOST** important consideration for an organization interacting with the media during a disaster?

A. Communicating specially drafted messages by an authorized person
B. Refusing to comment until recovery
C. Referring the media to the authorities
D. Reporting the losses and recovery strategy to the media

A is the correct answer.

Justification:

A. Proper messages need to be sent quickly through a specific identified person so that there are no rumors or statements made that may damage reputation.
B. Refusing to comment until recovery is recommended until the message to be communicated is made clear and the spokesperson has spoken to the media.
C. Referring the media to the authorities is not recommended.
D. Reporting the losses and recovery strategy to the media is not recommended.

S4-61 During the security review of organizational servers it was found that a file server containing confidential human resources (HR) data was accessible to all user IDs. What is the **FIRST** step the security manager should perform?

A. Copy sample files as evidence.
B. Remove access privileges to the folder containing the data.
C. Report this situation to the data owner.
D. Train the HR team on properly controlling file permissions.

C is the correct answer.

Justification:

A. Copying sample files as evidence is not advisable because it breaches confidentiality requirements on the file.
B. Removing access privileges to the folder containing the data should be done by the data owner or by the security manager in consultation with the data owner—frequently the security manager would not have this right; regardless, this would be done only after formally reporting the incident.
C. The data owner should be notified prior to any action being taken.
D. Training the human resources team on properly controlling file permissions is the method to prevent such incidents in the future, but this should take place after the incident reporting and investigation activities are completed.

S4-62 What is the **PRIMARY** focus if an organization considers taking legal action on a security incident?

A. Obtaining evidence as soon as possible
B. Preserving the integrity of the evidence
C. Disconnecting all IT equipment involved
D. Reconstructing the sequence of events

B is the correct answer.

Justification:

A. Obtaining evidence as soon as possible is part of the investigative procedure but is not as important as preserving the integrity of the evidence.
B. The integrity of evidence should be kept, following the appropriate forensic techniques to obtain the evidence and a chain of custody procedure to maintain the evidence (in order to be accepted in a court of law).
C. Disconnecting involved IT equipment is part of the investigative procedure but is not as important as preserving the integrity of the evidence.
D. Reconstructing the sequence of events is part of the investigative procedure but is not as important as preserving the integrity of the evidence.

S4-63 Which of the following has the highest priority when defining an emergency response plan?

A. Critical data
B. Critical infrastructure
C. Safety of personnel
D. Vital records

C is the correct answer.

Justification:

A. Critical data are secondary to safety of personnel.
B. Critical infrastructure is secondary to safety of personnel.
C. The safety of an organization's employees should be the most important consideration given human safety laws. Human safety is considered first in any process or management practice.
D. Vital records are secondary to safety of personnel.

X **S4-64** The **PRIMARY** purpose of involving third-party teams for carrying out postincident reviews of information security incidents is to:

A. enable independent and objective review of the root cause of the incidents.
B. obtain support for enhancing the expertise of the third-party teams.
C. identify lessons learned for further improving the information security management process.
D. obtain better buy-in for the information security program.

A is the correct answer.

Justification:

A. It is always desirable to avoid the conflict of interest involved in having the information security team carry out the postevent review.
B. Obtaining support for enhancing the expertise of the third-party teams is one of the advantages but is not the primary driver.
C. Identifying lessons learned for further improving the information security management process is the general purpose of carrying out the postevent review.
D. Obtaining better buy-in for the information security program is a secondary reason for involving third-party teams.

S4-65 What is the **MOST** important objective of a postincident review?

A. Capture lessons learned to improve the process.
B. Develop a process for continuous improvement.
C. Develop a business case for the security program budget.
D. Identify new incident management tools.

A is the correct answer.

Justification:
A. The main purpose of a postincident review is to identify areas of improvement in the process.
B. Developing a process for continuous improvement is not the objective of a postincident review.
C. Developing a business case for the security program budget may be supported by the analysis of the incident but is not the key objective.
D. Identifying new incident management tools may come from the analysis of the incident but is not the key objective.

S4-66 Which of the following is the **MOST** critical consideration when collecting and preserving admissible evidence during an incident response?

A. Unplugging the systems
B. Chain of custody
C. Segregation of duties
D. Clock synchronization

B is the correct answer.

Justification:
A. Unplugging the systems is generally the preferred option in preserving evidence but is just one step.
B. Admissible evidence must be collected and preserved by maintaining the chain of custody.
C. Segregation of duties is not necessary in evidence collection and preservation because the entire process can be done by a single person.
D. Clock synchronization is not as important for the collection and preservation of admissible evidence.

S4-67 In a forensic investigation, which of the following would be the **MOST** important factor?

A. Operation of a robust incident management process
B. Identification of areas of responsibility
C. Involvement of law enforcement
D. Expertise of resources

D is the correct answer.

Justification:
A. Operation of a robust incident management process should occur prior to an investigation.
B. The identification of areas of responsibility should occur prior to an investigation.
C. Involvement of law enforcement is dependent upon the nature of the investigation.
D. The most important factor in a forensic investigation is the expertise of the resources participating in the project due to the inherent complexity.

S4-68 When a major vulnerability in the security of a critical web server is discovered, immediate notification should be made to the:

A. system owner to take corrective action.
B. incident response team to investigate.
C. data owners to mitigate damage.
D. development team to remediate.

A is the correct answer.

Justification:

A. In order to correct the vulnerabilities, the system owner needs to be notified quickly before an incident can take place.
B. Sending the incident response team to investigate is not correct because the incident has not taken place and notification could delay implementation of the fix.
C. Data owners would be notified only if the vulnerability could have compromised data.
D. The development team may be called upon by the system owner to resolve the vulnerability.

S4-69 Different types of tests exist for testing the effectiveness of recovery plans. Which of the following choices occurs during a parallel test that does not occur during a simulation test?

A. The team members step through the individual recovery tasks.
B. The primary site operations are interrupted.
C. A fictitious scenario is used for the test.
D. The recovery site is brought to operational readiness.

D is the correct answer.

Justification:

A. A walk-through of all necessary recovery tasks is part of both tests.
B. Only a full interruption test includes interruption of primary site operations.
C. Both parallel tests and simulation tests rely on fictitious scenarios.
D. A parallel recovery test includes the test of the operational capabilities of the recovery site, while a simulation test focuses on role-playing.

S4-70 Which of the following choices is the **BEST** input for the definition of escalation guidelines?

A. Risk management issues
B. A risk and impact analysis
C. Assurance review reports
D. The effectiveness of resources

B is the correct answer.

Justification:

A. Risk management deals primarily with controls and is not a viable basis for the definition of escalation guidelines.
B. A risk and impact analysis will be a basis for determining what authority levels are needed to respond to particular incidents.
C. Assurance review reports and results are primarily suited for the monitoring of stakeholder communication such as the description of the assessment of reporting effectiveness.
D. The effectiveness of resources belongs to the description of reporting and communication and is not a viable basis for the definition of escalation guidelines.

S4-71 Which of the following choices is **MOST** important to verify to ensure the availability of key business processes at an alternate site?

A. Recovery time objective
B. Functional delegation matrix
C. Staff availability to the site
D. End-to-end transaction flow

D is the correct answer.

Justification:
A. Recovery time objective (RTO) may only address a part of requirements to ensure end-to-end business operations at the alternate site.
B. Functional delegation is of secondary importance to ensure the process availability at the alternate site.
C. Staff availability is important only to the extent that it impacts process availability at the alternate site.
D. Until end-to-end transaction flow is established, recovery is not complete. Whether or not the RTO has been met is less important than achieving full recovery.

S4-72 An organization has verified that its customer information was recently exposed. Which of the following is the **FIRST** step a security manager should take in this situation?

A. Inform senior management.
B. Determine the extent of the compromise.
C. Report the incident to the authorities.
D. Communicate with the affected customers.

B is the correct answer.

Justification:
A. Before reporting to senior management, the extent of the exposure needs to be assessed.
B. Before reporting to senior management, affected customers or the authorities, the extent of the exposure needs to be assessed.
C. Reporting the incident to authorities is a management decision and not up to the security manager.
D. Communication with affected customers is a management task and is not the responsibility of the security manager.

S4-73 Which of the following choices is the **BEST** method of determining the impact of a distributed denial-of-service attack on a business?

A. Identify the sources of the malicious traffic.
B. Interview the users and document their responses.
C. Determine the criticality of the affected services.
D. Review the logs of the firewalls and intrusion detection system.

C is the correct answer.

Justification:
A. Identifying the sources of the attack may be useful to stop the attack, but does not aid in determining impact.
B. The overall impact of a distributed denial-of-service attack may be beyond the comprehension of the users as servers, databases, routers, etc., may be affected.
C. Criticality of affected services will determine the impact on the business. If affected services are not critical, then there is no cause for alarm.
D. Logs may identify the nature of the attack rather than the impact.

S4-74 What is the **PRIMARY** consideration when defining recovery time objectives for information assets?

A. Regulatory requirements
B. Business requirements
C. Financial value
D. IT resource availability

B is the correct answer.

Justification:

A. Regulatory requirements may not be consistent with business requirements.
B. The criticality to business should always drive the decision.
C. The financial value of an asset may not correspond to its business value and is irrelevant.
D. While a consideration, IT resource availability is not a primary factor.

S4-75 What task should be performed after a security incident has been verified?

A. Identify the incident.
B. Contain the incident.
C. Determine the root cause of the incident.
D. Perform a vulnerability assessment.

B is the correct answer.

Justification:

A. Identifying the incident means verifying whether an incident has occurred and finding out more details about the incident.
B. After an incident has been confirmed (identified), the incident management team should limit further exposure.
C. Determining the root cause takes place after the incident has been contained.
D. Performing a vulnerability assessment takes place after the root cause of an incident has been determined to determine if the vulnerability has been addressed.

X **S4-76** The **PRIMARY** factor determining maximum tolerable outage is:

A. available resources.
B. operational capabilities.
C. long haul network diversity.
D. last mile protection.

A is the correct answer.

Justification:

A. The main variable affecting the ability to operate in the recovery site is adequate resource availability such as diesel fuel to operate generators. Although resources would be taken into account during initial calculation of the maximum tolerable outage (MTO), circumstances associated with disaster recovery frequently have unexpected impacts on availability of resources. As a result, the expectations may not be met during real-world events.
B. The operational capabilities of the recovery site would have been predetermined and factored into the MTO.
C. Long haul diversity does not affect MTO.
D. Last mile protection does not affect MTO.

S4-77 While defining incident response procedures, an information security manager must **PRIMARILY** focus on:

A. closing incident tickets in a predetermined time frame.
B. reducing the number of incidents.
C. minimizing operational interruptions.
D. meeting service delivery objectives.

D is the correct answer.

Justification:
A. Closing tickets is not a priority of incident response.
B. Reducing the number of incidents is the focus of overall incident management.
C. Minimizing the impact on operations is not necessarily the primary focus. Some disruption in operations may be within acceptable limits.
D. The primary focus of incident response is to ensure that business-defined service delivery objectives are met.

S4-78 Which of the following capabilities is **MOST** important for an effective incident management process? The organization's capability to:

A. detect the incident.
B. respond to the incident.
C. classify the incident.
D. record the incident.

A is the correct answer.

Justification:
A. An organization must be able to detect the incident to respond, record and classify the incident. Even if response is not possible, detection allows stakeholders to be informed.
B. Responding to an incident is an essential part of incident management, but it must be detected first.
C. Incidents detected are typically classified based on impact.
D. Incidents cannot be recorded unless detected.

S4-79 Which of the following would be **MOST** appropriate for collecting and preserving evidence?

A. Encrypted hard drives
B. Generic audit software
C. Proven forensic processes
D. Log correlation software

C is the correct answer.

Justification:
A. Whether hard drives are encrypted or not is not relevant to collecting and preserving evidence.
B. Audit software is not useful for collecting and preserving evidence.
C. When collecting evidence about a security incident, it is very important to follow appropriate forensic procedures to handle electronic evidence by a method approved by local jurisdictions.
D. Log correlation software may help when collecting data about the incident; however, these data might not be accepted as evidence in a court of law if they are not collected by a method approved by local jurisdictions.

S4-80 Which of the following is the **MOST** important aspect of forensic investigations that will potentially involve legal action?

A. The independence of the investigator
B. Timely intervention
C. Identifying the perpetrator
D. Chain of custody

D is the correct answer.

Justification:

A. The independence of the investigator may be important but is not the most important aspect.
B. Timely intervention is important for containing incidents but not important for forensic investigation.
C. Identifying the perpetrator is important, but maintaining the chain of custody is more important in order to have the perpetrator convicted in court.
D. Establishing the chain of custody is one of the most important steps in conducting forensic investigations because it preserves the evidence in a manner that is admissible in court.

S4-81 In the course of examining a computer system for forensic evidence, data on the suspect media were inadvertently altered. Which of the following should have been the **FIRST** course of action in the investigative process?

A. Perform a backup of the suspect media to new media.
B. Create a bit-by-bit image of the original media source onto new media.
C. Make a copy of all files that are relevant to the investigation.
D. Run an error-checking program on all logical drives to ensure that there are no disk errors.

B is the correct answer.

Justification:

A. A backup does not preserve 100 percent of the data, such as erased or deleted files and data in slack space—which may be critical to the investigative process.
B. The original hard drive or suspect media should never be used as the source for analysis. The source or original media should be physically secured and only used as the master to create a bit-by-bit image. The original should be stored using the appropriate chain of custody procedures, depending on location. The image created for forensic analysis should be used for analysis.
C. Once data from the source are altered, they may no longer be admissible in court.
D. Continuing the investigation, documenting the date, time and data altered, are actions that may not be admissible in legal proceedings. The organization would need to know the details of collecting and preserving forensic evidence relevant to their jurisdiction.

S4-82 Which of the following recovery strategies has the **GREATEST** chance of failure?

A. Hot site
B. Redundant site
C. Reciprocal arrangement
D. Cold site

C is the correct answer.

Justification:

A. A hot site is incorrect because it is a site kept fully equipped with processing capabilities and other services by the vendor.
B. A redundant site is incorrect because it is a site equipped and configured exactly like the primary site.
C. A reciprocal arrangement is an agreement that allows two organizations to back up each other during a disaster. This approach sounds desirable, but it has the greatest chance of failure due to problems in keeping agreements and plans up to date and providing adequate processing capacity.
D. A cold site is incorrect because it is a building that has a basic environment such as electrical wiring, air conditioning, flooring, etc., and is ready to receive equipment in order to operate.

S4-83 Recovery point objectives can be used to determine which of the following?

A. Maximum tolerable period of data loss
B. Maximum tolerable downtime
C. Baseline for operational resiliency
D. Time to restore backups

A is the correct answer.

Justification:

A. The recovery point objective (RPO) is determined based on the acceptable data loss in the case of disruption of operations. RPOs effectively quantify the permissible amount of data loss in the case of interruption.
B. RPO cannot be used to determine allowable down time.
C. RPO does not set the baseline for operational resiliency.
D. RPO will determine the required frequency and type of backup. The shorter the RPO, the more frequent the backups.

S4-84 Which of the following choices is the **PRIMARY** purpose of maintaining an information security incident history?

A. To provide evidence for forensic analysis
B. To record progress and document exceptions
C. To determine a severity classification of incidents
D. To track errors to assign accountability

B is the correct answer.

Justification:

A. Recording incidents helps in providing evidence of forensic analysis in case legal action is required. Providing evidence for forensic analysis may or may not be the primary requirement for all incidents.
B. Recording information security incidents helps in maintaining a record of events from detecting the incident to closure of the incident. This helps the incident management teams in ensuring that all related aspects required for resolving, closing and preventing recurrence of incidents are covered.
C. Recording incidents also helps in identifying all required parameters for determining a severity classification; however, incident management is focused on containment, prevention and recovery.
D. Tracking errors to assign accountability is not the primary purpose for recording details of information security incidents. Process improvement is the primary purpose.

S4-85 When electronically stored information is requested during a fraud investigation, which of the following should be the **FIRST** priority?

A. Assigning responsibility for acquiring the data
B. Locating the data and preserving the integrity of the data
C. Creating a forensically sound image
D. Issuing a litigation hold to all affected parties

B is the correct answer.

Justification:

A. While assigning responsibility for acquiring the data is a step that should be taken, it is not the first step or the highest priority.
B. Locating the data and preserving data integrity are the first priorities.
C. Creating a forensically sound image may or may not be a necessary step, depending on the type of investigation, but it would never be the first priority.
D. Issuing a litigation hold to all affected parties might be a necessary step early on in an investigation of certain types, but not the first priority.

S4-86 When creating a forensic image of a hard drive, which of the following should be the **FIRST** step?

A. Identify a recognized forensics software tool to create the image.
B. Establish a chain of custody log.
C. Connect the hard drive to a write blocker.
D. Generate a cryptographic hash of the hard drive contents.

B is the correct answer.

Justification:

A. Identifying a recognized forensics software tool to create the image is one of the important steps, but it should come after several of the other options.
B. The first step in any investigation requiring the creation of a forensic image should always be to maintain the chain of custody.
C. Connecting the hard drive to a write blocker is an important step, but it must be done after the chain of custody has been established.
D. Generating a cryptographic hash of the hard drive contents is another important subsequent step.

S4-87 When a significant security breach occurs, what should be reported **FIRST** to senior management?

A. A summary of the security logs that illustrates the sequence of events
B. An explanation of the incident and corrective action taken
C. An analysis of the impact of similar attacks at other organizations
D. A business case for implementing stronger logical access controls

B is the correct answer.

Justification:

A. A summary of security logs would be too technical to report to senior management.
B. When reporting an incident to senior management, the initial information to be communicated should include an explanation of what happened and how the breach was resolved.
C. An analysis of the impact of similar attacks would be desirable; however, this would be communicated later in the process.
D. A business case for improving controls may be appropriate subsequently as a result of investigating the cause of the breach.

S4-88 The **BEST** time to determine who should be responsible for declaring a disaster is:

A. during the establishment of the plan.
B. after an incident has been confirmed by operations staff.
C. after fully testing the incident management plan.
D. after the implementation details of the plan have been approved.

A is the correct answer.

Justification:

A. Roles and responsibilities for all involved in incident response should be established when the incident response plan is established.
B. Determining roles and responsibilities during a disaster is not the best time to make such decisions, unless it is absolutely necessary.
C. While testing the plan may drive some changes in roles based on test results, roles (including who declares the disaster) should have been established before testing and plan approval.
D. Roles and responsibilities for all involved in incident response should be established when the incident response plan is established, not after the details have been approved.

S4-89 The **PRIMARY** objective of incident response is to:

A. investigate and report results of the incident to management.
B. gather evidence.
C. minimize business disruptions.
D. assist law enforcement in investigations.

C is the correct answer.

Justification:
A. Investigating and reporting results of the incident is a responsibility of incident response teams but not the primary objective.
B. Gathering evidence is an activity that an incident response team may conduct, depending on circumstances, but not a primary objective.
C. The primary role of incident response is to detect, respond to and contain incidents so that impact to business operations is minimized.
D. Assisting law enforcement is an activity that an incident response team may conduct, depending on circumstances, but not a primary objective.

S4-90 An information security manager is in the process of investigating a network intrusion. One of the enterprise's employees is a suspect. The manager has just obtained the suspect's computer and hard drive. Which of the following is the **BEST** next step?

A. Create an image of the hard drive.
B. Encrypt the data on the hard drive.
C. Examine the original hard drive.
D. Create a logical copy of the hard drive.

A is the correct answer.

Justification:
A. One of the first steps in an investigation is to create an image of the original hard drive. A physical copy will copy the data, block by block, including any hidden data blocks and hidden partitions that can be used to conceal evidence.
B. Encryption is not required.
C. Examining the hard drive is not good practice because it risks destroying or corrupting evidence.
D. A logical copy will only copy the files and folders and may not copy other necessary data to properly examine the hard drive for forensic evidence.

S4-91 The factor that is **MOST** likely to result in identification of security incidents is:

A. effective communication and reporting processes.
B. clear policies detailing incident severity levels.
C. intrusion detection system capabilities.
D. security awareness training.

D is the correct answer.

Justification:
A. Timely communication and reporting is only useful after identification of an incident has occurred.
B. Understanding how to establish severity levels is important, but it is not the essential element of ensuring that the information security manager is aware of anomalous events that might signal an incident.
C. Intrusion detection systems are useful for detecting IT-related incidents but not useful in identifying other types of incidents such as social engineering or physical intrusion.
D. Ensuring that employees have the knowledge to recognize and report a suspected incident is most likely to result in identification of security incidents.

S4-92 Which of the following functions is responsible for determining the members of the enterprise's response teams?

A. Governance
B. Risk management
C. Compliance
D. Information security

D is the correct answer.

Justification:
A. The governance function will determine the strategy and policies that will set the scope and charter for incident management and response capabilities.
B. While response is a component of managing risk, the basis for risk management is determined by governance and strategy requirements.
C. Compliance would not be directly related to this activity, although this function may have representation on the incident response team.
D. The information security manager, or designated manager for incident response, should select the team members required to ensure that all required disciplines are represented on the team.

X **S4-93** The typical requirement for security incidents to be resolved quickly and service restored is:

A. always the best option for an enterprise.
B. often in conflict with effective problem management.
C. the basis for enterprise risk management activities.
D. a component of forensics training.

B is the correct answer.

Justification:

A. Quickly restoring service will not always be the best option such as in cases of criminal activity, which requires preservation of evidence precluding use of the systems involved.
B. Problem management is focused on investigating and uncovering the root cause of incidents, which will often be a problem when restoring service compromises the evidence needed.
C. Managing risk goes beyond the quick restoration of services (e.g., if doing so increased some other risk disproportionately).
D. Forensics is concerned with legally adequate collection and preservation of evidence, not with service continuity.

X **S4-94** Which of the following should be the **FIRST** action to take when a fire spreads throughout the building?

A. Check the facility access logs.
B. Call together the crisis management team.
C. Launch the disaster recovery plan.
D. Launch the business continuity plan.

A is the correct answer.

Justification:

A. Safety of people always comes first; therefore, verifying access logs of personnel to the facility should be the first action in order to ensure that all staff can be accounted for.
B. Calling the crisis management team together should be done after the initial emergency response (i.e., evacuation of people).
C. Launching the disaster recovery plan is not the first action.
D. Launching the business continuity plan is not the first action.

S4-95 Which of the following tests gives the **MOST** assurance that a business continuity plan works, without potentially impacting business operations?

A. Checklist tests
B. Simulation tests
C. Walk-through tests
D. Full operational tests

B is the correct answer.

Justification:

A. With checklist tests, copies of the business continuity plan are distributed to various persons for review. In these tests, people do not exercise a plan.

B. Business continuity coordinators come together to practice executing a plan based on a specific scenario. This does not interrupt normal operations and provides the most assurance of the given nonintrusive methods.

C. In walk-through tests, representatives come together to go over the plan (one or more scenarios) and ensure the plan's accuracy. The plan itself is not executed.

D. Full operational tests are the most intrusive to regular operations and business productivity. The original site is actually shut down and processing is performed at another site, thus providing the most assurance, but interrupting normal business productivity.

S4-96 An employee's computer has been infected with a new virus. What should be the **FIRST** action?

A. Execute the virus scan.
B. Report the incident to senior management.
C. Format the hard disk.
D. Disconnect the computer from the network.

D is the correct answer.

Justification:

A. The virus may start infecting other computers while the virus scan is running.

B. Only when the impact to the IT environment is significant should it be reported to senior management.

C. A case of virus infection does not warrant the action. Formatting the hard disk is the last resort.

D. The first action should be containing the risk (i.e., disconnecting the computer so that it will not infect other computers on the network).

X

S4-97 The **PRIMARY** reason for senior management review of information security incidents is to:

A. ensure adequate corrective actions were implemented.
B. demonstrate management commitment to the information security process.
C. evaluate the incident response process for deficiencies.
D. evaluate the ability of the security team.

A is the correct answer.

Justification:

A. Although some corrective actions are being taken by the security team and the incident response team, management review will ensure whether there are any other corrective actions that need to be taken. Sometimes this will result in improvements to information security policies.
B. Management will not review information security incidents merely to demonstrate management commitment.
C. Management will not perform a review for fault findings such as examining the incident response process for deficiencies.
D. Management will not perform a review for fault findings such as evaluating the ability of the security team.

S4-98 Observations made by staff during a disaster recovery test are **PRIMARILY** reviewed to:

A. identify people who have not followed the process.
B. determine lessons learned.
C. identify equipment that is needed.
D. maintain evidence of review.

B is the correct answer.

Justification:

A. It is not the aim of observation to identify people who have not followed the process.
B. After a test, results should be reviewed to ensure that lessons learned are applied.
C. Identifying equipment that is needed may be part of the lessons learned but is not the sole reason for the review.
D. Review is conducted not only to maintain evidence but to make improvements.

S4-99 The **PRIMARY** selection criterion for an offsite media storage facility is:

A. that the primary and offsite facilities are not subject to the same environmental disasters.
B. that the offsite storage facility is in close proximity to the primary site.
C. the overall storage and maintenance costs of the offsite facility.
D. the availability of cost-effective media transportation services.

A is the correct answer.

Justification:

A. It is important to prevent a disaster that could affect both sites, and ensuring that the primary and offsite facilities are not subject to the same environmental disasters addresses this concern.
B. The distance between sites may be important in cases of widespread disasters; however, this is covered by ensuring that the same environmental disasters do not affect the primary and offsite facilities.
C. The costs are a secondary criterion to selection.
D. A cost-effective media transport service may be a consideration but is not the main concern.

S4-100 The recovery time objective is reached at which of the following milestones?

A. Disaster declaration
B. Recovery of the backups
C. Restoration of the system
D. Return to business as usual processing

C is the correct answer.

Justification:
A. Disaster declaration occurs at the beginning of this period.
B. Recovery of the backups occurs shortly after the beginning of this period.
C. The recovery time objective (RTO) is based on the amount of time required to restore a system.
D. Return to business as usual processing occurs significantly later than the RTO. RTO is an "objective," and full restoration may or may not coincide with the RTO. RTO can be the minimum acceptable operational level, far short of normal operations.

S4-101 The recovery point objective requires which of the following?

A. Disaster declaration
B. Before-image restoration
C. System restoration
D. After-image processing

B is the correct answer.

Justification:
A. Disaster declaration is independent of this processing checkpoint.
B. The recovery point objective is the point in the processing flow at which system recovery should occur. This is the predetermined state of the application processing and data used to restore the system and to continue the processing flow.

C. Restoration of the system can occur at a later date.
D. After-image processing can occur at a later date.

S4-102 Who would be in the **BEST** position to determine the recovery point objective for business applications?

A. Business continuity coordinator
B. Chief operations officer
C. Information security manager
D. Internal audit

B is the correct answer.

Justification:
A. It would be inappropriate for a business continuity coordinator to determine the recovery point objective (RPO) because he/she is not directly responsible for the data or the operation.
B. The RPO is the processing checkpoint to which systems are recovered. In addition to data owners, the chief operations officer is the most knowledgeable person to make this decision.
C. It would be inappropriate for the information security manager to determine the RPO because he/she is not directly responsible for the data or the operation.
D. It would be inappropriate for internal audit to determine the RPO because they are not responsible for the data or the operation.

S4-103 When the computer incident response team finds clear evidence that a hacker has penetrated the corporate network and modified customer information, an information security manager should **FIRST** notify:

A. the information security steering committee.
B. customers who may be impacted.
C. data owners who may be impacted.
D. regulatory agencies overseeing privacy.

C is the correct answer.

Justification:

A. The information security steering committee will be notified later as required by corporate policy requirements.
B. Customers will be notified later as required by corporate policy and regulatory requirements.
C. The data owners should be notified first so they can take steps to determine the extent of the damage and coordinate a plan for corrective action with the computer incident response team.
D. Regulatory agencies will be notified later as required by corporate policy and regulatory requirements.

S4-104 The systems administrator forgot to immediately notify the security officer about a malicious attack. An information security manager could prevent this situation by:

A. periodically testing the incident response plans.
B. regularly testing the intrusion detection system.
C. establishing mandatory training of all personnel.
D. periodically reviewing incident response procedures.

A is the correct answer.

Justification:

A. Security incident response plans should be tested to find any deficiencies and improve existing processes.
B. Testing the intrusion detection system is a good practice but would not have prevented this situation.
C. All personnel need to go through formal training to ensure that they understand the process, tools and methodology involved in handling security incidents. However, testing of the actual plans is more effective in ensuring that the process works as intended.
D. Reviewing the response procedures is not enough; the security response plan needs to be tested on a regular basis.

S4-105 Which of the following would a security manager establish to determine the target for restoration of normal processing?

A. Recovery time objective
B. Maximum tolerable outage
C. Recovery point objectives
D. Service delivery objectives

A is the correct answer.

Justification:

A. **Recovery time objective is the length of time from the moment of an interruption until the time the process must be functioning at a service level sufficient to limit financial and operational impacts to an acceptable level.**
B. Maximum tolerable outage is the maximum time for which an organization can operate in alternate mode.
C. Recovery point objectives relate to the age of the data required for recovery.
D. Service delivery objectives are the levels of service required for acceptable operations.

S4-106 Which of the following should be the **PRIMARY** basis for making a decision to establish an alternate site for disaster recovery?

A. A business impact analysis, which identifies the requirements for availability of critical business processes
B. Adequate distance between the primary site and the alternate site so that the same disaster does not simultaneously impact both sites
C. A benchmarking analysis of similarly situated enterprises in the same geographic region to demonstrate due diligence
D. Differences between the regulatory requirements applicable at the primary site and those at the alternate site

A is the correct answer.

Justification:

A. **The business impact analysis will help determine the recovery time objective and recovery point objective for the enterprise. This information will drive the decision on the requirements for an alternate site.**
B. Natural disasters are just one of many factors that an enterprise must consider when it decides whether to pursue an alternate site for disaster recovery.
C. While a benchmark could provide useful information, the decision should be based on a BIA, which considers factors specific to the enterprise.
D. Regulatory requirements are just one of many factors that an enterprise must consider when it decides whether to pursue an alternate site for disaster recovery.

S4-107 During a business continuity plan test, one department discovered that its new software application was not going to be restored soon enough to meet the needs of the business. This situation can be avoided in the future by:

A. conducting a periodic and event-driven business impact analysis to determine the needs of the business during a recovery.
B. assigning new applications a higher degree of importance and scheduling them for recovery first.
C. developing a help desk ticket process that allows departments to request recovery of software during a disaster.
D. conducting a thorough risk assessment prior to purchasing the software.

A is the correct answer.

Justification:
A. A periodic business impact analysis (BIA) can help compensate for changes in the needs of the business for recovery during a disaster.
B. Assigning new applications a higher degree of importance and scheduling them for recovery first is an incorrect assumption regarding the automatic importance of a new program.
C. Developing a help desk ticket process that allows departments to request recovery of software during a disaster is not an appropriate recovery procedure because it allows individual business units to make unilateral decisions without consideration of broader implications.
D. The risk assessment may not include the BIA.

S4-108 The main mail server of a financial institution has been compromised at the superuser level; the only way to ensure that the system is secure would be to:

A. change the root password of the system.
B. implement multifactor authentication.
C. rebuild the system from the original installation medium.
D. disconnect the mail server from the network.

C is the correct answer.

Justification:
A. Changing the root password of the system does not ensure the integrity of the mail server.
B. Implementing multifactor authentication is an aftermeasure and does not clear existing security threats.
C. Rebuilding the system from the original installation medium is the only way to ensure all security vulnerabilities and potential stealth malicious programs have been destroyed.
D. Disconnecting the mail server from the network is an initial step but does not guarantee security.

S4-109 Which of the following would present the **GREATEST** risk to information security?

A. Virus signature files updates are applied to all servers every day.
B. Security access logs are reviewed within five business days.
C. Critical patches are applied within 24 hours of their release.
D. Security incidents are investigated within five business days.

D is the correct answer.

Justification:
A. Virus signature files updated every day do not pose a great risk.
B. Reviewing security access logs within five days is not the greatest risk.
C. Patches applied within 24 hours is not a significant risk.
D. Waiting to investigate security incidents can pose a major risk.

S4-110 The effectiveness of an incident response team is **BEST** measured by the:

A. percentage of incidents resolved within previously agreed-on time limits.
B. number of change requests submitted as a result of reported incidents.
C. percentage of unresolved events still open at the end of any given month.
D. number of incidents originating from external sources.

A is the correct answer.

Justification:

A. The goal of incident response is to resolve incidents within agreed-on time limits.
B. The number of change requests related to infrastructure changes simply indicates that there have been required changes to the internal architecture. Those change requests may or may not have anything to do with found vulnerabilities or reported incidents.
C. The end of the month is an arbitrary time, unrelated to agreed-on time limits for incident resolution.
D. The source of incidents does not provide input concerning the effectiveness of incident management.

S4-111 Prioritization of incident response activities is driven primarily by a:

A. recovery point objective.
B. quantitative risk assessment.
C. business continuity plan.
D. business impact analysis.

D is the correct answer.

Justification:

A. A recovery point objective identifies the maximum acceptable data loss associated with successful recovery. It does not prioritize the order of incident response.
B. Risk assessment (both qualitative and quantitative) examines sources of threat, associated vulnerability and probability of occurrence. At the point that an incident occurs, the probability aspect of risk is no longer unknown, so the degree of impact drives the prioritization of incident response, captured in the specialized business impact analysis.
C. Business continuity plans define procedures to follow when business functions are impacted. They do not prioritize the order of incident response.
D. Business impact analysis is a systematic activity designed to assess the effect upon an organization associated with impairment or loss of a function. At the point that an incident occurs, its probability is no longer unknown, so it is the potential impact on the organization that determines prioritization of response activities.

S4-112 Which of the following is the **BEST** indicator that operational risk is effectively managed in an enterprise?

A. A tested business continuity plan/disaster recovery plan
B. An increase in timely reporting of incidents by employees
C. Extent of risk management education
D. Regular review of risk by senior management

A is the correct answer.

Justification:

A. **A tested business continuity plan/disaster recovery plan is the best indicator that operational risk is managed effectively in the enterprise.**
B. Reporting incidents by employees is an indicator but not the best choice because it is dependent upon the knowledge of the employees.
C. Extent of risk management education is not correct because this may not necessarily indicate that risk is effectively managed in the enterprise. A high level of risk management education would help but would not necessarily mean that risk is managed effectively.
D. Regular review of risk by senior management is not correct because this may not necessarily indicate that risk is effectively managed in the enterprise. Top management involvement would greatly help but would not necessarily mean that risk is managed effectively.

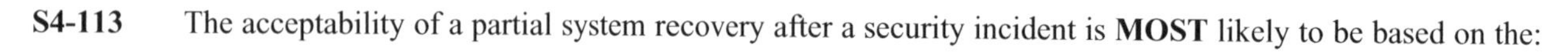

S4-113 The acceptability of a partial system recovery after a security incident is **MOST** likely to be based on the:

A. ability to resume normal operations.
B. maximum tolerable outage.
C. service delivery objective.
D. acceptable interruption window.

C is the correct answer.

Justification:

A. The ability to resume normal operations is situational and would not be a standard for acceptability.
B. While the maximum tolerable outage, in addition to many other factors, is part of a service delivery objective (SDO), by itself, it does not address the acceptability of a specific level of operational recovery.
C. **A prior determination of acceptable levels of operation in the event of an outage is the SDO. The SDO may be set at less than normal operation levels, but sufficient to sustain essential business functions.**
D. While the acceptable interruption window, in addition to many other factors, is part of an SDO, by itself, it does not address the acceptability of a specific level of operational recovery.

S4-114 A password hacking tool was used to capture detailed bank account information and personal identification numbers. Upon confirming the incident, the **NEXT** step is to:

A. notify law enforcement.
B. start containment.
C. make an image copy of the media.
D. isolate affected servers.

B is the correct answer.

Justification:
A. Notifying law enforcement should be performed after the containment plan has been executed.
B. After an incident has been confirmed, containment is the first priority of incident response because it will generally mitigate further impact.
C. Making an image copy of the media should be performed after the containment plan has been executed.
D. Isolating affected servers is part of containment.

S4-115 An employee has found a suspicious file on a server. The employee thinks the file is a virus and contacts the information security manager. What is the **FIRST** step to take?

A. Contain the file.
B. Delete the file.
C. Verify whether the file is malicious.
D. Report the suspicious file to management.

C is the correct answer.

Justification:
A. Containment is the next step in the incident response cycle.
B. Deleting the file could be part of the containment process after it has been determined that it is safe to do so.
C. The first step in incident response is to verify whether the file is malicious.
D. Reporting to management would be a later step in the incident handling cycle and will vary based on policy, but it would not come before verification or general containment.

✗ **S4-116** Which of the following **MOST** effectively reduces false-positive alerts generated by a security information and event management process?

A. Building use cases
B. Conducting a network traffic analysis
C. Performing an asset-based risk assessment
D. The quality of the logs

A is the correct answer.

Justification:

A. Implementing a security information and event management (SIEM) process helps ensure that incidents are correctly identified and handled appropriately. Because an SIEM process depends on log analysis based on predefined rules, the most effective way to reduce false-positive alerts is to develop use cases for known threats to identified critical systems. The use cases would then be used to develop appropriate rules for the SIEM solution.

B. Although security monitoring requires traffic analysis, only properly defined use cases can ensure that the rules are accurately defined and that events are properly identified, thereby reducing false-positive alerts.

C. A risk assessment will not reduce false positive alerts.

D. The quality of the logs can affect alerts but is usually a minor consideration.

S4-117 A forensic team was commissioned to perform an analysis of unrecognized processes running on a desktop personal computer. The lead investigator advised the team against disconnecting the power in order to:

A. prevent disk corruption.
B. conduct a hot-swap of the main disk drive.
C. avoid loss of data in server logs.
D. avoid loss of data stored in volatile memory.

D is the correct answer.

Justification:

A. Preventing disk corruption does not address the capture of the data that exist in volatile memory.

B. Conducing a hot-swap of the main disk drive does not address the capture of the data that exist in volatile memory.

C. Avoiding loss of data in server logs does not address the capture of the data that exist in volatile memory.

D. Disconnecting power from a system results in loss of data stored in volatile memory. Those data could be vital for the investigation and for understanding the extent of the impact of the event. Disconnecting power is not recommended where analysis of running processes or the content of volatile memory is required.

X

S4-118 When attempting data recovery of a specific file during forensic analysis, an investigator would be challenged the **MOST** when:

A. all files in the directory have been deleted.
B. the partition table on the disk has been deleted.
C. the file content has been overwritten.
D. high-level disk formatting has been performed.

C is the correct answer.

Justification:
A. Deleted files that have not been physically overwritten can generally be retrieved using commonly available forensic tools.
B. Partition tables can generally be retrieved using commonly available forensic tools.
C. When the actual file content on the disk is overwritten, it generally cannot be recovered without significant resources and highly specialized tools, and frequently, it cannot be recovered at all.
D. Drives that have been high-level formatted can generally be retrieved using commonly available forensic tools.

S4-119 Major security events with serious legal implications should be communicated to:

A. appropriate civil authorities when there has been a crime committed.
B. management after the incident has been verified and the severity determined.
C. all affected stakeholders, including legal and the insurance carrier.
D. only to human resources and the legal department for appropriate action.

B is the correct answer.

Justification:
A. There are few, if any, circumstances where the information security manager should contact external authorities directly.
B. Communication regarding security events, particularly ones that have legal implications, is a business decision that is the responsibility of management.
C. It is the decision of management to determine which stakeholders and external entities should be informed. This process should be detailed in the enterprise's incident response communication plan.
D. Human resources and legal would not be the only departments communicated with in this situation.

S4-120 Which of the following is the **MOST** effective method to ensure that a business continuity plan (BCP) meets an organization's needs?

A. Require quarterly updating of the BCP.
B. Automate the survey of plan owners to obtain input to the plan.
C. Periodically test the cross-departmental plan with varied scenarios.
D. Conduct face-to-face meetings with management for discussion and analysis.

C is the correct answer.

Justification:
A. Quarterly updates do not establish that a plan meets the organization's needs.
B. Automated surveys is a method that could be used during testing but, on its own, is not sufficient.
C. Cross-departmental testing of a plan with varied scenarios is most effective in determining the validity of a business continuity plan (BCP).
D. Face-to-face meetings is a method that could be used during testing but, on its own, is not sufficient.

S4-121 After a significant security breach has occurred, what is the **MOST** important item to report to the chief information officer?

A. A summary of the security logs that illustrates the sequence of events
B. An analysis of the impact of similar attacks at other organizations
C. A business case for implementing stronger logical access controls
D. The impact of the incident and corrective actions taken

D is the correct answer.

Justification:
A. A summary of security logs would be too technical to report to the chief information officer (CIO).
B. An analysis of the impact of similar attacks would be helpful but is not the most important item to report.
C. A business case for implementing stronger controls would be helpful to report to management, but it is not the most important item to report and would be subsequent to reporting impact and corrective actions.
D. The actual impact to the organization and corrective actions taken would be the most important item to share with the CIO.

X **S4-122** In a large organization, effective management of security incidents will be **MOST** dependent on:

A. clear policies detailing incident severity levels.
B. broadly dispersed intrusion detection capabilities.
C. training employees to recognize security incidents.
D. effective communication and reporting processes.

D is the correct answer.

Justification:
A. Understanding severity levels is important but, on its own is, not sufficient to ensure that the information security manager is able to manage the incident effectively.
B. Intrusion detection is useful for detecting potential network security incidents, but without robust communication and reporting processes, the tool is less effective.
C. Conducting awareness training so individuals can recognize potential incidents is important, but not effective unless the information is communicated to the right people in a timely manner.
D. Timely communication and reporting is most likely to ensure that the information security manager receives the information necessary to effectively manage a security incident. Effective communication will also help ensure that the correct resources are engaged at the appropriate time.

S4-123 Which of the following choices includes the activity of evaluating the computing infrastructure by performing proactive security assessment and evaluation?

A. A disaster recovery plan
B. A business continuity plan
C. An incident management plan
D. A continuity of operations plan

C is the correct answer.

Justification:
A. A disaster recovery plan is a set of human, physical, technical and procedural resources to recover, within a defined time and cost, an activity interrupted by an emergency.
B. A business continuity plan is a plan used by an organization to respond to disruption of critical business processes. It depends on the contingency plan for restoration of critical systems.
C. This activity is part of the "protect" phase of the incident management planning process flow.
D. A continuity of operations plan is an effort within individual executive departments and agencies to ensure that primary mission-essential functions continue to be performed during a wide range of emergencies, including localized acts of nature, accidents and technological or attack-related emergencies.

S4-124 Proximity factors must be considered when:

A. conducting a business impact analysis.
B. conducting a table-top business continuity test.
C. developing disaster recovery metrics.
D. selecting an alternate recovery site.

D is the correct answer.

Justification:
A. Proximity to hazards is not a primary consideration in conducting a business impact analysis.
B. Proximity to hazards is not a primary consideration in conducting a table-top business continuity test.
C. Proximity to hazards is not a primary consideration in developing disaster recovery metrics.
D. Proximity to the primary site, the scope of potential hazards, and their possible impact on the recovery site are important considerations when selecting the location of a recovery site.

S4-125 An organization's chief information security officer would like to ensure that operations are prioritized correctly for recovery in case of a disaster. Which of the following would be the **BEST** to use?

A. A business impact analysis
B. An organization risk assessment
C. A business process map
D. A threat statement

A is the correct answer.

Justification:
A. A business impact analysis (BIA) ensures that operations are prioritized correctly for recovery in case of a disaster.
B. An organization risk assessment would not support prioritization of system recovery.
C. A business process map would not support prioritization of system recovery.
D. A threat statement would not support prioritization of system recovery.

S4-126 The purpose of incident management and response is to:

A. recover an activity interrupted by an emergency or disaster, within a defined time and cost.
B. perform a walk-through of the steps required to recover from an adverse event.
C. reduce business disruption insurance premiums for the business.
D. address disruptive events with the objective of controlling impacts within acceptable levels.

D is the correct answer.

Justification:

A. This is the definition of a disaster recovery plan (DRP). The incident response process is sequentially the first response to an adverse event with aims of preventing the incident from escalating to a disaster.
B. A DRP table-top test or walk-through is performed to exercise the DRP in a test scenario to determine whether the steps that the organization needs to take to recover are reliably documented.
C. Business disruption insurance is an instrument of the risk management strategy to diversify and distribute the costs associated with an adverse event to a third party. Business insurance premiums are not dependent on incident management and response.
D. Incident management and response is a component of business continuity planning. As a first response to adverse events, the objective of incident management and response is to prevent incidents from becoming problems and to prevent problems from becoming disasters.

S4-127 For global organizations, which of the following is **MOST** essential to the continuity of operations in an emergency situation?

A. A documented succession plan
B. Distribution of key process documents
C. A reciprocal agreement with an alternate site
D. Strong senior management leadership

B is the correct answer.

Justification:

A. During contingency situations, contact with one or more senior managers may be lost. In such cases, a documented succession plan is important as a means of establishing who is empowered to make decisions on behalf of the organization. However, if an organization experiencing a contingency situation has only a succession plan and no distributed key process documentation, the effectiveness of the empowered decision maker will be limited. A succession plan is, therefore, worthwhile but less important than process documentation.
B. Many factors come into play during contingency situations, but continuity is possible only when personnel who are able to resume key processes have the knowledge of how to do so. When key process documentation is distributed to contingency locations, it is available for the use of any staff who report to these locations during contingencies, and so long as that documentation is up to date, it may be used even by those who may not typically be involved in performing those functions.
C. Reciprocal agreements are established when contingency sites are shared among multiple business partners. There are business justifications for establishing these relationships, but having them established is generally not going to ensure continuity of operations.
D. Strong leadership by senior management drives the preparation that goes into continuity of operations planning before a contingency situation arises. Assuming that this preparation has been adequate, however, the continuity functions should be carried out by organization personnel even if leadership during the contingency is interrupted or lacking in strength.

S4-128 The **PRIMARY** objective of continuous monitoring is to:

A. minimize the magnitude of impact.
B. align the security program with IT goals.
C. identify critical information assets.
D. reduce the number of policy exceptions.

A is the correct answer.

Justification:
A. Continuous monitoring helps an organization identify adverse events in a timely manner. The reduced lag time to take steps to contain damage results in minimizing the impact.
B. Aligning the security program with IT goals is a derived benefit of continuous monitoring rather than the primary objective.
C. Identifying critical information assets is a prerequisite for implementing continuous monitoring.
D. Reduction of policy exceptions is not a direct benefit of continuous monitoring.

S4-129 Which of the following is the **FIRST** step in developing an incident response plan?

A. Set the minimum time required to respond to incidents.
B. Establish a process to report incidents to senior management.
C. Ensure the availability of skilled resources.
D. Categorize incidents based on likelihood and impact.

D is the correct answer.

Justification:
A. Determining response time is based on the categorization of incidents.
B. The process for reporting depends on the categorization. Management may want only high-severity incidents to be reported.
C. The resources required depend on the categorization of the incident and the established response time.
D. Incidents with higher likelihood and impact warrant more attention.

S4-130 A security operations center detected an attempted structured query language injection, but could not determine if it was successful. Which of the following resources should the information security manager approach to assess the possible impact?

A. Application support team
B. Business process owner
C. Network management team
D. System administrator

A is the correct answer.

Justification:
A. Structured query language (SQL) injection is an application-based attack. Because the security operations center has detected an attempt of SQL injection and could not determine if it was successful, the information security manager should approach the application support group that has access to data in order to identify the impact.
B. The business process owner may help the application support group determine the overall impact, after it has been determined if the attack has been successful.
C. Because SQL injection is an application-based attack, the network management team is not the best resource to assess the possible impact.
D. The system administrator is not the best resource to assess the possible impact. However, he/she may assist the application support team and assist with incident response activities, should the attack have been successful.

S4-131 Which of the following is the **FIRST** step after the intrusion detection system sends out an alert about a possible attack?

A. Assess the type and severity of the attack.
B. Determine whether it is an actual incident.
C. Contain the damage to minimize the risk.
D. Minimize the disruption of computer resources.

B is the correct answer.

Justification:
A. The type and severity of the attack should be studied after it is concluded that the incident is valid.
B. An administrator conducting regular maintenance activities may trigger a false-positive alarm from the intrusion detection system. One must validate a real incident before taking any action.
C. Damage should be contained and risk minimized after confirming a valid incident, thus discovering the type and severity of the attack.
D. One of the goals of incident response is to minimize the disruption of computer resources.

S4-132 After a service interruption of a critical system, the incident response team finds that it needs to activate the warm recovery site. Discovering that throughput is only half of the primary site, the team nevertheless notifies management that it has restored the critical system. This is **MOST** likely because it has achieved the:

A. recovery point objective.
B. recovery time objective.
C. service delivery objective.
D. maximum tolerable outage.

C is the correct answer.

Justification:
A. The recovery point objective (RPO) is determined based on the acceptable data loss in case of a disruption of operations. It indicates the earliest point in time that is acceptable to recover the data. The RPO effectively quantifies the permissible amount of data loss in case of interruption.
B. The recovery time objective is the target time to restore services to either the service delivery objective (SDO) or normal operations.
C. The SDO is the agreed-on level of service required to resume acceptable operations.
D. Maximum tolerable outage is the maximum length of time that the organization can operate at the recovery site.

S4-133 The **MOST** timely and effective approach to detecting nontechnical security violations in an organization is: 

A. the development of organizationwide communication channels.
B. periodic third-party auditing of incident reporting logs.
C. an automated policy compliance monitoring system.
D. deployment of suggestion boxes throughout the organization.

A is the correct answer.

Justification:

A. Timely reporting of all security-related activities provides the information needed to monitor and respond to information security governance issues. Effective communication channels also are important for disseminating security-related information to the organization.
B. Audits are one form of periodic reporting, but they are too infrequent for effective day-to-day information security management.
C. Automated policy compliance monitoring is useful for reporting IT-related processes but, by itself, is insufficient for overall information security governance monitoring.
D. Even if personnel could be persuaded to leave notes on policy violations in suggestion boxes, it would not be effective and is unlikely to be timely unless suggestions are collected daily.

S4-134 The **MOST** important purpose of implementing an incident response plan is to:

A. prevent the occurrence of incidents.
B. ensure business continuity.
C. train users on resolution of incidents.
D. promote business resiliency.

D is the correct answer.

Justification:

A. The incident response plan is a means to respond to an event but does not prevent the occurrence.
B. Business continuity plans, not incident response plans, are designed to restore business operations after a disaster; they cannot assure the actual outcome.
C. The incident management plan may address training users, but the incident response plan does not.
D. Business resilience refers to the ability of the business to withstand disruption. An effective incident response plan minimizes the impact of an incident to the level that it ideally is transparent to end users and business partners.

S4-135 Which of the following gives the **MOST** assurance of the effectiveness of an organization's disaster recovery plan?

A. Checklist test
B. Table-top exercise
C. Full interruption test
D. Simulation test

C is the correct answer.

Justification:

A. A checklist test does not provide more assurance than a full interruption test. Checklist tests are a preliminary step to a real test. Recovery checklists are distributed to all members of a recovery team to review and ensure that the checklist is current.

B. A table-top exercise does not provide more assurance than a full interruption test. Table-top exercises may consist of virtual walk-throughs of the disaster recovery plan (DRP), or they may involve virtual walk-throughs of the DRP based on different scenarios.

C. A full interruption test gives the organization the best assurance because it is the closest test to an actual disaster. It generally involves shutting down operations at the primary site and shifting them to the recovery site in accordance with the recovery plan; this is the most rigorous form of testing.

D. A simulation test does not provide more assurance than a full interruption test. During simulation testing, the recovery team role plays a prepared disaster scenario without activating processing at the recovery site.

S4-136 An information security manager has been notified that a server that is utilized within the entire organization has been breached. What is the **FIRST** step to take?

A. Inform management.
B. Notify users.
C. Isolate the server.
D. Verify the information.

D is the correct answer.

Justification:

A. The information security manager should inform management but not before verifying the information.

B. Users should be notified after the information security manager has verified the information and informed management.

C. Isolating the server is not the first step that the information security manager should take.

D. Before any action is taken, the information security manager should verify that there has been a breach.

S4-137 Which of the following is **MOST** likely to improve the effectiveness of the incident response team?

A. Briefing team members on the nature of new threats to information systems (IS) security
B. Periodic testing and updates to incorporate lessons learned
C. Ensuring that all members have a good understanding of IS technology
D. A nonhierarchical structure to ensure that team members can share ideas

B is the correct answer.

Justification:

A. The fact that threats can materialize into an incident requires the presence of system vulnerabilities. It is the vulnerabilities that should be the focus of analysis when considering incident management procedures.
B. Periodic testing and updates to incorporate lessons learned will ensure that implementation of the incident management response plan is aligned and kept current with the business priorities set by business management.
C. All of the members of the incident management response team do not need to have IS skills. Members who take charge of implementing the incident management response plan should be able to utilize different skills to ensure alignment with the organization's procedures and policies.
D. It is important that someone take ownership of implementing the incident management plan (e.g., to formally declare that such a plan needs to be put into place after an incident). A nonhierarchical structure can introduce ambiguity as to who is responsible for what aspects of the incident management response plan.

S4-138 Which of the following is the **BEST** way to confirm that disaster recovery planning is current?

A. Audits of the business process changes
B. Maintenance of the latest configurations
C. Regular testing of the disaster recovery plan
D. Maintenance of the personnel contact list

C is the correct answer.

Justification:

A. Auditing business process changes will not necessarily enable maintenance of the disaster recovery plan (DRP).
B. Maintenance of the latest configuration will not show how current the process is, which is vital for disaster recovery planning.
C. When a DRP is properly tested, the results of the tests will reveal shortcomings and opportunities for improvement.
D. The maintenance of the personnel contact list is an indication of the personnel to be involved in the DRP. Although indicative of how current the DRP is, the DRP also should include the suppliers, customers and vendors needed for its success.

X

S4-139 Which of the following activities **MOST** increases the probability that an organization will be able to resume operations after a disaster?

A. Restoration testing
B. Establishment of a "warm site"
C. Daily data backups
D. An incident response plan

A is the correct answer.

Justification:

A. **A demonstrated ability to restore data is the best way to ensure that data can be restored after a disaster, and data drive the majority of business processes. If an organization is unable to restore its data, it will be of little value to have other considerations in place. On the other hand, if data can be restored, the organization can likely find work-arounds for other challenges that it may face.**

B. Having a "warm site" speeds up the process of disaster recovery by providing the facilities and equipment where data can be restored and operations reconstituted. However, if the data themselves cannot be restored, having the facilities and equipment will not be nearly as useful.

C. Performing data backups on a daily or other periodic basis is a good practice, but it is not until recovery is attempted that an organization gains knowledge of whether these backups are effective. Should the organization diligently perform backups for months or years and then discover that it cannot restore the data, all of the time and expense of the backup program will have been wasted.

D. Recovery procedures are documented in the disaster recovery plan rather than in the incident response plan.

S4-140 Which of the following **BEST** contributes to the design of data restoration plans?

A. Transaction turnaround time
B. Mean time between failures
C. Service delivery objectives
D. The duration of the data restoration job

C is the correct answer.

Justification:

A. Transaction turnaround time may be a concern when the effectiveness of an application system is evaluated. Normally it is not the main agenda in the restoration stage.

B. Mean time between failures (MTBF) is the predicted elapsed time between inherent failures of a system during operation. MTBF is not a factor in determining restoration of data.

C. **The service delivery objective (SDO) relates directly to the business needs; SDO is the level of services to be reached during the alternate process mode until the normal situation is restored.**

D. The duration of a data restoration job may be of secondary importance. The strategic importance of data should be considered first.

S4-141 Which of the following contributes **MOST** to incident response team efficiency?

A. Security policies and procedures
B. Defined roles and responsibilities
C. Digital forensic analysis skills
D. Reporting line structure

B is the correct answer.

Justification:

A. Knowing about security policies and procedures may be important. However, this knowledge is not a "must" item for an incident response team to work efficiently.
B. Incident response team members need to work in a disrupted environment; therefore, it is essential that they be clearly aware of roles and responsibility prior to engagement.
C. There could be an instance when a digital forensic analyst is needed. In such a case, assigning a qualified professional may be the best solution rather than having the response team learn the skills.
D. The reporting structure is a mandatory component for a team to operate. However, in the case of incident response, team size is usually small and the reporting line may be flat. Thus, it may not be a major contributor to efficiency.

S4-142 Which of the following needs to be **MOST** seriously considered when designing a risk-based incident response management program?

A. The chance of collusion among staff
B. Degradation of investigation quality
C. Minimization of false-positive alerts
D. Monitoring repeated low-risk events

D is the correct answer.

Justification:

A. In general, any control practice is vulnerable to collusion, and if an incident is carefully crafted among a number of staff, it is hard to detect. However, successful collusion is not common.
B. As long as it is well defined, it is unlikely that the quality of incident investigation will fall short.
C. A risk-based approach may not guarantee the minimization of false-positive alerts.
D. A risk-based approach focuses on high-risk items. Those attempting to commit fraud may take advantage of its weaknesses. When risk-based monitoring is in place, there is a higher chance of overlooking low-risk activities. Even though the impact of a low-risk event is small, it may not be possible to ignore the accumulated damage from its repeated occurrence. Therefore, it also is essential to review the chance of the repeated occurrence of low-risk events.

X **S4-143** What is the **MOST** appropriate IT incident response management approach for an organization that has outsourced its IT and incident management function?

A. A tested plan and a team to provide oversight
B. An individual to serve as the liaison between the parties
C. Clear notification and reporting channels
D. A periodic audit of the provider's capabilities

A is the correct answer.

Justification:

A. An approved and tested plan will provide assurance of the provider's ability to address incidents within an acceptable recovery time and an internal team to provide oversight and liaison functions to ensure that the response is according to plan.
B. Identifying a liaison is not sufficient by itself to provide assurance of adequate incident response performance.
C. Notification and reporting is not a sufficient assurance of suitable response activities and provides no capability for input, participation or addressing related issues in a timely manner.
D. Audits provide a periodic snapshot of the sufficiency of the provider's plans and capabilities, but are not adequate to manage collateral and consequential issues in the event of a significant incident.

S4-144 Which of the following choices is **MOST** important to ensure the admissibility of forensic evidence?

A. Adequacy of the retention period
B. Storage on read-only media
C. Review by an independent authority
D. Traceability of control

D is the correct answer.

Justification:

A. While evidence must necessarily be retained long enough to be submitted, the length of the retention period does not itself affect the admissibility of evidence.
B. Read-only media reduces the possibility of tampering as a technical capability, but maintenance of a chain of custody serves as an adequate safeguard in any case.
C. Review by an independent authority does not guarantee admissibility.
D. Evidence is inadmissible without a clear chain of custody, which is a tracing of who had control of the evidence throughout the process.

S4-145 What action should an incident response team take if the investigation of an incident response event cannot be completed in the time allocated?

A. Continue to work the current action.
B. Escalate to the next level for resolution.
C. Skip on to the next action in the plan.
D. Declare a disaster.

B is the correct answer.

Justification:

A. Every unsuccessful action simply wastes time; escalate and move on.
B. Because the investigation process must have time constraints, if the initial team cannot find resolution in the plan time allotted, they should escalate the resolution to the next level and move on to system recovery.
C. The activity in an incident response event should not stop until the root cause has been determined, but other teams may need to be called in to divide the work and complete the response plan.
D. A disaster should not be declared until the event root cause has been determined or senior management has determined that the resolution will take longer than acceptable for a system outage.

S4-146 Which of the following choices is a characteristic of security information and event management (SIEM) technology?

A. SIEM promotes compliance with security policies.
B. SIEM is primarily a means of managing residual risk.
C. SIEM replaces the need to install a firewall.
D. SIEM provides a full range of compensating controls.

A is the correct answer.

Justification:

A. Security information and event management (SIEM) can provide information on policy compliance as well as incident monitoring and other capabilities, if properly deployed, configured and tuned.
B. SIEM is not used to manage residual risk.
C. SIEM is an automated review of logs through aggregation and correlation, and does not replace the need for firewalls.
D. SIEM provides a series of detective controls, not compensating controls.

S4-147 Which of the following items is **MOST** important to determine the recovery point objective for a critical process in an enterprise?

A. The number of hours of acceptable downtime
B. The total cost of recovering critical systems
C. The acceptable reduction in the level of service
D. The extent of data loss that is acceptable

D is the correct answer.

Justification:

A. The recovery time objective (RTO) is the amount of time allowed for the recovery of a business function or resource after a disaster. The RTO is not a factor in determining the recovery point objective (RPO).
B. The determination of the RPO would have already taken cost into consideration.
C. The service delivery level is directly related to the business needs. It is the level of services to be reached during the alternate process mode until the normal situation is restored.
D. The RPO is determined based on the acceptable data loss in case of a disruption of operations. It indicates the earliest point in time that is acceptable to recover the data. The RPO effectively quantifies the permissible amount of data loss in case of interruption.

S4-148 What action should the security manager take **FIRST** when incident reports from different organizational units are inconsistent and highly inaccurate?

A. Ensure that a clear organizational incident definition and severity hierarchy exists.
B. Initiate a companywide incident identification training and awareness program.
C. Escalate the issue to the security steering committee for appropriate action.
D. Involve human resources in implementing a reporting enforcement program.

A is the correct answer.

Justification:

A. The first action is to validate that clear incident definition and severity criteria are established and communicated throughout the organization.
B. A training program will not be effective until clear incident identification and severity criteria have been established.
C. The steering committee may become involved after incident criteria have been clearly established and communicated.
D. Enforcement activities will not be effective unless incident criteria have been clearly established and communicated.

S4-149 Which of the following measurements is integrated into the incident response plan by the statement "If the database is corrupted by an incident, the backup at the close of work on the previous day should be restored"?

A. The recovery time objective
B. The recovery point objective
C. The service delivery objective
D. The maximum tolerable outage

B is the correct answer.

Justification:

A. The recovery time objective (RTO) is the amount of time allowed for the recovery of a business function or resource after a disaster occurs. The statement did not mention the time for the restoration to be concluded.

B. The recovery point objective (RPO) is determined based on the acceptable data loss in case of a disruption of operations. It indicates the earliest point in time that is acceptable to recover the data. The RPO effectively quantifies the permissible amount of data loss in case of interruption. The statement allows for the loss of current day's data.

C. Directly related to the business needs, the service delivery objective (SDO) is the level of services to be reached during the alternate process mode until the normal situation is restored. The SDO is the acceptable level of service within the RTO.

D. The maximum tolerable outage is the maximum time that an enterprise can support processing in alternate mode.

S4-150 In following up on a security incident, the system administrator is to copy data from one hard disk to another. From a forensic perspective, which of the following tasks must be ensured?

A. Copy to the same disk model as the original.
B. Make a dual backup of the original disk.
C. Keep the digital hash from both hard disks.
D. Perform a restoration test after replication.

C is the correct answer.

Justification:

A. To make a copy, the target hard disk does not necessarily have to be the one to maintain the same specification as the original disk; therefore, this is not the best choice.

B. It is a good practice to make a dual backup; however, it does not prove that data are not modified by anyone.

C. In order to prove that data are not modified before and after the copy, it is best to keep a digital hash from both hard disks. The hashes alone are not adequate to meet the standards of evidence admissibility, but these will support other aspects of integrity in the context of data forensics.

D. It is a good practice to perform a restoration test. This is to ensure availability rather than to maintain evidential capability.

S4-151 While a disaster recovery exercise in the organization's hot site successfully restored all essential services, the test was deemed a failure. Which of the following circumstances would be the **MOST** likely cause?

A. The maximum tolerable outage exceeded the acceptable interruption window (AIW).
B. The recovery plans specified outdated operating system versions.
C. Some restored systems exceeded service delivery objectives.
D. Aggregate recovery activities exceeded the AIW.

all or overall

D is the correct answer.

Justification:

A. The maximum tolerable outage, the amount of time the organization can operate in alternate mode, would normally exceed the acceptable interruption window (AIW).
B. While a difference in operating system versions might cause a delay, it would probably be minor.
C. Service delivery objectives (SDOs) are directly related to the business needs. The SDO is the level of services to be reached during the alternate process mode until the normal situation is restored. Not meeting SDOs on some systems might be a concern, but would not necessarily lead to the conclusion that the test was a failure.
D. Exceeding the AIW would cause the organization significant damage and must be avoided. The acceptable interruption window is the maximum period of time that a system can be unavailable before compromising the achievement of the enterprise's business objectives.

S4-152 Which of the following choices is the **MOST** important incident response resource for timely identification of an information security incident?

A. A fully updated intrusion detection system
B. Multiple channels for distribution of information
C. A well-defined and structured communication plan
D. A regular schedule for review of network device logs

C is the correct answer.

Justification:

A. Not all information security incidents originate from the network; an intrusion detection system will provide no detection value for a variety of incident types.
B. Diversifying the means of communication increases the odds that information reaches the people to whom it is sent, but it does nothing to ensure that the correct people receive the correct information at the correct time.
C. An incident is not identified within an organization until it is declared, which is a business responsibility beyond the scope of the technical staff. A well-defined and structured communication plan ensures that information flows from the technical staff to decision makers in a timely fashion, allowing incidents to be recognized, declared and appropriately addressed.
D. Reviewing logs provides an opportunity to identify irregular traffic patterns that may indicate an information security incident, but these logs provide insight into only a subset of attack vectors (e.g., external penetration would generally be covered, but insider threats may not). Additionally, if analysts who identify potentially revealing information do not have mechanisms in place to share those revelations with others in the organization, an effective response is less likely.

S4-153 Which of the following actions is the **BEST** to ensure that incident response activities are consistent with the requirements of business continuity?

A. Develop a scenario and perform a structured walk-through.
B. Draft and publish a clear practice for enterprise-level incident response.
C. Establish a cross-departmental working group to share perspectives.
D. Develop a project plan for end-to-end testing of disaster recovery.

A is the correct answer.

Justification:

A. A structured walk-through including both incident response and business continuity personnel provides the best opportunity to identify gaps or misalignments between the plans.
B. Publishing an enterprise-level incident response plan would be effective only if business continuity aligned itself to incident response. Incident response supports business continuity, not the other way around.
C. Sharing perspectives is valuable, but a working group does not necessarily lead to action ensuring that the interface between plans is workable.
D. A project plan developed for disaster recovery will not necessarily address deficiencies in business continuity or incident response.

S4-154 A newly-hired information security manager examines the 10-year old business continuity plan and notes that the maximum tolerable outage (MTO) is much shorter than the allowable interruption window (AIW). What action should be taken as a result of this information?

X

A. Reassess the MTO.
B. Conduct a business impact analysis and update the plan.
C. Increase the service delivery objective.
D. Take no action; MTO is not related to AIW.

B is the correct answer.

Justification:

A. Performing a business impact analysis (BIA) will include reassessment of the maximum tolerable outage (MTO); until that time, there is no way to determine whether it is the MTO or the allowable interruption window (AIW) that is incorrect.
B. The first issue is to determine whether the plan is current and then update requirements as necessary. The BIA will most likely be a collaborative effort with the business process owners.
C. The service delivery objective will need to be updated by performing a BIA.
D. The MTO should always be at least equal to the AIW and is generally longer.

S4-155 Which of the following benefits that the enterprise receives from employing a systematic incident management program with a formal methodology is **MOST** important?

A. A formal methodology makes incident management more flexible.
B. A formal methodology is more reliant on business continuity activities.
C. Each incident responder is able to get broad-based experience.
D. Evidence of due diligence supports legal and liability claims.

D is the correct answer.

Justification:
A. The more formalized that something becomes, the less flexible it is.
B. A formal methodology is actually able to more easily operate as a stand-alone function, with less reliance on business continuity activities.
C. Having a formal methodology means that duties are generally assigned based on competence and availability of time.
D. Legal and liability claims are most credible when the mechanisms used to collect them are formally documented, repeatable and regularly practiced.

S4-156 Untested response plans:

A. depend on up-to-date contact information.
B. pose an unacceptable risk to the organization.
C. pose a risk that the plan will not work when needed.
D. are quickly distinguished from tested plans.

C is the correct answer.

Justification:
A. While up-to-date contact information is important, it is no more important in an untested plan than would be true for a tested plan.
B. Whether a risk is acceptable or not is a business determination and is not a function of testing.
C. A response plan may prove unworkable upon testing despite appearing to cover all areas as written.
D. Whether a plan has or has not been tested is not quickly apparent from inspection.

S4-157 What makes an incident management program effective?

A. It identifies, assesses and prevents recurrence of incidents.
B. It detects and documents incidents.
C. It includes a risk management strategy.
D. It reflects the capabilities of the organization.

A is the correct answer.

Justification:
A. Incident management identifies and assesses incidents as they happen. Then it implements improvements to prevent future occurrences.
B. Detecting and documenting incidents is only part of the process; future occurrences need to be addressed and prevented.
C. Risk management occurs outside of the incident management program.
D. Objectives are set based on business need and capabilities are built to meet those objectives.

S4-158 An organization is primarily concerned with the financial impact of downtime associated with an information security incident. Which of the following items would be the **MOST** appropriate compensating control to have in place?

A. An offsite media storage contract
B. Business interruption insurance
C. A real-time failover architecture
D. A disaster recovery plan

B is the correct answer.

Justification:

A. Storing backup media offsite improves the odds that they will be available to use for recovery activities, but it also increases the amount of time needed to complete the recovery. In a situation where the primary concern is the financial impact of downtime, an offsite media storage contract is not helpful.
B. Business interruption insurance does not help restore operations, but it does compensate a business for the financial impact associated with interruption. In this scenario, the financial impact of downtime is the primary concern; therefore, insurance is an appropriate compensating control.
C. An architecture that provides for real-time failover prevents financial impact from downtime, but it does so at significant cost. An organization that is primarily concerned with financial impact (rather than operational efficiency or other concerns) is unlikely to accept this higher cost because the other benefits associated with real-time failover are not seen as justified.
D. A disaster recovery plan aids an organization in performing the steps needed to return to normal operations after a disaster, but even a clearly drafted and tested plan does not compensate for the financial impact of downtime, and many information security incidents have impacts that do not meet the disaster threshold.

S4-159 If an organization has a requirement for continuous operations, which of the following approaches would be **BEST** to test response and recovery?

A. A full interruption test
B. A simulation test
C. A parallel test
D. A structured walk-through

C is the correct answer.

Justification:

A. A full interruption test, in which operations are shut down at the primary site and shifted to the recovery site, is the most stringent form of response and recovery testing, but it is potentially disruptive. Even though the organization in this scenario might accept the cost of such a test, the need for continuous operations makes it inappropriate.
B. Simulation testing tests people and processes but does not go so far as to start up recovery-site operations; therefore, it provides a lower level of assurance than what would be provided by a parallel test.
C. The organization in this scenario requires continuous operations. A parallel test, in which operations are brought online at the recovery site alongside primary-site operations, is the closest that an organization can come to full testing without risking a business impact; therefore, it is the best fit for the requirement.
D. Structured walk-throughs are pen-and-paper activities. A walk-through may help identify constraints, deficiencies and opportunities for enhancement, but the level of assurance that it provides is low relative to a parallel test.

X **S4-160** When establishing a new incident management team whose members will serve on a part-time basis, which of the following means of training is **MOST** effective?

A. Formal training
B. Mentoring
C. On-the-job training
D. Induction

A is the correct answer.

Justification:

A. Formal training is a good choice when everyone is new because it does not assume any prior knowledge and ensures that everyone covers the same material.
B. Mentoring is most effective when senior members of an established team can be paired with new members. It does not work well when everyone is new.
C. On-the-job training is a suitable choice when the material to be learned is part of the participants' everyday duties. For an incident management team comprised of part-time members, there will be limited opportunities to train in the course of regular, day-to-day activities.
D. Induction provides a basic overview of incident management team activities and serves as a basis for further training. By itself, it is not an effective means of training.

S4-161 The **PRIMARY** business objective of incident management is:

A. containment.
B. root-cause analysis.
C. eradication.
D. impact control.

D is the correct answer.

Justification:

A. Containment is one of the steps of the standard incident management process, not the primary objective. Depending on the nature of the incident and its potential impact on the organization, containment may or may not be a priority.
B. Root-cause analysis facilitates long-term remediation of vulnerabilities to prevent the recurrence of a given type of incident, but it is not the purpose of incident management.
C. Eradication is one of the steps of the standard incident management process, not the primary objective.
D. The purpose of incident management is to identify and respond to unexpected disruptive events with the objective of controlling impacts within acceptable levels.

S4-162 Which of the following technologies is likely to be the **MOST** useful in countering advanced persistent threats?

A. Anomaly-based intrusion detection system
B. Security information and event management system
C. Automated vulnerability scanning tools
D. Integrated network management system

B is the correct answer.

Justification:

A. Intrusion detection systems can detect and notify of a potential attack but provide no information on subsequent breaches, making these systems less effective at identifying persistent threats than system information and event management (SIEM) systems.

B. SIEM systems can identify incidents or potential incidents, prioritize according to potential impact, track incidents until they are closed and provide substantial trend analysis over time.

C. Vulnerability scanning tools identify weaknesses in systems and networks that correspond to known paradigms. In general, advanced persistent threats (APTs) involve exploits that are outside the scope of published vulnerabilities, making vulnerability scanning a limited countermeasure with regards to the APTs.

D. Integrated network management typically provides a limited subset of the capabilities of fully implemented SIEM.

S4-163 When establishing effective incident escalation processes for the incident response team, it is **PRIMARILY** necessary to state how:

A. long a member should wait for a response and what to do if no response occurs.
B. critical the incident is and which business units are directly impacted.
C. the incident is communicated to senior managers and other affected stakeholders.
D. incident response team managers are informed quickly about high-risk incidents.

A is the correct answer.

Justification:

A. When defining and establishing effective incident escalation processes, it is primarily relevant to state how long a team member should wait for an incident response and what to do if no such response occurs. This is the necessary (initial) platform for all further steps of an effective escalation process.

B. It is relevant to know how critical an incident is and which business units are impacted, but when establishing escalation processes, it is much more relevant to state how long a person should wait for a response and what to do if no response occurs.

C. Communication to stakeholders is part of the incident response process, but it is more important to establish waiting times and alternative responses because time is of the essence.

D. It is relevant to inform incident response team managers quickly, but as an initial aspect, it is more relevant in this connection to state how long a person should wait for a response and what to do if no response occurs.

X

S4-164 While developing incident response procedures an information security manager must ensure that the procedure is **PRIMARILY** aimed at:

A. containing incidents to minimize damage.
B. identifying root causes of incidents.
C. implementing solutions to prevent recurrence.
D. recording and closing incident tickets.

A is the correct answer.

Justification:

A. Incident response procedures primarily focus on containing the incident and minimizing damage.
B. Root cause analysis is a component of the overall incident management process rather than the incident response procedure.
C. Implementing solutions is possible only after a cause has been determined.
D. Recording and closing tickets is part of the subsequent documentation process but is not the primary focus of incident response.

S4-165 Which of the following would be the **BEST** indicator of the readiness of the incident response team in the context of the overall incident management program?

A. Amount of time for incident detection
B. Time between incident detection and severity determination
C. Time between detection and response
D. Amount of time between incident occurrence and its resolution

C is the correct answer.

Justification:

A. The time to detect is a measure of detection capability, which is typically provided by automated controls.
B. Time between detection and determining severity is a part of response.
C. Readiness is the time it takes from detection to initiate a response. The first time that the incident response team typically becomes aware of an event is when an alert is provided by monitoring mechanisms.
D. Time between incident and resolution is a function of response capability.

S4-166 An organization decides its old recovery facility is no longer adequate because it is not capable of operation for an extended period. The organization decides to build a new facility in another location that would address the major shortcomings of the old site and provide more space for possible future expansion. Until the new facility is completed, which of the following objectives for recovery will have to be changed?

A. Maximum tolerable outage
B. Recovery point objective
C. Service delivery objective
D. Allowable interruption window

C is the correct answer.

Justification:

A. Although the current recovery facility cannot satisfy the maximum tolerable outage (MTO), that does not change the MTO. The organization should document an inability to meet the MTO and continue developing a new facility that will satisfy the objective.
B. The recovery point objective (RPO) is not affected by the stated deficiencies in the current recovery facility.
C. The service delivery objective (SDO) reflects a commitment to internal customers to meet certain performance standards. To be realistic, the objective must be changed to reflect the operating capabilities of the current recovery facility.
D. The MTO must be at least as great as the allowable interruption window (AIW). Therefore, it is possible that exceeding the MTO will result in not being able to meet the AIW, which will result in unacceptable damage to the organization. However, as with the MTO, the inability to meet the AIW does make the associated damage acceptable, so changing the AIW would not be appropriate.

S4-167 Which of the following poses the **GREATEST** challenge to establishing effective security incident management processes?

A. Security technologies are not kept up to date.
B. Stakeholders are not defined within security policies.
C. Incidents are not controlled by process owners.
D. Escalation paths are insufficiently defined.

D is the correct answer.

Justification:

A. Security technologies are not typically the cause of substantial challenges in building effective security processes.
B. Security policies rarely define all stakeholders and notification to stakeholders is typically outside the scope of initial incident management, making the definition of escalation paths a greater concern. Escalations processes are typically procedures.
C. Control of incidents by process owners is not a primary requirement of effective security incident management.
D. Inadequately defined escalation paths may result lack of adequate authority, substantial delays, lack of notification of the appropriate individuals, and other significant negative impacts.

X **S4-168** An organization determined that in a worst case situation it was not feasible to recreate all of the data lost in a system crash in the time available. Various constraints prevent increasing the frequency of backups. What other solutions to this issue could the information security manager suggest?

A. Increase the recovery time objective
B. Decrease the service delivery objective
C. Adjust the maximum tolerable outage
D. Increase the allowable interruption window

A is the correct answer.

Justification:

A. Because the original recovery time objective (RTO) cannot be met due to the time required to restore data, the RTO could be increased.
B. Decreasing the service delivery objective (SDO) would increase the problem and is not a solution.
C. Adjusting the maximum tolerable outage (MTO) will not have any effect on the situation.
D. Increasing the allowable interruption window (AIW) is based on the maximum time the organization can be down before major financial impacts occur.

S4-169 What is the **FIRST** step in investigating an information security incident for which the organization may want to file criminal charges?

A. Notify law enforcement and senior management
B. Prevent contamination of evidence
C. Activate the incident response team
D. Contain the scope of impact

B is the correct answer.

Justification:

A. Notification to law enforcement or senior management may occur in tandem with other activities, but preventing contamination of evidence takes priority. In many organizations, the decision to notify law enforcement is made by senior management.
B. If criminal charges may be filed, preventing contamination of evidence is the foremost concern to facilitate prosecution.
C. Activation of the incident response team must be delayed until after steps have been taken to prevent contamination of evidence in situations where criminal charges may be filed.
D. Containment is part of an effective incident response strategy, but preventing contamination of evidence takes priority over containment in situation where criminal charges may be filed.

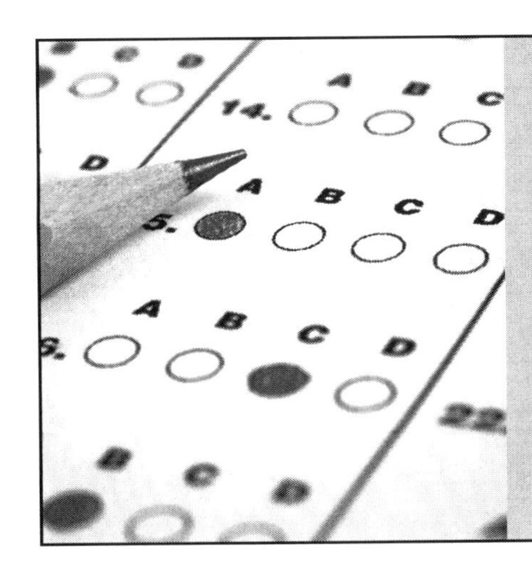

POSTTEST

If you wish to take a posttest to determine strengths and weaknesses, the Sample Exam begins on page 409 and the posttest answer sheet begins on page 435. You can score your posttest with the Sample Exam Answer and Reference Key on page 431.

Page intentionally left blank

SAMPLE EXAM

1. Which of the following is an indicator of effective governance?

   A. A defined information security architecture
   B. Compliance with international security standards
   C. Periodic external audits
   D. An established risk management program

2. Which of the following is the **PRIMARY** prerequisite to implementing data classification within an organization?

   A. Defining job roles
   B. Performing a risk assessment
   C. Identifying data owners
   D. Establishing data retention policies

3. What is the **PRIMARY** factor that should be taken into consideration when designing the technical solution for a disaster recovery site?

   A. Service delivery objective
   B. Recovery time objective
   C. Allowable interruption window
   D. Maximum tolerable outage

4. Which of the following is the **MOST** effective way to treat a risk such as a natural disaster that has a low probability and a high impact level?

   A. Implement countermeasures.
   B. Eliminate the risk.
   C. Transfer the risk.
   D. Accept the risk.

5. Due to limited storage media, an IT operations employee has requested permission to overwrite data stored on a magnetic tape. The decision of the authorizing manager will **MOST** likely be influenced by the data:

   A. classification policy.
   B. retention policy.
   C. creation policy.
   D. leakage protection.

6. Which of the following is **MOST** effective in protecting against the attack technique known as phishing?

   A. Firewall blocking rules
   B. Up-to-date signature files
   C. Security awareness training
   D. Intrusion detection system monitoring

7. Which of the following is the **MOST** cost-effective approach to achieve strategic alignment?

   A. Periodically survey management
   B. Implement a governance framework
   C. Ensure that controls meet objectives
   D. Develop an enterprise architecture

8. An internal review of a web-based application system reveals that it is possible to gain access to all employees' accounts by changing the employee's ID used for accessing the account on the uniform resource locator. The vulnerability identified is:

A. broken authentication.
B. unvalidated input.
C. cross-site scripting.
D. structured query language injection.

9. Which of the following is the **MOST** effective at preventing an unauthorized individual from following an authorized person through a secured entrance (tailgating or piggybacking)?

A. Card key door locks
B. Photo identification
C. Biometric scanners
D. Awareness training

10. Which of the following is the **MOST** important factor to be considered in the loss of mobile equipment with unencrypted data?

A. Disclosure of personal information
B. Sufficient coverage of the insurance policy for accidental losses
C. Potential impact of the data loss
D. Replacement cost of the equipment

11. Recovery point objectives can be used to determine which of the following?

A. Maximum tolerable period of data loss
B. Maximum tolerable downtime
C. Baseline for operational resiliency
D. Time to restore backups

12. An information security manager has implemented procedures for monitoring specific activities on the network. The system administrator has been trained to analyze the network events, take appropriate action and provide reports to the information security manager. What additional monitoring should be implemented to give a more accurate, risk-based view of network activity?

A. The system administrator should be monitored by a separate reviewer.
B. All activity on the network should be monitored.
C. No additional monitoring is needed in this situation.
D. Monitoring should be done only by the information security manager.

13. Which of the following is **MOST** likely to be discretionary?

A. Policies
B. Procedures
C. Guidelines
D. Standards

14. Which of the following is **MOST** important in developing a security strategy?

A. Creating a positive security environment
B. Understanding key business objectives
C. Having a reporting line to senior management
D. Allocating sufficient resources to information security

15. Which of the following roles is responsible for legal and regulatory liability?

A. Chief security officer
B. Chief legal counsel
C. Board of directors and senior management
D. Information security steering group

16. What is the **PRIMARY** basis for a detailed business continuity plan?

A. Consideration of different alternatives
B. The solution that is least expensive
C. Strategies that cover all applications
D. Strategies validated by senior management

17. Which of the following is the **MOST** appropriate frequency for updating antivirus signature files for antivirus software on production servers?

A. Daily
B. Weekly
C. Concurrently with operating system patch updates
D. During scheduled change control updates

18. Why is public key infrastructure the preferred model when providing encryption keys to a large number of individuals?

A. It is computationally more efficient.
B. It is more scalable than a symmetric key.
C. It is less costly to maintain than a symmetric key approach.
D. It provides greater encryption strength than a secret key model.

19. Which of the following **BEST** ensures that modifications made to in-house developed business applications do not introduce new security exposures?

A. Stress testing
B. Patch management
C. Change management
D. Security baselines

20. What is the **MOST** important item to be included in an information security policy?

A. The definition of roles and responsibilities
B. The scope of the security program
C. The key objectives of the security program
D. Reference to procedures and standards of the security program

21. What is the **PRIMARY** objective of a postincident review in incident response?

A. To adjust budget provisioning
B. To preserve forensic data
C. To improve the response process
D. To ensure the incident is fully documented

22. Which of the following actions is **MOST** important when a server is infected with a virus?

A. Isolate the infected server(s) from the network.
B. Identify all potential damage caused by the infection.
C. Ensure that the virus database files are current.
D. Establish security weaknesses in the firewall.

23. Who should determine the appropriate classification of accounting ledger data located on a database server and maintained by a database administrator in the IT department?

A. Database administrator
B. Finance department management
C. Information security manager
D. IT department management

24. When a major vulnerability in the security of a critical web server is discovered, immediate notification should be made to the:

A. system owner to take corrective action.
B. incident response team to investigate.
C. data owners to mitigate damage.
D. development team to remediate.

25. What should documented standards/procedures for the use of cryptography across the enterprise achieve?

A. They should define the circumstances where cryptography should be used.
B. They should define cryptographic algorithms and key lengths.
C. They should describe handling procedures of cryptographic keys.
D. They should establish the use of cryptographic solutions.

26. Which of the following would be the **MOST** relevant factor when defining the information classification policy?

A. Quantity of information
B. Available IT infrastructure
C. Benchmarking
D. Requirements of data owners

27. A certificate authority is required for a public key infrastructure:

A. in cases where confidentiality is an issue.
B. when challenge/response authentication is used.
C. except where users attest to each other's identity.
D. in role-based access control deployments.

28. For risk management purposes, the value of a physical asset should be based on:

A. original cost.
B. net cash flow.
C. net present value.
D. replacement cost.

29. An enterprise is transferring its IT operations to an offshore location. An information security manager should **PRIMARILY** focus on:

A. reviewing new laws and regulations.
B. updating operational procedures.
C. validating staff qualifications.
D. conducting a risk assessment.

30. An organization's board of directors is concerned about recent fraud attempts that originated over the Internet. What action should the board take to address this concern?

A. Direct information security regarding specific resolutions that are needed to address the risk.
B. Research solutions to determine appropriate actions for the organization.
C. Take no action; information security does not report to the board.
D. Direct management to assess the risk and to report the results to the board.

31. Which of the following roles is **MOST** responsible for ensuring that information protection policies are consistent with applicable laws and regulations?

A. Executive management
B. The quality manager
C. The board of directors
D. The auditor

32. The **PRIMARY** goal of a corporate risk management program is to ensure that an organization's:

A. IT assets in key business functions are protected.
B. business risk is addressed by preventive controls.
C. stated objectives are achieved.
D. IT facilities and systems are always available.

33. Of the following, retention of business records should be **PRIMARILY** based on:

A. periodic vulnerability assessment.
B. business requirements.
C. device storage capacity and longevity.
D. legal requirements.

34. Who should generally determine the classification of an information asset?

A. The asset custodian
B. The security manager
C. Senior management
D. The asset owner

35. An information security manager mapping a job description to types of data access is **MOST** likely to adhere to which of the following information security principles?

A. Ethics
B. Proportionality
C. Integration
D. Accountability

36. What activity needs to be performed for previously accepted risk?

A. Risk should be reassessed periodically because risk changes over time.
B. Accepted risk should be flagged to avoid future reassessment efforts.
C. Risk should be avoided next time to optimize the risk profile.
D. Risk should be removed from the risk log after it is accepted.

37. An enterprise has been recently subject to a series of denial-of-service attacks due to a weakness in security. The information security manager needs to present a business case for increasing the investment in security. The **MOST** significant challenge in obtaining approval from senior management for the proposal is:

A. explaining technology issues of security.
B. demonstrating value and benefits.
C. simulating various risk scenarios.
D. obtaining benchmarking data for comparison.

38. When creating a forensic image of a hard drive, which of the following should be the **FIRST** step?

A. Identify a recognized forensics software tool to create the image.
B. Establish a chain of custody log.
C. Connect the hard drive to a write blocker.
D. Generate a cryptographic hash of the hard drive contents.

39. When an organization is using an automated tool to manage and house its business continuity plans, which of the following is the **PRIMARY** concern?

A. Ensuring accessibility should a disaster occur
B. Versioning control as plans are modified
C. Broken hyperlinks to resources stored elsewhere
D. Tracking changes in personnel and plan assets

40. Security risk assessments are **MOST** cost-effective to a software development organization when they are performed:

A. before system development begins.
B. at system deployment.
C. before developing a business case.
D. at each stage of the system development life cycle.

41. Which of the following is **MOST** effective in preventing disruptions to production systems?

A. Patch management
B. Security baselines
C. Virus detection
D. Change management

42. The segregation of duties principle is violated if which of the following individuals has update rights to the database access control list?

A. Data owner
B. Data custodian
C. Systems programmer
D. Security administrator

43. During which phase of development is it **MOST** appropriate to begin assessing the risk of a new application system?

    A. Feasibility
    B. Design
    C. Development
    D. Testing

44. When a proposed system change violates an existing security standard, the conflict would be **BEST** resolved by:

    A. calculating the risk.
    B. enforcing the security standard.
    C. redesigning the system change.
    D. implementing mitigating controls.

45. Which of the following technologies is utilized to ensure that an individual connecting to a corporate internal network over the Internet is not an intruder masquerading as an authorized user?

    A. Intrusion detection system
    B. IP address packet filtering
    C. Two-factor authentication
    D. Embedded digital signature

46. Which of the following are seldom changed in response to technological changes?

    A. Standards
    B. Procedures
    C. Policies
    D. Guidelines

47. While implementing information security governance, an organization should **FIRST**:

    A. adopt security standards.
    B. determine security baselines.
    C. define the security strategy.
    D. establish security policies.

48. Who should be assigned as data owner for sensitive customer data that is used only by the sales department and stored in a central database?

    A. The sales department
    B. The database administrator
    C. The chief information officer
    D. The head of the sales department

49. Which of the following actions should be taken when an online trading company discovers a network attack in progress?

    A. Shut off all network access points
    B. Dump all event logs to removable media
    C. Isolate the affected network segment
    D. Enable trace logging on all events

50. Which of the following is the **MOST** important consideration when performing a risk assessment?

A. Management supports risk mitigation efforts.
B. Annual loss expectancies have been calculated for critical assets.
C. Assets have been identified and appropriately valued.
D. Attack motives, means and opportunities are understood.

51. Which program element should be implemented **FIRST** in asset classification and control?

A. Risk assessment
B. Classification
C. Valuation
D. Risk mitigation

52. Information security governance is **PRIMARILY** driven by:

A. technology constraints.
B. regulatory requirements.
C. litigation potential.
D. business strategy.

53. What is the **BEST** method for mitigating against network denial-of-service (DoS) attacks?

A. Ensure all servers are up to date on OS patches.
B. Employ packet filtering to drop suspect packets.
C. Implement network address translation to make internal addresses nonroutable.
D. Implement load balancing for Internet facing devices.

54. Which of the following would be **MOST** appropriate for collecting and preserving evidence?

A. Encrypted hard drives
B. Generic audit software
C. Proven forensic processes
D. Log correlation software

55. What is the **BEST** means to standardize security configurations in similar devices?

A. Policies
B. Procedures
C. Technical guides
D. Baselines

56. Serious security incidents typically lead to renewed focus by management on information security that then usually fades over time. What opportunity should the information security manager seize to **BEST** use this renewed focus?

A. To improve the integration of business and information security processes
B. To increase information security budgets and staffing levels
C. To develop tighter controls and stronger compliance efforts
D. To acquire better supplemental technical security controls

57. The use of insurance is an example of which of the following?

A. Risk mitigation
B. Risk acceptance
C. Risk elimination
D. Risk transfer

58. Information security frameworks can be **MOST** useful for the information security manager because they:

A. provide detailed processes and methods.
B. are designed to achieve specific outcomes.
C. provide structure and guidance.
D. provide policy and procedure.

59. What activity **BEST** helps ensure that contract personnel do not obtain unauthorized access to sensitive information?

A. Set accounts to pre-expire.
B. Avoid granting system administration roles.
C. Ensure they successfully pass background checks.
D. Ensure their access is approved by the data owner.

60. When should risk assessments be performed for optimum effectiveness?

A. At the beginning of security program development
B. On a continuous basis
C. While developing the business case for the security program
D. During the business change management process

61. Under what circumstances is it **MOST** appropriate to reduce control strength?

A. Assessed risk is below acceptable levels.
B. Risk cannot be determined.
C. The control cost is high.
D. The control is not effective.

62. Information security policy enforcement is the responsibility of the:

A. security steering committee.
B. chief information officer.
C. chief information security officer.
D. chief compliance officer.

63. What is the **PRIMARY** role of the information security manager related to the data classification and handling process within an organization?

A. Defining and ratifying the organization's data classification structure
B. Assigning the classification levels to the information assets
C. Securing information assets in accordance with their data classification
D. Confirming that information assets have been properly classified

64. Who should **PRIMARILY** provide direction on the impact of new regulatory requirements that may lead to major application system changes?

A. The internal audit department
B. System developers/analysts
C. Key business process owners
D. Corporate legal counsel

65. Maturity levels are an approach to determine the extent that sound practices have been implemented in an organization based on outcomes. Another approach that has been developed to achieve essentially the same result is:

A. controls applicability statements.
B. process performance and capabilities.
C. probabilistic risk assessment.
D. factor analysis of information risk.

66. Which of the following is the **MOST** important action to take when engaging third-party consultants to conduct an attack and penetration test?

A. Request a list of the software to be used.
B. Provide clear directions to IT staff.
C. Monitor intrusion detection system and firewall logs closely.
D. Establish clear rules of engagement.

67. When securing wireless access points, which of the following controls would **BEST** assure confidentiality?

A. Implementing wireless intrusion prevention systems
B. Not broadcasting the service set identifier
C. Implementing wired equivalent privacy authentication
D. Enforcing a virtual private network over wireless

68. Which of the following should be performed **EXCLUSIVELY** by the information security department?

A. Monitoring unauthorized access to operating systems
B. Configuring user access to operating systems
C. Approving operating system access standards
D. Configuring the firewall to protect operating systems

69. An organization has to comply with recently published industry regulatory requirements—compliance that potentially has high implementation costs. What should the information security manager do **FIRST**?

A. Consult the security committee.
B. Perform a gap analysis.
C. Implement compensating controls.
D. Demand immediate compliance.

70. After a service interruption of a critical system, the incident response team finds that it needs to activate the warm recovery site. Discovering that throughput is only half of the primary site, the team nevertheless notifies management that it has restored the critical system. This is **MOST** likely because it has achieved the:

A. recovery point objective.
B. recovery time objective.
C. service delivery objective.
D. maximum tolerable outage.

71. Which of the following is the **MOST** important step before implementing a security policy?

A. Communicating to employees
B. Training IT staff
C. Identifying relevant technologies for automation
D. Obtaining sign-off from stakeholders

72. The acceptability of a partial system recovery after a security incident is **MOST** likely to be based on the:

A. ability to resume normal operations.
B. maximum tolerable outage.
C. service delivery objective.
D. acceptable interruption window.

73. Which of the following is the **MOST** important objective of an information security strategy review?

A. Ensuring that risk is identified, analyzed and mitigated to acceptable levels
B. Ensuring the information security strategy is aligned with organizational goals
C. Ensuring the best return on information security investments
D. Ensuring the efficient utilization of information security resources

74. Which of the following choices is the **MOST** important consideration when developing the security strategy of a company operating in different countries?

A. Diverse attitudes toward security by employees and management
B. Time differences and the ability to reach security officers
C. A coherent implementation of security policies and procedures in all countries
D. Compliance with diverse laws and governmental regulations

75. Temporarily deactivating some monitoring processes, even if supported by an acceptance of operational risk, may not be acceptable to the information security manager if:

A. it implies compliance risk.
B. short-term impact cannot be determined.
C. it violates industry security practices.
D. changes in the roles matrix cannot be detected.

76. An information security manager is performing a security review and determines that not all employees comply with the access control policy for the data center. The **FIRST** step to address this issue should be to:

A. assess the risk of noncompliance.
B. initiate security awareness training.
C. prepare a status report for management.
D. increase compliance enforcement.

77. Which of the following has the highest priority when defining an emergency response plan?

A. Critical data
B. Critical infrastructure
C. Safety of personnel
D. Vital records

78. A mission-critical system has been identified as having an administrative system account with attributes that prevent locking and change of privileges and name. Which would be the **BEST** approach to prevent a successful brute force attack of the account?

A. Prevent the system from being accessed remotely.
B Create a strong random password.
C. Ask for a vendor patch.
D. Track usage of the account by audit trails.

79. A regulatory authority has just introduced a new regulation pertaining to the release of quarterly financial results. The **FIRST** task that the security officer should perform is to:

A. identify whether current controls are adequate.
B. communicate the new requirement to audit.
C. implement the requirements of the new regulation.
D. conduct a cost-benefit analysis of implementing the control.

80. What does the following statement reflect: "All desktops are required to use Windows 7 Service Pack 1, and all servers are required to use Windows Server 2008 R2 Service Pack 1."

A. The statement is a policy.
B. The statement is a guideline.
C. The statement is a standard.
D. The statement is a procedure.

81. What is the **FIRST** step of performing an information risk analysis?

A. Establish the ownership of assets.
B. Evaluate the risk to the assets.
C. Take an asset inventory.
D. Categorize the assets.

82. Which of the following **BEST** helps calculate the impact of losing frame relay network connectivity for 18 to 24 hours?

A. Hourly billing rate charged by the carrier
B. Value of the data transmitted over the network
C. Aggregate compensation of all affected business users
D. Financial losses incurred by affected business units

83. Which of the following would be the **FIRST** step when developing a business case for an information security investment?

A. Defining the objectives
B. Calculating the cost
C. Defining the need
D. Analyzing the cost-effectiveness

84. Which of the following **BEST** assists the information security manager in identifying new threats to information security?

A. Performing more frequent reviews of the organization's risk factors
B. Developing more realistic information security risk scenarios
C. Understanding the flow and classification of information used by the organization
D. A process to monitor postincident review reports prepared by IT staff

85. A cost-benefit analysis is performed on any proposed control to:

A. define budget limitations.
B. demonstrate due diligence to the budget committee.
C. verify that the cost of implementing the control is within the security budget.
D. demonstrate the costs are justified by the reduction in risk.

86. Which of the following functions is responsible for determining the members of the enterprise's response teams?

A. Governance
B. Risk management
C. Compliance
D. Information security

87. A database was compromised by guessing the password for a shared administrative account and confidential customer information was stolen. The information security manager was able to detect this breach by analyzing which of the following?

A. Invalid logon attempts
B. Write access violations
C. Concurrent logons
D. Firewall logs

88. An information security manager at a global organization has to ensure that the local information security program will initially be in compliance with the:

A. corporate data privacy policy.
B. data privacy policy where data are collected.
C. data privacy policy of the headquarters' country.
D. data privacy directive applicable globally.

89. In controlling information leakage, management should **FIRST** establish:

A. a data leak prevention program.
B. user awareness training.
C. an information classification process.
D. a network intrusion detection system.

90. To determine how a security breach occurred on the corporate network, a security manager looks at the logs of various devices. Which of the following **BEST** facilitates the correlation and review of these logs?

A. Database server
B. Domain name server
C. Time server
D. Proxy server

91. Who in an organization has the responsibility for classifying information?

A. Data custodian
B. Database administrator
C. Information security officer
D. Data owner

92. Which of the following attacks is **BEST** mitigated by using strong passwords?

A. Man-in-the-middle attack
B. Brute force attack
C. Remote buffer overflow
D. Root kit

93. Which of the following choices is the **BEST** method of determining the impact of a distributed denial-of-service attack on a business?

A. Identify the sources of the malicious traffic.
B. Interview the users and document their responses.
C. Determine the criticality of the affected services.
D. Review the logs of the firewalls and intrusion detection system.

94. Which of the following is **MOST** important in determining whether a disaster recovery test is successful?

A. Only business data files from offsite storage are used.
B. IT staff fully recovers the processing infrastructure.
C. Critical business processes are duplicated.
D. All systems are restored within recovery time objectives.

95. IT-related risk management activities are **MOST** effective when they are:

A. treated as a distinct process.
B. conducted by the IT department.
C. integrated within business processes.
D. communicated to all employees.

96. Investments in information security technologies should be based on:

A. vulnerability assessments.
B. value analysis.
C. business climate.
D. audit recommendations.

97. Which of the following is the **MOST** appropriate position to sponsor the design and implementation of a new security infrastructure in a large global enterprise?

A. Chief security officer
B. Chief operating officer
C. Chief privacy officer
D. Chief legal counsel

98. Simple Network Management Protocol v2 (SNMP v2) is used frequently to monitor networks. Which of the following vulnerabilities does it always introduce?

A. Remote buffer overflow
B. Cross-site scripting
C. Cleartext authentication
D. Man-in-the-middle attack

99. The **PRIMARY** objective of incident response is to:

A. investigate and report results of the incident to management.
B. gather evidence.
C. minimize business disruptions.
D. assist law enforcement in investigations.

100. To **BEST** improve the alignment of the information security objectives in an organization, the chief information security officer should:

A. revise the information security program.
B. evaluate a business balanced scorecard.
C. conduct regular user awareness sessions.
D. perform penetration tests.

101. Who is ultimately responsible for ensuring that information is categorized and that protective measures are taken?

A. Information security officer
B. Security steering committee
C. Data owner
D. Data custodian

102. A project manager is developing a developer portal and requests that the security manager assign a public Internet Protocol address so that it can be accessed by in-house staff and by external consultants outside the organization's local area network. What should the security manager do **FIRST**?

A. Understand the business requirements of the developer portal.
B. Perform a vulnerability assessment of the developer portal.
C. Install an intrusion detection system.
D. Obtain a signed nondisclosure agreement from the external consultants before allowing external access to the server.

103. In conducting an initial technical vulnerability assessment, which of the following choices should receive top priority?

A. Systems impacting legal or regulatory standing
B. Externally facing systems or applications
C. Resources subject to performance contracts
D. Systems covered by business interruption insurance

104. An information security manager is in the process of investigating a network intrusion. One of the enterprise's employees is a suspect. The manager has just obtained the suspect's computer and hard drive. Which of the following is the **BEST** next step?

A. Create an image of the hard drive.
B. Encrypt the data on the hard drive.
C. Examine the original hard drive.
D. Create a logical copy of the hard drive.

105. The decision as to whether an IT risk has been reduced to an acceptable level should be determined by:

A. organizational requirements.
B. information systems requirements.
C. information security requirements.
D. international standards.

106. A third party was engaged to develop a business application. Which of the following is the **BEST** test for the existence of back doors?

A. System monitoring for traffic on network ports
B. Security code reviews for the entire application
C. Reverse engineering the application binaries
D. Running the application from a high-privileged account on a test system

107. A contract has just been signed with a new vendor to manage IT support services. Which of the following tasks should the information security manager ensure is performed **NEXT**?

A. Establish vendor monitoring.
B. Define reporting relationships.
C. Create a service level agreement.
D. Have the vendor sign a nondisclosure agreement.

108. Which of the following is the **MOST** effective security measure to protect data held on mobile computing devices?

A. Biometric access control
B. Encryption of stored data
C. Power-on passwords
D. Protection of data being transmitted

109. An information security manager can **BEST** attain senior management commitment and support by emphasizing:

A. organizational risk.
B. performance metrics.
C. security needs.
D. the responsibilities of organizational units.

110. Which of the following is an inherent weakness of signature-based intrusion detection systems?

A. A higher number of false positives
B. New attack methods will be missed
C. Long duration probing will be missed
D. Attack profiles can be easily spoofed

111. Why is it important to develop an information security baseline? The security baseline helps define:

A. critical information resources needing protection.
B. a security policy for the entire organization.
C. the minimum acceptable security to be implemented.
D. required physical and logical access controls.

112. To achieve effective strategic alignment of information security initiatives, it is important that:

A. steering committee leadership rotates among members.
B. major organizational units provide input and reach a consensus.
C. the business strategy is updated periodically.
D. procedures and standards are approved by all departmental heads.

113. Which of the following is the **MOST** important item to consider when evaluating products to monitor security across the enterprise?

A. Ease of installation
B. Product documentation
C. Available support
D. System overhead

114. Which of the following is the **MOST** serious exposure of automatically updating virus signature files on every desktop each Friday at 11:00 p.m. (2300 hours)?

A. Most new viruses' signatures are identified over weekends.
B. Technical personnel are not available to support the operation.
C. Systems are vulnerable to new viruses during the intervening week.
D. The update's success or failure is not known until Monday.

115. Which of the following should an information security manager **PRIMARILY** use when proposing the implementation of a security solution?

A. Risk assessment report
B. Technical evaluation report
C. Business case
D. Budgetary requirements

116. Which of the following choices is the **MOST** significant single point of failure in a public key infrastructure?

A. A certificate authority's (CA) public key
B. A relying party's private key
C. A CA's private key
D. A relying party's public key

117. Which of the following is the **BEST** method for ensuring that temporary employees do not receive excessive access rights?

A. Mandatory access controls
B. Discretionary access controls
C. Lattice-based access controls
D. Role-based access controls

118. What is the **BEST** way to determine if an anomaly-based intrusion detection system (IDS) is properly installed?

A. Simulate an attack and review IDS performance.
B. Use a honeypot to check for unusual activity.
C. Audit the configuration of the IDS.
D. Benchmark the IDS against a peer site.

119. A serious vulnerability is reported in the firewall software used by an organization. Which of the following should be the immediate action of the information security manager?

A. Ensure that all operating system patches are up to date.
B. Block inbound traffic until a suitable solution is found.
C. Obtain guidance from the firewall manufacturer.
D. Commission a penetration test.

120. Which of the following should be determined **FIRST** when establishing a business continuity program?

A. Cost to rebuild information processing facilities
B. Incremental daily cost of the unavailability of systems
C. Location and cost of offsite recovery facilities
D. Composition and mission of individual recovery teams

121. What responsibility do data owners normally have?

A. Applying emergency changes to application data
B. Administering security over database records
C. Migrating application code changes to production
D. Determining the level of application security required

122. The **PRIMARY** reason for senior management review of information security incidents is to:

A. ensure adequate corrective actions were implemented.
B. demonstrate management commitment to the information security process.
C. evaluate the incident response process for deficiencies.
D. evaluate the ability of the security team.

123. Which of the following choices would be the **MOST** significant key risk indicator?

A. A deviation in employee turnover
B. The number of packets dropped by the firewall
C. The number of viruses detected
D. The reporting relationship of IT

124. A virtual desktop infrastructure enables remote access. The benefit of this approach from a security perspective is to:

A. optimize the IT resource budget by reducing physical maintenance to remote personal computers (PCs).
B. establish segregation of personal and organizational data while using a remote PC.
C. enable the execution of data wipe operations into a remote PC environment.
D. terminate the update of the approved antivirus software list for remote PCs.

125. What is the **PRIMARY** purpose of segregation of duties?

A. Employee monitoring
B. Reduced supervisory requirements
C. Fraud prevention
D. Enhanced compliance

126. The **MOST** effective approach to address issues that arise between IT management, business units and security management when implementing a new security strategy is for the information security manager to:

A. escalate issues to an external third party for resolution.
B. ensure that senior management provide authority for security to address the issues.
C. insist that managers or units not in agreement with the security solution accept the risk.
D. refer the issues to senior management along with any security recommendations.

127. Which of the following will require the **MOST** effort when supporting an operational information security program?

A. Reviewing and modifying procedures
B. Modifying policies to address changing technologies
C. Writing additional policies to address new regulations
D. Drafting standards to address regional differences

128. Which of the following would be the **BEST** way to improve employee attitude toward, and commitment to, information security?

A. Implement restrictive controls.
B. Customize methods training to the audience.
C. Apply administrative penalties.
D. Initiate stronger supervision.

129. Which of the following processes is **CRITICAL** for deciding prioritization of actions in a business continuity plan?

A. Business impact analysis
B. Risk assessment
C. Vulnerability assessment
D. Business process mapping

130. In a large enterprise, what makes an information security awareness program **MOST** effective?

A. The program is developed by a professional training company.
B. The program is embedded into the orientation process.
C. The program is customized to the audience using the appropriate delivery channel.
D. The program is required by the information security policy.

131. Which of the following choices is the **BEST** input for the definition of escalation guidelines?

A. Risk management issues
B. A risk and impact analysis
C. Assurance review reports
D. The effectiveness of resources

132. After a risk assessment, it is determined that the cost to mitigate the risk is much greater than the benefit to be derived. The information security manager should recommend to business management that the risk be:

A. transferred.
B. treated.
C. accepted.
D. terminated.

133. The **FIRST** step in developing a business case is to:

A. determine the probability of success.
B. calculate the return on investment.
C. analyze the cost-effectiveness.
D. define the issues to be addressed.

134. Which of the following is the **BEST** way to erase confidential information stored on magnetic tapes?

A. Performing a low-level format
B. Rewriting with zeros
C. Burning them
D. Degaussing them

135. Effective governance of enterprise security is **BEST** ensured by:

A. using a bottom-up approach.
B. management by the IT department.
C. referring the matter to the organization's legal department.
D. using a top-down approach.

136. Which of the following is a key component of an incident response policy?

A. Updated call trees
B. Escalation criteria
C. Press release templates
D. Critical backup files inventory

137. Which of the following is the **MOST** important aspect of forensic investigations that will potentially involve legal action?

A. The independence of the investigator
B. Timely intervention
C. Identifying the perpetrator
D. Chain of custody

138. Which of the following risk scenarios would **BEST** be assessed using qualitative risk assessment techniques?

A. Theft of purchased software
B. Power outage lasting 24 hours
C. Permanent decline in customer confidence
D. Temporary loss of email services

139. Who is in the **BEST** position to determine the level of information security needed for a specific business application?

A. The system developer
B. The information security manager
C. The system custodian
D. The data owner

140. Which of the following roles is responsible for ensuring that information is classified?

A. Senior management
B. The security manager
C. The data owner
D. The data custodian

141. Which of the following is the **BEST** way to mitigate the risk of the database administrator reading sensitive data from the database?

A. Log all access to sensitive data.
B. Employ application-level encryption.
C. Install a database monitoring solution.
D. Develop a data security policy.

142. An organization's information security strategy should be based on:

A. managing risk relative to business objectives.
B. managing risk to a zero level and minimizing insurance premiums.
C. avoiding occurrence of risk so that insurance is not required.
D. transferring most risk to insurers and saving on control costs.

143. Risk acceptance is a component of which of the following?

A. Risk assessment
B. Risk mitigation
C. Risk identification
D. Risk monitoring

144. What is the **PRIMARY** objective of security awareness?

A. Ensure that security policies are understood.
B. Influence employee behavior.
C. Ensure legal and regulatory compliance.
D. Notify of actions for noncompliance.

145. Which of the following steps in conducting a risk assessment should be performed **FIRST**?

A. Identify business assets
B. Identify business risk
C. Assess vulnerabilities
D. Evaluate key controls

146. Successful implementation of information security governance will **FIRST** require:

A. security awareness training.
B. updated security policies.
C. a computer incident management team.
D. a security architecture.

147. A control policy is **MOST** likely to address which of the following implementation requirements?

A. Specific metrics
B. Operational capabilities
C. Training requirements
D. Failure modes

148. The director of auditing has recommended a specific information security monitoring solution to the information security manager. What should the information security manager do **FIRST**?

A. Obtain comparative pricing bids and complete the transaction with the vendor offering the best deal.
B. Add the purchase to the budget during the next budget preparation cycle to account for costs.
C. Perform an assessment to determine correlation with business goals and objectives.
D. Form a project team to plan the implementation.

149. An enterprise has a network of suppliers that it allows to remotely access an important database that contains critical supply chain data. What is the **BEST** control to ensure that the individual supplier representatives who have access to the system do not improperly access or modify information within this system?

A. User access rights
B. Biometric access controls
C. Password authentication
D. Two-factor authentication

150. Which of the following should be included in an annual information security budget that is submitted for management approval?

A. A cost–benefit analysis of budgeted resources
B. All of the resources that are recommended by the business
C. Total cost of ownership
D. Baseline comparisons

CISM® Review Questions, Answers & Explanations Manual 9th Edition

SAMPLE EXAM ANSWER AND REFERENCE KEY

Exam Question #	Key	Ref. #	Exam Question #	Key	Ref. #	Exam Question #	Key	Ref. #	Exam Question #	Key	Ref. #
1	D	S1-104	39	A	S4-22	77	C	S4-63	115	C	S1-58
2	C	S2-46	40	D	S2-131	78	B	S2-50	116	C	S3-136
3	C	S4-42	41	D	S2-149	79	A	S1-133	117	D	S3-106
4	C	S2-84	42	C	S3-177	80	C	S3-224	118	A	S3-109
5	B	S2-174	43	A	S2-8	81	C	S2-33	119	C	S4-56
6	C	S3-40	44	A	S2-107	82	D	S2-22	120	B	S4-1
7	A	S1-160	45	C	S3-28	83	C	S1-102	121	D	S3-133
8	A	S2-113	46	C	S1-21	84	C	S2-175	122	A	S4-97
9	D	S3-187	47	C	S1-39	85	D	S2-158	123	A	S1-129
10	C	S2-63	48	D	S2-72	86	D	S4-92	124	B	S3-204
11	A	S4-83	49	C	S4-3	87	A	S4-52	125	C	S2-90
12	A	S3-260	50	C	S2-60	88	B	S1-42	126	D	S1-67
13	C	S1-19	51	C	S2-59	89	C	S2-124	127	A	S1-146
14	B	S1-50	52	D	S1-5	90	C	S4-54	128	B	S3-226
15	C	S1-38	53	B	S4-50	91	D	S1-48	129	A	S2-125
16	D	S4-36	54	C	S4-79	92	B	S2-36	130	C	S3-236
17	A	S3-16	55	D	S2-92	93	C	S4-73	131	B	S4-70
18	B	S3-318	56	A	S3-320	94	C	S4-13	132	C	S2-69
19	C	S3-37	57	D	S2-96	95	C	S2-89	133	D	S1-131
20	C	S1-117	58	C	S1-44	96	B	S1-7	134	D	S3-247
21	C	S4-35	59	B	S3-124	97	B	S1-24	135	D	S1-86
22	A	S4-44	60	B	S2-74	98	C	S3-215	136	B	S4-9
23	B	S3-163	61	A	S2-140	99	C	S4-89	137	D	S4-80
24	A	S4-68	62	C	S1-41	100	B	S3-56	138	C	S2-13
25	A	S3-144	63	A	S1-49	101	B	S3-147	139	D	S3-92
26	D	S2-71	64	C	S2-111	102	A	S2-49	140	C	S2-68
27	C	S3-50	65	B	S1-107	103	D	S2-42	141	B	S3-235
28	D	S2-5	66	D	S3-140	104	A	S4-90	142	A	S1-81
29	D	S2-94	67	D	S3-243	105	A	S2-18	143	B	S2-10
30	D	S1-127	68	C	S3-228	106	B	S3-184	144	B	S3-120
31	C	S3-173	69	B	S2-64	107	A	S3-249	145	A	S2-39
32	C	S2-29	70	C	S4-132	108	B	S3-75	146	B	S1-10
33	B	S2-115	71	D	S3-82	109	A	S1-15	147	D	S3-21
34	D	S2-100	72	C	S4-113	110	B	S3-128	148	C	S1-97
35	B	S1-52	73	B	S1-110	111	C	S3-53	149	A	S3-241
36	A	S2-85	74	D	S1-73	112	B	S1-61	150	A	S1-82
37	B	S1-78	75	A	S2-105	113	D	S3-35			
38	B	S4-86	76	A	S2-98	114	C	S4-26			

Reference example: S1-132 = See domain 1, question 132 for explanation of answer.

Page intentionally left blank

CISM® Review Questions, Answers & Explanations Manual 9th Edition

SAMPLE EXAM ANSWER SHEET (PRETEST)

(side 1)

Please use this answer sheet to take the sample exam as a pretest to determine strengths and weaknesses. The answer key/reference grid is on page 431.

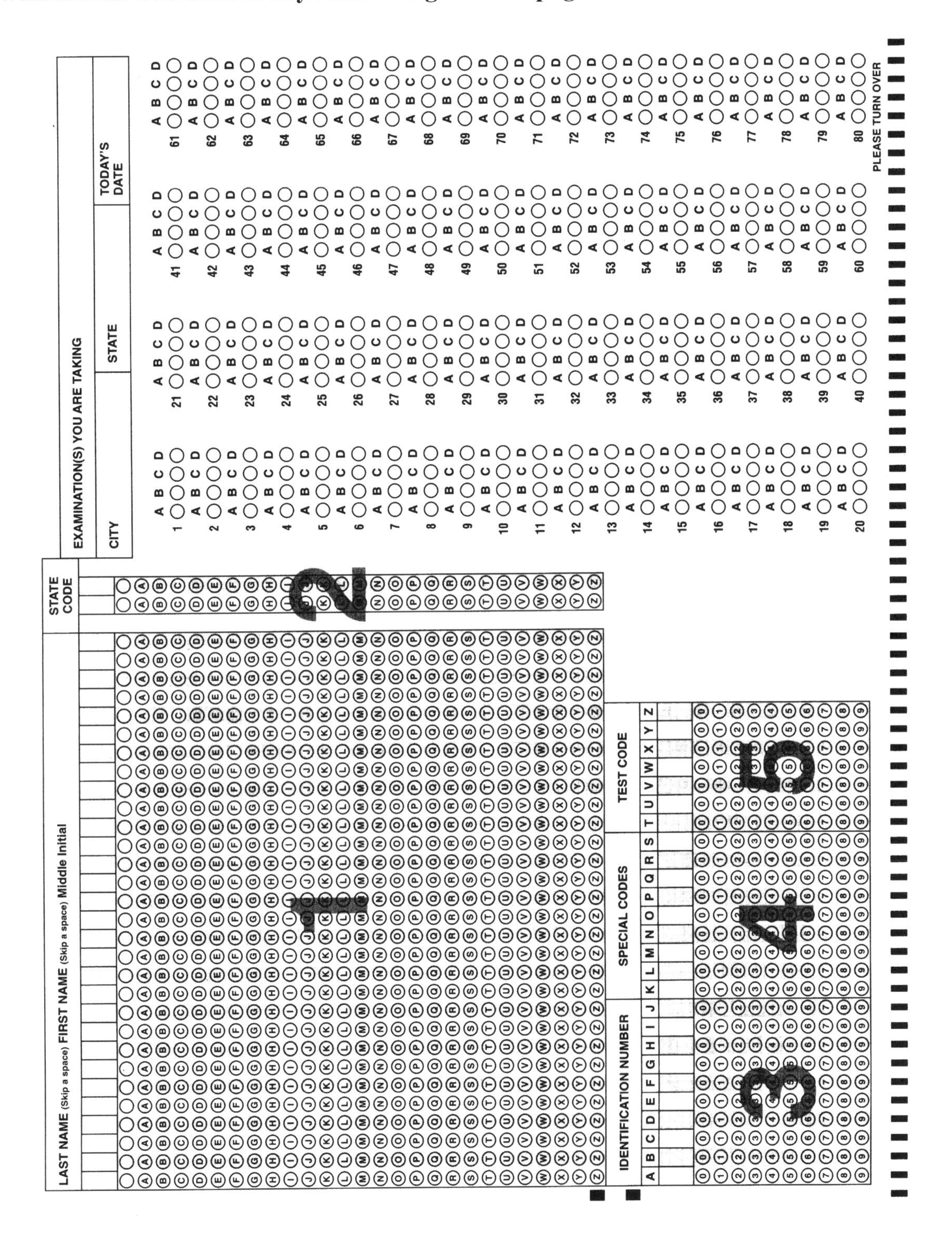

(side 2)

Please use this answer sheet to take the sample exam as a pretest to determine strengths and weaknesses. The answer key/reference grid is on page 431.

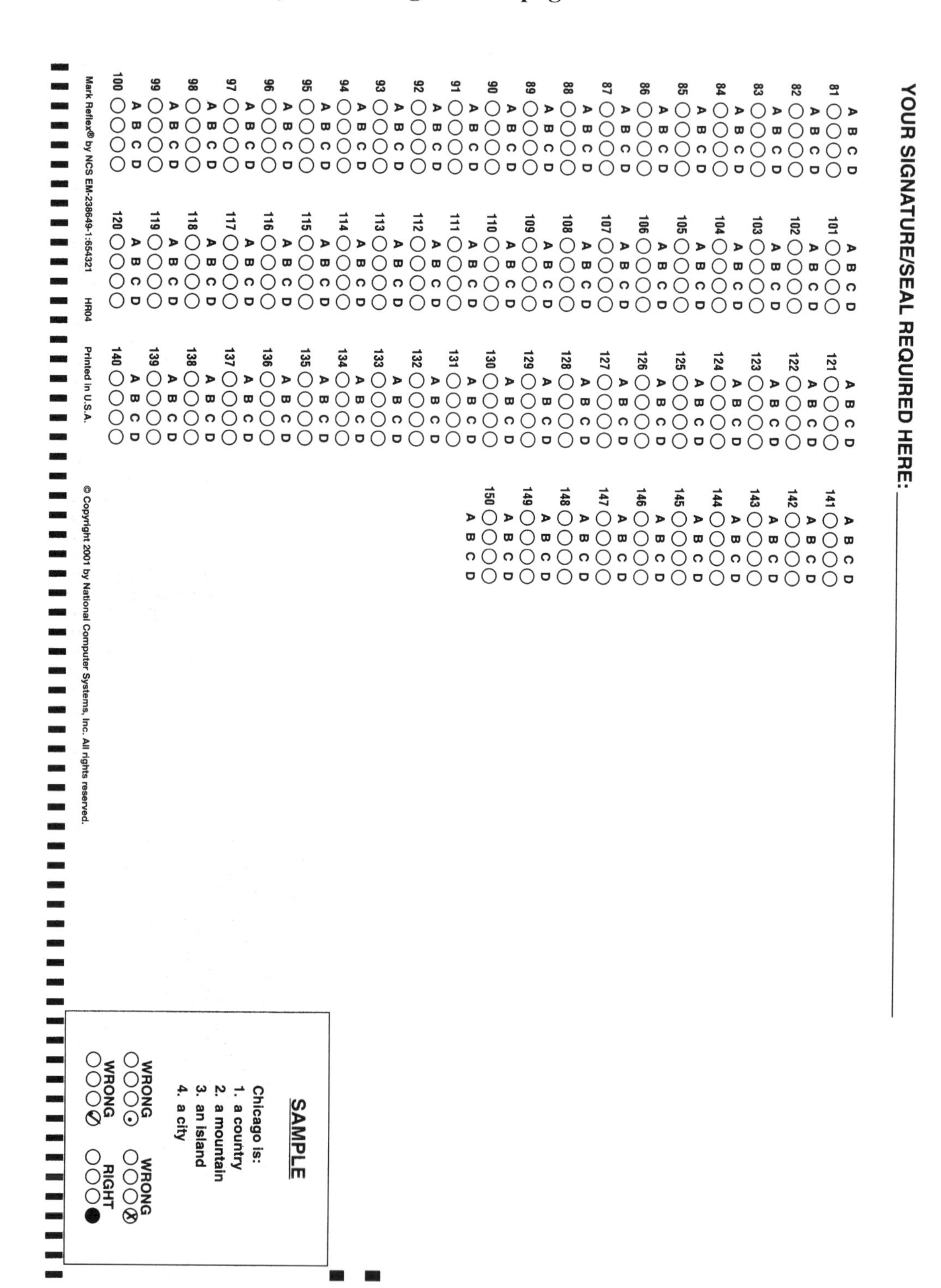

YOUR SIGNATURE/SEAL REQUIRED HERE: ______________________

81 A B C D
82 A B C D
83 A B C D
84 A B C D
85 A B C D
86 A B C D
87 A B C D
88 A B C D
89 A B C D
90 A B C D
91 A B C D
92 A B C D
93 A B C D
94 A B C D
95 A B C D
96 A B C D
97 A B C D
98 A B C D
99 A B C D
100 A B C D

101 A B C D
102 A B C D
103 A B C D
104 A B C D
105 A B C D
106 A B C D
107 A B C D
108 A B C D
109 A B C D
110 A B C D
111 A B C D
112 A B C D
113 A B C D
114 A B C D
115 A B C D
116 A B C D
117 A B C D
118 A B C D
119 A B C D
120 A B C D

121 A B C D
122 A B C D
123 A B C D
124 A B C D
125 A B C D
126 A B C D
127 A B C D
128 A B C D
129 A B C D
130 A B C D
131 A B C D
132 A B C D
133 A B C D
134 A B C D
135 A B C D
136 A B C D
137 A B C D
138 A B C D
139 A B C D
140 A B C D

141 A B C D
142 A B C D
143 A B C D
144 A B C D
145 A B C D
146 A B C D
147 A B C D
148 A B C D
149 A B C D
150 A B C D

SAMPLE

Chicago is:
1. a country
2. a mountain
3. an island
4. a city

WRONG WRONG
WRONG RIGHT

Mark Reflex® by NCS EM-238649-1:654321 HR04 Printed in U.S.A.

CISM® Review Questions, Answers & Explanations Manual 9th Edition

SAMPLE EXAM ANSWER SHEET (POSTTEST)

(side 1)

Please use this answer sheet to take the sample exam as a posttest to determine strengths and weaknesses. The answer key/reference grid is on page 431.

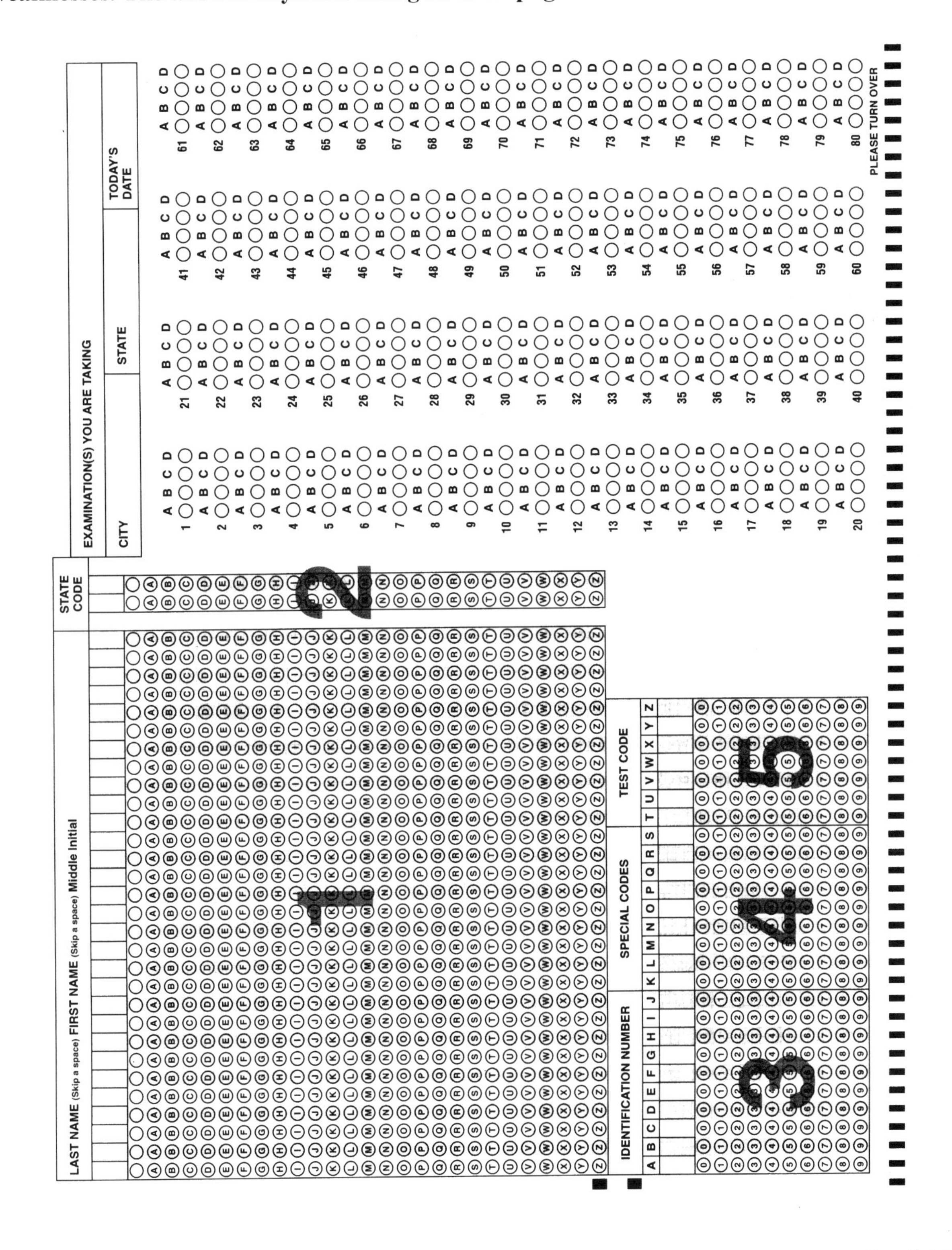

(side 2)

Please use this answer sheet to take the sample exam as a posttest to determine strengths and weaknesses. The answer key/reference grid is on page 431.

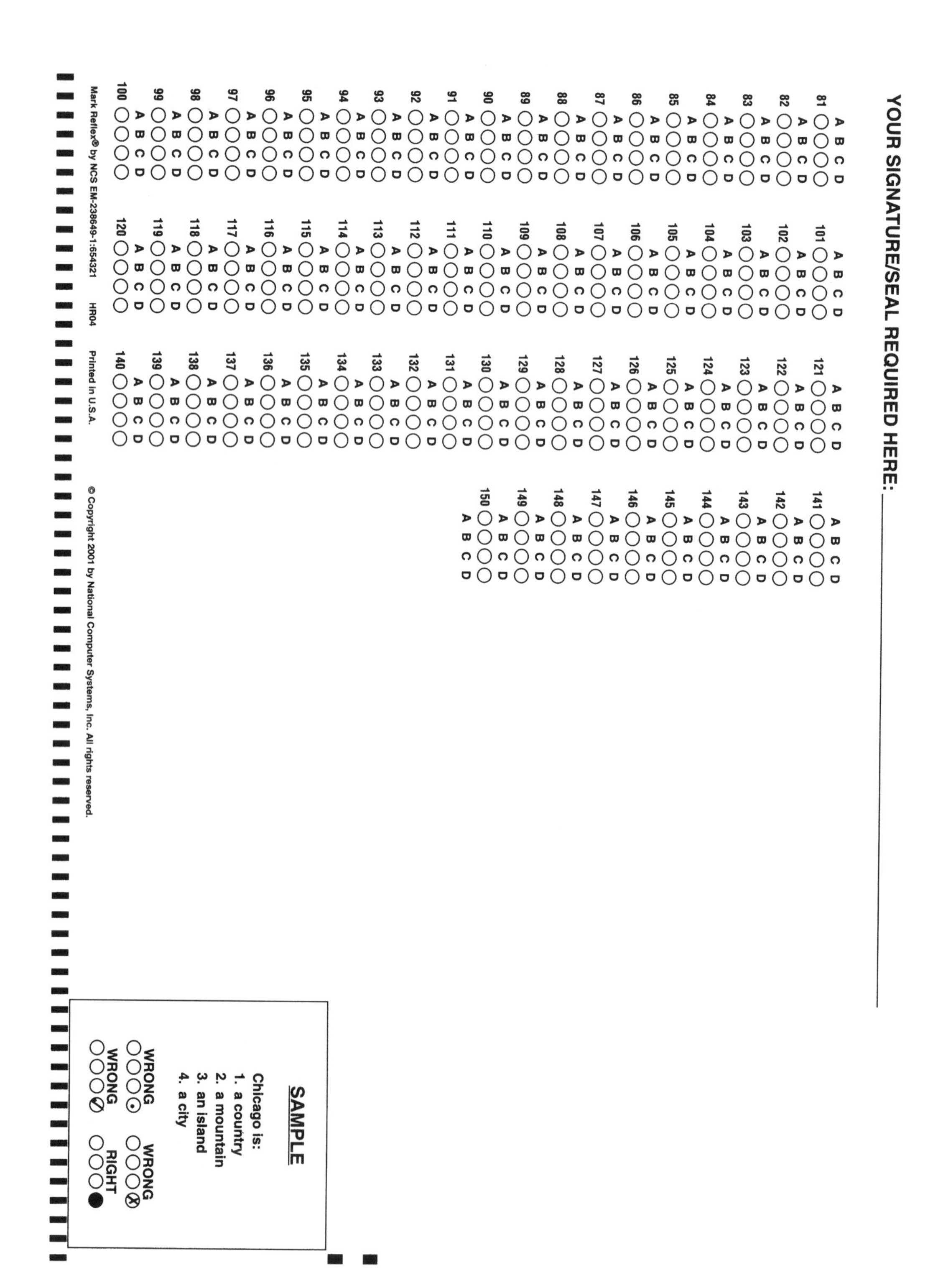

Page intentionally left blank